ORGANIZATION AND MANAGEMENT OF ADVANCED MANUFACTURING

ORGANIZATION AND MANAGEMENT OF ADVANCED MANUFACTURING

Edited by

WALDEMAR KARWOWSKI

GAVRIEL SALVENDY

A Wiley-Interscience Publication

JOHN WILEY & SONS, INC.

New York / Chichester / Brisbane / Toronto / Singapore

Library of Congress Cataloging in Publication Data:

Organization and management of advanced manufacturing / edited by Waldemar Karwowski, Gavriel Salvendy.

p. cm.

Includes index.

ISBN: 978-0-471-55508-7

1. Human engineering. 2. Computer integrated manufacturing systems. 3. Personnel management. 4. Manufacturing resource planning. I. Karwowski, Waldemar, 1953– . II. Salvendy, Gavriel, 1938–

TA166.067 1993

670.42—dc20 92-36731

10 9 8 7 6 5 4 3 2 1

CONTRIBUTORS

R. BADHAM, University of Wollongong, Australia

A. BADIRU, University of Oklahoma, USA

R. D. BRETZ, JR., Cornell University, USA

H. BULLINGER, Frauhofer-Inst. fur Arbeitswirtschaft, Germany

T. CHARLES, Commission of the European Communities, Belgium

J. A. EDOSOMWAN, Johnson & Johnson Associates, Inc., USA

K.-P. FÄHNRICH, Frauhofer-Inst. fur Arbeitswirtschaft, Germany

B. GERHART, Cornell University, USA

W. GOLOMSKI, Golomski & Associates, USA

E. HAVN, Technical University of Denmark, Denmark

M. HUESER, Institute for Manufacturing Engineering and Automation (IPA), Germany

W. KARWOWSKI, University of Louisville, USA

P. T. KIDD, Cheshire Henbury Research and Consultancy, United Kingdom

L. KLEIN, The Bayswater Institute, United Kingdom

S. KOLLI, University of Louisville, USA

R. J. KOUBEK, Purdue University, USA

J. LAMARSH, LaMarsh & Associates, USA

J. K. LIKER, University of Michigan, USA

D. LILES, University of Texas at Arlington, USA

A. MAJCHRZAK, University of Southern California, USA

J. NIEMEIER, Frauhofer-Inst. fur Arbeitswirtschaft, Germany

H. R. PARSAEI, University of Louisville, USA

J. RANTA, Technical Research Centre of Finland, Finland

G. SALVENDY, Purdue University, USA

B. SCHLEGEL, University of Oklahoma, USA

J. SHARIT, State University of New York—Buffalo, USA

H.J. WARNECKE, Institute for Manufacturing Engineering and Automation (IPA), Germany

J. R. WILSON, University of Nottingham, United Kingdom

W. WOBBE, Commission of the European Communities FAST Programme, Belgium

CONTENTS

PREFACE

In the foreword to *Handbook of Human Factors*, edited by Gavriel Salvendy and published in 1987 by John Wiley and Sons, E. M. Estes, retired president of General Motors Corporation, stated that: "We are now, along with our worldwide competitors, in a race to take advantage of the technological breakthroughs that have occurred, mainly in computer science and communications. These changes are occurring particularly in manufacturing, service industries, and communication and are requiring major changes in the role of people. This revolution has increased the necessity for more emphasis on the relationship of all workers from the top executive down to the manufacturing service and sales operators. Although the difference between success and failure in any endeavor has always been the people involved, today the human factor is even more important." These statements together with the findings of the National Research Council's *Committee on the Effectiveness of Implementation of Advanced Manufacturing Technology* concluded that the human resource practices for implementing a vast manufacturing technologies are the key elements to success. The senior editor was privileged to participate in this committee, together with key leaders from trade unions and top executives from industry. The document of this report, published by the U.S. National Academy Press in 1986, concluded that such factors as the mode of planning plant organization, job design, compensation and raises, selection, training and education of personnel, and labor–management relations prevailing in the organization were the key elements for the effective implementation of advanced technologies in the workplace. In view of these developments, it was believed essential to prepare/develop a comprehensive book fully dedicated to the human factors issues in advanced manufacturing, the effective use of which would make the difference between the failure and success of industrial corporations.

As a result of this belief, two companion books were edited. This book consists of 16 chapters written by leading international authorities in the subject areas from academia, industry, and government agencies. Each contributing author was guided by a set of objectives, including the following:

Purpose The purpose of this project is to provide a comprehensive review of the human factors relevant to agile manufacturing. Advanced manufacturing systems aim to implement integration of manufacturing resources planning, computer-aided process planning, computer-aided manufacturing (including robots, FMS, CNC/DNC, etc.), computer-aided design and engineering, and concurrent engineering. Such integration will involve fundamental changes in, for example, human resource practices, cognitive task design, employee training and education, organizational design and management, and explicit definition of human roles and human–machine function allocation schedules.

This book, along with the companion book entitled *Design of Work and Development of Personnel in Advanced Manufacturing* (Salvendy and Karwowski, eds.), also published by John Wiley (1994), address and examine all of these as well as other issues relevant to agile manufacturing philosophy.

Intended Audience This text will be of interest to both practitioners and researchers of advanced manufacturing systems in academia, business, and industry, including industrial and manufacturing engineers, managers, plant supervisors, human factors specialists, cognitive and work psychologists, sociologists, and human–computer communication specialists.

We thank all the contributing authors most sincerely for agreeing so willingly to create this book with us. Each chapter was carefully reviewed by independent authorities in the subject area and the editors. The following individuals have kindly contributed to the review process:

R. Badham, Australia
A. Badiru, USA
R. Bretz, USA
P. Brodner, USA
H. Bullinger, Germany
W. Compton, USA
C. Drury, USA
J. Edosomwan, USA
G. Evans, USA
B. Gerhart, USA
W. Golomski, USA
E. Havn, Denmark
J. Hedge, USA
J. Jacobs, USA
P. Kidd, United Kingdom
L. Klein, United Kingdom
R. Koubek, USA
R. Kuivanen, Finland

A. Kusiak, USA
J. LaMarsh, USA
G. Lay, Germany
J. Liker, USA
H. Luczak, Germany
A. Majchrzak, USA
S. Miller, USA
L. Mills, USA
D. Morrison, Australia
S. Nakajima, Japan
N. Nakamura, Japan
H. Neerland, Norway
H. Parsaei, USA
J. Ranta, Finland
J. Rasmussen, Denmark
J. Reason, United Kingdom
P. Sanderson, USA
J. Sharit, USA
T. Sheridan, USA
R. Shultz-Wild, Germany
T. J. Smith, USA
G. Spur, Germany
K. Stecke, USA
M. Tachibana, Japan
E. VonWeizacher, Germany
H. Warnecke, Germany
W. Wobbe, Belgium

The editing of this book was made possible through the excellent work of Laura Abell, administrative assistant to Waldemar Karwowski. Our sincere thanks and appreciation goes to her. It was a pleasure working on this project with Frank Cerra, the John Wiley editor, who is a truly outstanding facilitator and editor.

WALDEMAR KARWOWSKI
GAVRIEL SALVENDY

Louisville, Kentucky
West Lafayette, Indiana
September 1993

ORGANIZATION AND MANAGEMENT OF ADVANCED MANUFACTURING

CHAPTER 1

TECHNOLOGIES OF ADVANCED MANUFACTURING

H.J. WARNECKE and M. HUESER
Fraunhofer-Institut für Produktionstechnik und Automatisierung
Stuttgart, Germany

As a result of intensified competition due to the transition from a seller's to a buyer's market, it is becoming increasingly important for companies to reduce costs or increase utility for the customer. An advantage achieved through ongoing product and process innovation cannot be maintained forever. The know-how will ultimately become common knowledge and will seep through to competitors or threshold countries. Companies in the industrial world can only pursue two strategies in order to maintain competitiveness:

- The services and products offered must contain so much generally unavailable know-how and competence that competitors are discouraged either through industrial property rights and licence payments or through the expense involved in research and development.
- Production itself involves a high degree of know-how.

Organization and Management of Advanced Manufacturing. Edited by Waldemar Karwowski and Gavriel Salvendy.
ISBN 0-471-55508-8

1.1 NO RULES FOR A COMPLEX FACTORY

In technical science it is assumed that all phenomena are based on the principle of reason and effect: small reason—small effect, large reason—large effect. Parts examined according to this attitude are subsequently put together again under the illusion that this is the way to obtain an exact image of the whole. However, this is a fundamental misconception. In real systems the smallest reasons can build up to large effects due to complex feedback mechanisms. The original order is replaced by an irregular unforeseeable behavior. If these chains of effects are ignored in model creation, as usual, the look at the whole system gets lost. So only suboptimal solutions are found that often deviate considerably from the overall optimum.

Although the linear and monocausal worldview has already been questioned with the emergence of quantum theory at the beginning of our century, we are still looking for the "world formula." For manufacturing engineering, this means, for example, the search for the "factory of the future."

Neither physicists nor engineers will find the world formula. It was only recently that scientists accepted chaos indeed as one of the fundamental patterns of our existence, but they are still far from being familiar with it. When coping with our daily tasks in this environment, we will be most successful if we do not look for "the truth" or "the solution" as the only superordinate value. It is much more important, however, to emphasize the importance of constant efforts and improvements in an environment that is rather unpredictable and subject to quick changes.

A production system is exposed to more decisive influences than those three or four we consider to be important and that we are able to relate to according to certain rules. Nevertheless, we are faced with the necessity to elaborate guidelines and rules for technology management. (See Fig. 1.1.)

1.2 QUANTITY AND DIVERSITY IN THE PRODUCTION STRATEGY

The significant feature of our economic system is its multiformity. The superiority to all alternatives based on planned economy supports the fact that the interplay of all elements is unpredictable. The failure of planned economy coincides with the paradigmatic change in science mentioned earlier.

Thus, it can be stated that the system performance "factory" tends toward chaos. Even models describing it, which are created with highest efforts, for example, in the field of production control, are inadequate on principle. As a consequence, the competition strategies of enterprises differ considerably.

Structures based on the division of labor derived by Taylor as a seemingly optimal solution, aim at the use of quantity effects and progressive cost reduction with increasing production quantity. Such learning curves can be proved even for high-value capital goods and small piece numbers (Fig. 1.2). In contrast, there is the characteristic of manufacturing systems that intend

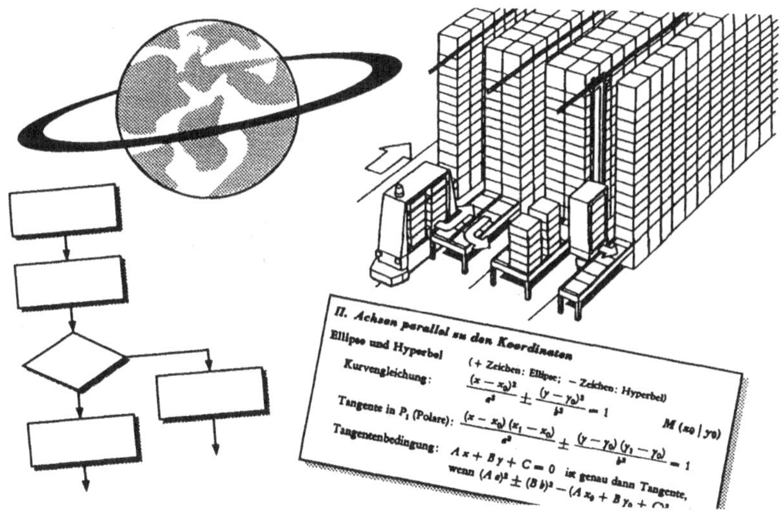

(a)

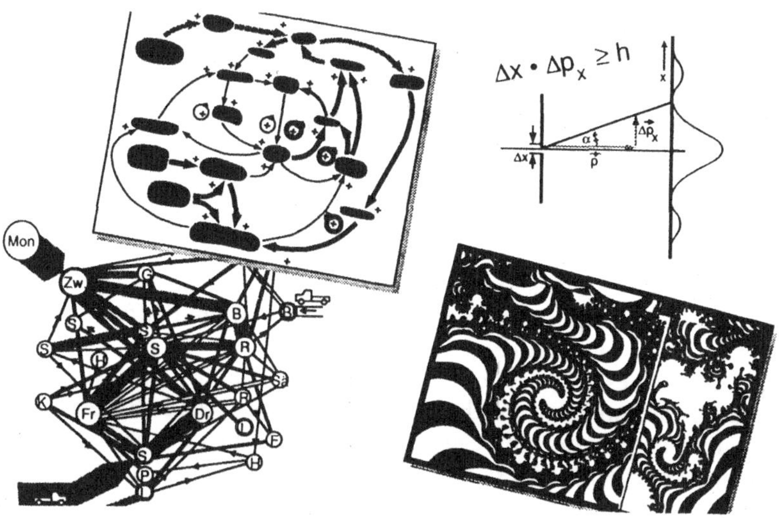

(b)

Fig. 1.1. Deterministic and chaotic world outlook.

to satisfy the individual needs of customers. The special benefits for the customer can lie in the product itself (e.g., great diversity of types) or in the service. The results of a software market study were (Stalk and Hout, 1990):

- Of the users questioned 85 percent were willing to pay an extra charge of 10 percent for service on the same day, 60 percent would pay a 20

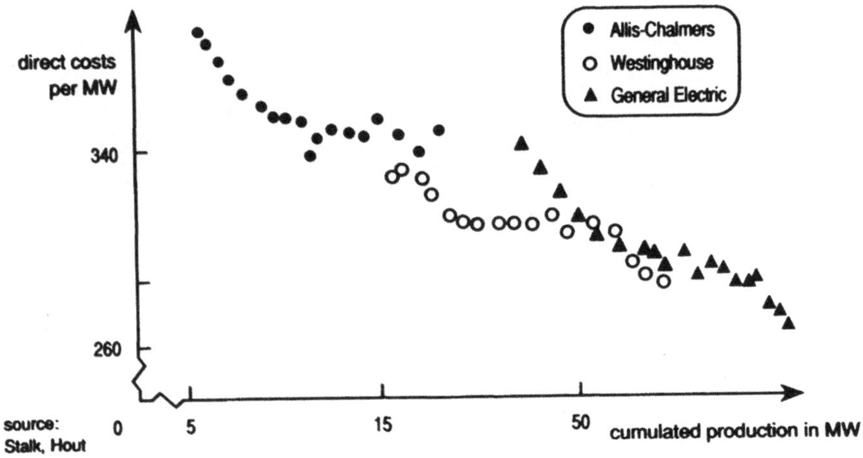

Fig. 1.2. Progressive cost reduction, for example, steam turbines (Stalk and Hout, 1990).

percent extra charge, and 40 percent would even pay a 30 percent extra charge.

- Brand name and dealer reputation were worth half the extra charge for quick service.
- Technological features only had 25 percent of the significance of the service offered.

Each enterprise occupies the individual position in the market that seems most favorable (Porter, 1985). Over a long period of time different production strategies can be pursued (Fig. 1.3).

The advantages of production focusing on quantity or diversity can be better used the more one of the extremes is approached. Based on this discovery areas of responsibility, product areas, and process areas with responsible teams or even business units are increasingly formed. If there are diversifications, there cannot be any solution covering the whole spectrum of products equally.

1.3 FACTORY STRUCTURES AS A FUNCTION OF ENVIRONMENT

The production strategies described lead to specific problem areas for structuring a factory. An enterprise aiming at distinctiveness from its competitors is facing increased uncertainty. In order to minimize the risks associated with that, the reaction speed with regard to changing boundary conditions has to be as high as possible. This can easily be realized when focusing on diversity, since in that case flexible organizational structures are existing anyway.

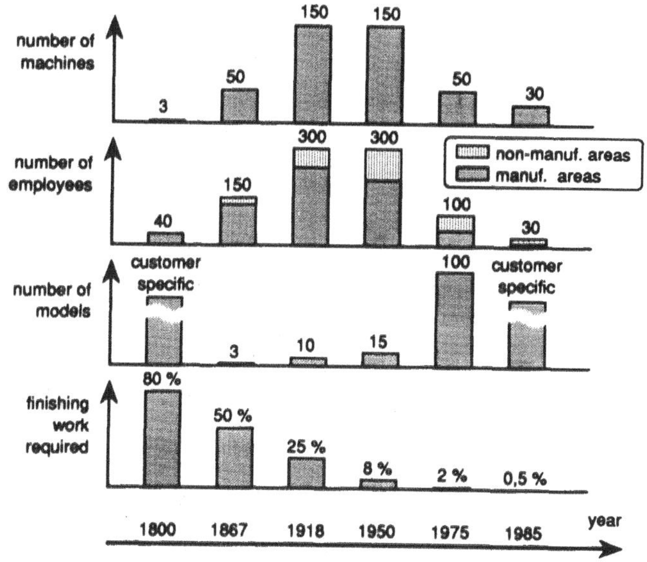

Fig. 1.3. Long-term development of enterprise organization, for example, Beretta (according to Jaikumar, 1986).

However, enterprises that consider themselves price competitors and that now tend more and more toward a differentiation strategy are suddenly facing great difficulties. In particular this applies to the automotive industry. Organizational structures that benefit from quantity advantages react on changes very slowly. There are different possibilities to cope with that situation (Womack et al., 1990).

Steps of product development following logical sequences are largely processed simultaneously (simultaneous engineering). Here coordination efforts are high, since all activities are interconnected. The communication efforts between all participants increase correspondingly. An increase in development costs, however, is much less significant for business success than exceeded deadlines or problems of quality when the product is introduced (Fig. 1.4). In extremely innovative branches, as for example, in the semiconductor industry, early presence on the market had become a question of life and death because adequate prices can only be obtained within a time period measured on the basis of months. For that reason, a stable production, impeccable under quality aspects, has to be realized as soon as possible after the start of the production run.

The problems arising from a lack of communication can be solved by creating small autonomously acting units. This type of organization, which is either product-oriented or process-oriented, promotes the identification of the involved employees with their tasks, and it allows autonomous and responsible actions. Thus, scopes are created concerning working hours, wages

Factor	Profitability loss
6 month late to market	34 %
9 % material cost overrun	27 %
50 % development cost overrun	3,5 %

Automobile with a 5 year life cycle
from job 1 to end of production

Source: Case study by Ford / McKinsey, Nov. 1988

Fig. 1.4. Effects of development deficits (Ford and McKinsey, 1988).

and salaries, and qualification. This applies to all hierarchical levels and to all functions of an enterprise.

The acquisition of time data (storage times, waiting times, rest periods, etc.) as an objective data basis as well as subsequent reduction of this unproductive time has proved to be an efficient instrument for the detection of potentials, reserves, and possible improvements. A lasting effect, however, can only be achieved if this procedure becomes a routine matter in practice.

The establishment of flat hierarchies tends to promote the generalization of staff members and expresses the existing and challenging complexity within an organization. Nevertheless, it is undisputed that specialization will remain indispensable particularly concerning technological issues. Therefore, staff has to consist of specialists *and* generalists. Otherwise important know-how might be lost within a short time. Changes in the organizational structure therefore have to be made very carefully, since drastic about-turns may hold dangers that must not be underestimated. This is even more so since the number of laws and official orders by which firms are constrained is continuously growing. The same applies to the reduction of overhead and contract services. In the case of nonkey functions of the enterprise, the process expenditure (presumably higher) on the one hand, and the expenditure for logistics (certainly increasing) for the integration of better and faster suppliers on the other hand, need to be weighted.

1.4 THE DILEMMA OF VARIANTS

Mastering the diversity of variants or maintaining the positive quantity effects has become a key task in production. The approaches can be found both in the fields of organization and technics. They pursue the strategy of *first avoiding variants, then mastering variants*.

A intentional technical overdimensioning of variants that are subjected to

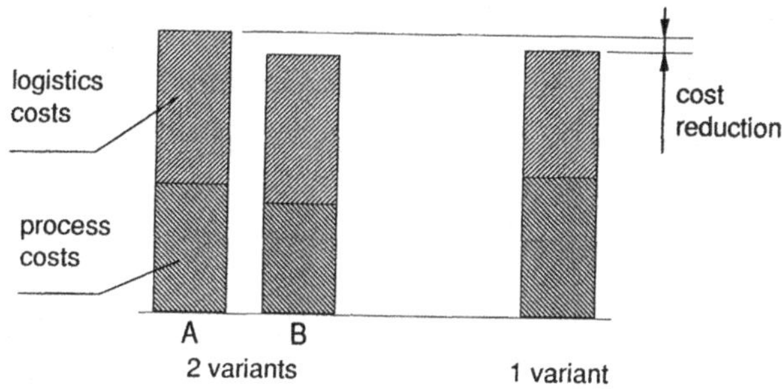

Fig. 1.5. Cost reduction by avoiding variants.

less stress can increase the number of identical parts. The insignificant increase of material required and, if necessary, of processing efforts can often be neglected compared with the saving potential (Fig. 1.5). For the realization of this strategy it is necessary to estimate the costs of different alternatives already in the design state. For these tasks generalists are required.

The creation of variants is not totally avoided, but it is at least shifted by the reduction of the production depth concerning those components that determine the variants. In principle, avoiding variants is preferred to controlling diversity. In reality, however, there is a tendency toward the opposite.

Often the problem is considerably alleviated when the variant is created in the production process as late as possible. Here it is again the design department that is to take suitable measures. Under certain circumstances it may be reasonable to increase the number of parts and to get into conflict with traditional cost calculation: A drill hole, for example, is standardized on the largest diameter required. Smaller diameters are realized by suitable drill sockets shortly before the part is finished.

A further alleviation, especially in distribution, can be achieved by having the customer or a member of a specialist shop determine the variant. This is always possible if only simple assembly tasks have to be carried out.

1.5 INFORMATION PROCESSING IN FUTURE MANUFACTURING

In spite of missing consideration in industrial management we must understand the following:

- An enterprise is a system that processes information.
- The information expenditure is decisive for production design.

Two main areas of expenditure can be distinguished:

- Technical information flow (product-oriented and process-oriented) from the idea to the product

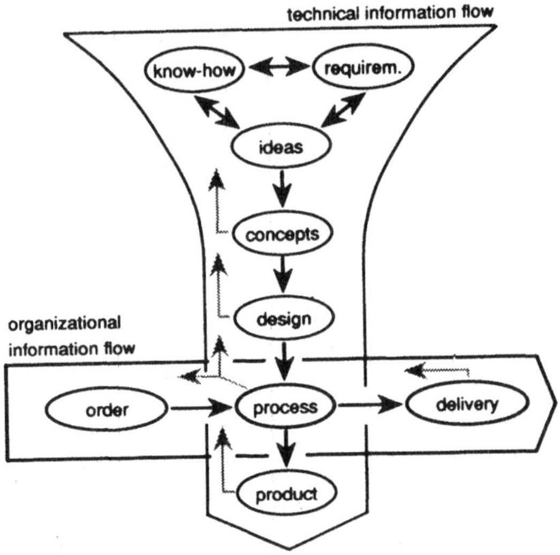

Fig. 1.6. Technical and organizational information flow in an enterprise.

- Logistical information flow (flow-oriented) from the order to delivery and service

Production means the conversion of information into shaped substance. Therefore, computer-aided design (CAD) systems are the basis and starting point for an efficient technical information flow. In contrast to that, the organizational information flow is mainly determined by Production Planning and Control (PPC) systems (Fig. 1.6).

Structures that are heavily based on the division of labor with the advantages of quantity and learning effects are characterized by small process expenditures. The information as such is hardly recognized anymore since it was fixed once: the arrangement of a manufacturing line (transfer line, belt conveyor) as an expression of the operation sequence.

From the viewpoint of information technology, the expenditure for logistics or communication rises in the case of production focused on diversity, since, for example, neither the shape of the workpiece nor the processing sequence is stored in the tool or processing equipment. Sequences, processing times, and dates have to be determined again and again. The process expenditure is of secondary importance.

When the process tolerances can be reliably kept within the permissible limits, the expenditure for information technology in the area of quality assurance is minimized as well. In this way total quality can be achieved economically. Total quality has always been and will always be a precondition for economically efficient automation.

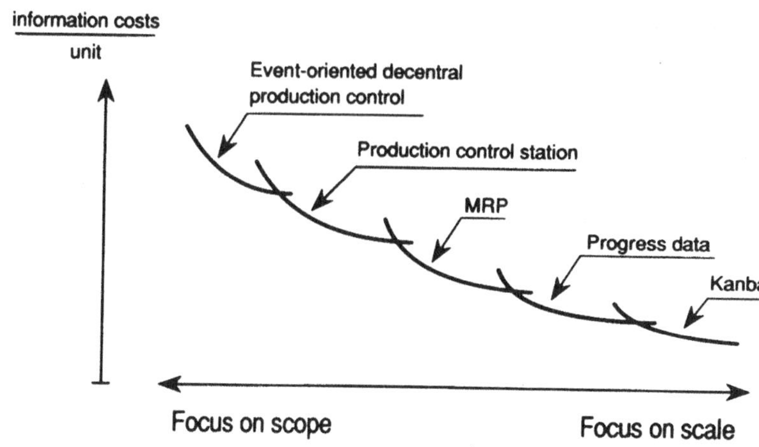

Fig. 1.7. Adapted methods for process control.

The rule for the creation of focused production units is as follows:

Process expenditures + expenditures for logistics = minimum value

Today a modular system of different methods for the configuration of a production system is available. All these different methods are to be combined and applied sensibly. Each method has a limited range of application and requires specific selection and substitution when boundary conditions change (Fig. 1.7).

This method involves the danger that isolated solutions are created. An integrated system of methods guarantees consistently available data allowing all components access to the required information (Fig. 1.8). This is also the

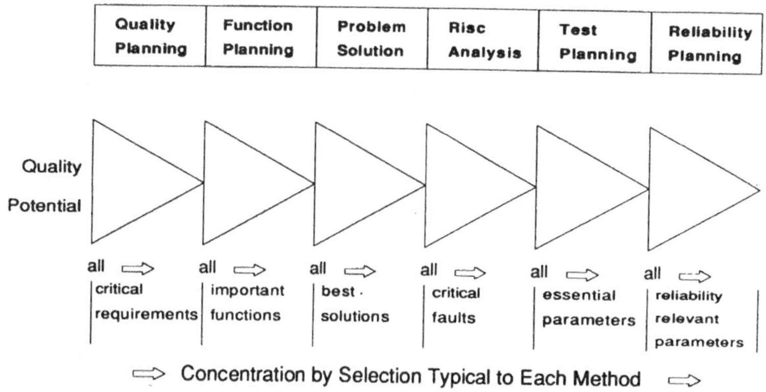

Fig. 1.8. Amplification effects of integrated methods system (IMS) (Kersten, 1991).

fundamental idea of computer-integrated manufacturing (CIM): Open modular systems with standardized interfaces.

Development of software and hardware supporting this concept has reached different levels. Computer-aided design systems as stand-alone solutions can be regarded as highly developed. Currently 60 percent of them are based on two-dimensional elements. Three-dimensional modeling, however, will gain more importance, since its range of application increases exponently. Future development will combine surface-oriented and volume-oriented description systems to hybrid models. Besides geometrical data, they will be able to process technological and organizational data. Demand for these data arises with additional applications, for example, planning functions.

While software for administration of parts lists and process plans is widespread, there is a strong demand for software generating process plans. The fault of current solutions lies in their limitation to special geometrical and technological conditions, for example, rotational parts. Specific systems for process planning can be based on decision tables, but changes in manufacturing machinery result in a huge demand for program adjustments. A breakthrough toward generally applicable process planning systems will be achieved by employing expert system techniques combined with an integrated product model (Warnecke, 1991).

Interactive creation of numerical control (NC) programs will be characterized by user-friendly dialog systems. Consequently, requirements for system know-how will diminish. Similar to process planning, the automatic generation of NC data will gain more importance.

Computer-aided manufacturing (CAM), that is, control and coordination of all actions on the shop floor, is by nature an interface problem. Therefore, development of nonspecific software systems depends on communication standards.

Development of computer-aided quality assurance is characterized by high dynamic. Quality data will be used more and more as feedback information to adjust manufacturing processes. General availability for further tasks like diagnostics and maintenance will enlarge their field of application considerably.

The horizontal integration of planning, supervision, and control of all production activities is still a major problem. Up to now, a lack of powerful methods for production control and resources management can be observed. As different planning techniques are necessary to cope with all situations, modular systems will be developed that can be adjusted to current conditions.

To maximize the benefits of computer-aided technologies in manufacturing, a strong demand for communication arises among the single modules (Fig. 1.9). Therefore, standardization activities will dominate future development of CIM.

System integration needs a user's view. Evaluating the alternatives and making the right choices to produce usable standards needs a tight link with final system users and with integrators who daily build and modify industrial

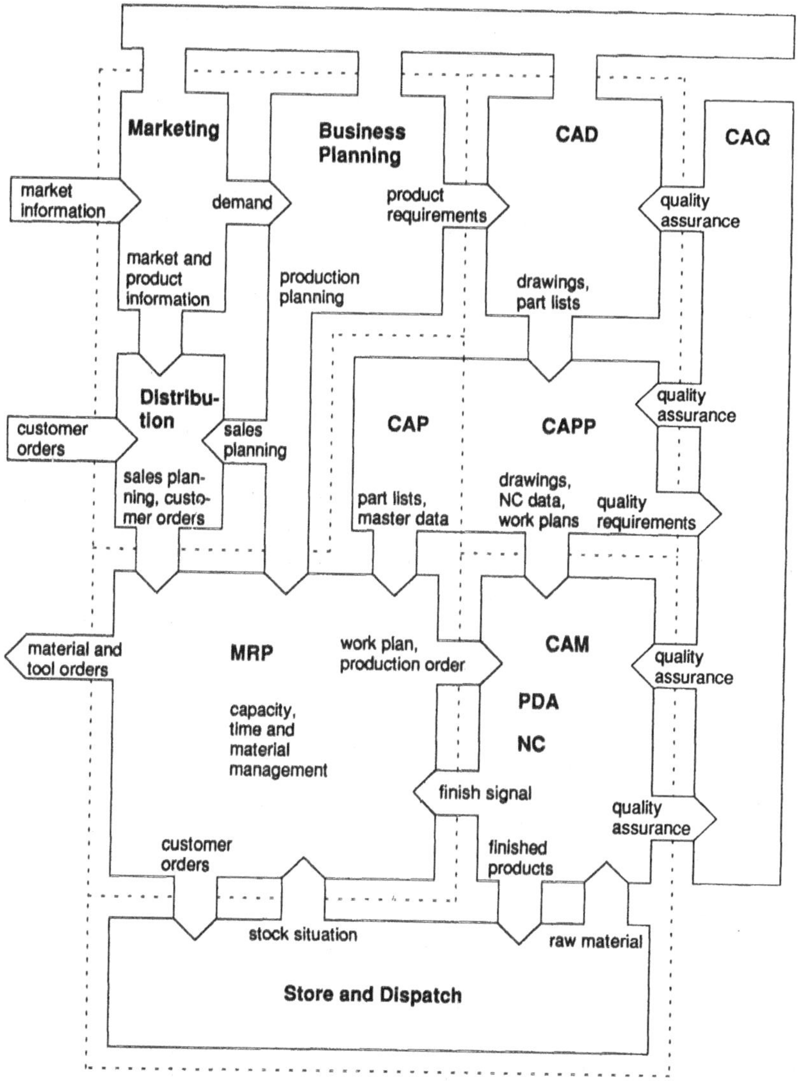

Fig. 1.9. Exchange of information among individual production areas.

applications. The best way to fulfill the user's requirements is to incorporate them in the process of integrating all the tasks in a global view. To be able to manage this in reality (with respect to time and money), industry should choose an open way, thereby inducing a strong need for standardization of both methods/languages and data.

One of the prerequisites for implementation of CIM concepts is the data integration at all levels of the control and planning concept. The production

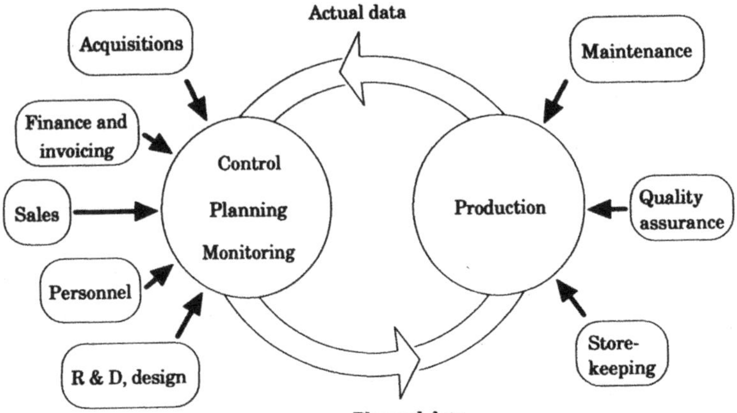

Fig. 1.10. Production data acquisition system.

control segment, in particular, requires feedback from the factory floor in order to fulfill its tasks (Kaempf, 1985).

The collection of current production data and supervision of the current production status are the tasks of a production data acquisition (PDA) system. Production data are the data giving information about the status and current procedures of the production. Production data acquisition includes all measures necessary to provide these data in a machine-usable form at the processing station.

The actual data acquired cover various elements in the relevant production process (Fig. 1.10). This includes:

- Actual time required for processing a part
- Actual utilization time of a production device
- Actual material requirement per period or per order
- Actual order start and end

The data provided for the production control system must be up to date. They are required as input variables for improved scheduling or as a basis for personnel and capacity planning. Additionally, other areas (e.g., recalculation, wage accounting, statistics) require the same data. These applications are less dependent on latest updates but more on completeness, accuracy, and condensation of data. The updated information from production also form a database for control loops within the production process. These loops can be used to optimize planning. It is also possible that control data are involved in preventive maintenance or product quality improvement. Up-to-date quality data might also be utilized for optimizing existing NC programs.

The PDA system represents the information interface between both planning and controlling elements of factory control. Technical (NC data) as well as forward-planning information (theoretical part quantities) can be transferred automatically to the production control computer.

1.6 ORGANIZATIONAL INTEGRATION OF FUNCTIONAL PRODUCTION AREAS

The integration of the different functional areas within a CAM system is contingent on the synchronization of the different technical, organizational, and planning processes.

The dictionary defines synchronization as "occurring simultaneously" or the "coordinated action of two processes, appliances or appliance parts." "Simultaneous" refers to time while "coordinated" refers to functional coincidence within an operation. In practice, this means that the technical functions (production, storage, assembly, and transport) must be coordinated with the organizational functions—CAD, computer-aided process planning (CAP), material requirements planning (MRP), and computer-aided quality assurance (CAQ)—in accordance with the required objectives (increased supply availability, reduced stocks on hand). In MRP, for example, the management of time and materials (methods, data, etc.) must be coordinated in order to reduce throughput times.

As mentioned, synchronization also stands for coincidence in time. Apart from the different types of information and production machines, the time parameter also depends on the production levels to be considered. At the implementation level (production devices) directly measurable variables are controlled or regulated. At the production control level these devices are monitored, while the planning and control level is responsible for optimization of the entire process.

The predominant hierarchical structures prevailing in companies determine the nature of the synchronization device. If, for example, time or function elements are classified in a hierarchy, vertical synchronization is applied. Elements at the same hierarchical level are linked by horizontal synchronization.

The production tasks and subtasks distribution, development/design, material management, production, assembly, quality assurance, and shipping are coordinated horizontally. Vertical synchronization mainly involves organizational functions such as planning, control, and monitoring. Method applications and data processing are the basic elements within these functions.

The second vertical linking aspect in this matrix is derived from the assignment of tasks to the various function levels. Within the organizational and technical control of material flow, for example, transport control data have to be coordinated with material flow control data and production control data.

Because of the vast amount of information exchanged among the individual production areas, the information must be structured suitably. A distinction

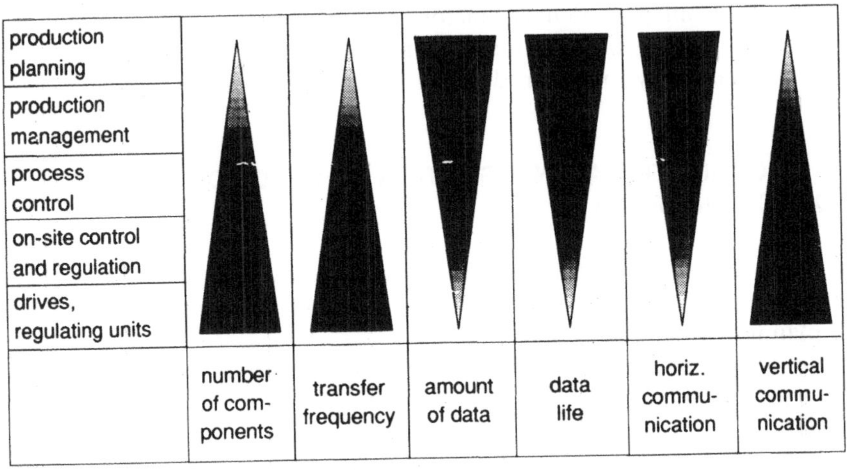

Fig. 1.11. Communications requirements.

is made between vertical and horizontal communication, depending on the direction of information paths and the way in which functions are linked with the communication system.

Vertical communication means integration of the corporate decision hierarchy in an information system permitting an exchange of information from top management to the factory floor. Horizontal communication, on the other hand, involves data transfer within one hierarchical or functional level: Data is exchanged among employees within a single department, among machines in the production process, and between the warehouse and transport systems.

In order to link the computers and control systems used for the individual function units, a universal communication system can provide information at all times at each point it is required. This communication system can be set up with the aid of local area networks (LANs) as offered on the market today. Terminals that can be connected to a LAN include independent data processing units, minicomputers, micros, dedicated word processors, computer terminals, workstation systems, data acquisition devices, "severs" (print server, file server), and other communicating systems.

The tasks to be performed by a communication system depend on the different requirements, as shown in Fig. 1.11. As far as automation is concerned, control systems for NC machines, robot controllers (RC), transport and storage systems, programmable logic controllers (PLC), sensor systems, and so forth must be able to communicate with each other and with higher level control computers. The distinctive features of this type of communication are frequent, rapid and short control commands (start, stop, acknowledge), alarms, and irregularity reports on the one hand and less frequent transmission of large data blocks (files) of program or production data on the other.

1.7 PRODUCTION PROCESSES

Progress in the application of production processes can be achieved as follows:

- Increase of performance
- Application of new processes
- Integration of processes

To improve the performance of existing tools and processes is an important task of research and development (R&D) in manufacturing technology. The better we understand the parameters that influence the process the better we can choose the optimum conditions.

Performance and typical application fields of cutting tools have shown an impressive development during the last few decades. Permanent development of existing materials and introduction of new cutting materials led to an exponential increase in maximum cutting speed.

Apart from maximum cutting speed as an indicator of high-temperature stability, the decisive characteristics of cutting tools are hardness and ductility. Up to now high-speed steel is the best choice for extreme cutting conditions like interrupted cutting or treatment of titanium alloy. Deficits in resistance to abrasion can be reduced by coatings such as titanium nitrid. Chemical vapor deposition (CVD) and physical vapor deposition (PVD) are state-of-the-art techniques to combine the best characteristics of different materials. Actually, Cermets, cubic boron nitride (CBN), cemented carbide, ceramics, and diamonds are considered as materials with the highest application potential.

Besides these developments, new production principles will dominate future manufacturing. *Generative manufacturing* follows the idea that geometrical information directly and exclusively leads to the actual workpiece. The first application of this principle is stereo lithography: Controlled by geometrical data supplied by a CAD system, an ultraviolet (UV) laser hardens the surface of liquid plastics. Layer by layer the laser beam copies the shape of the workpiece. In this way even complicated parts can be produced within a few hours. This technique enables "rapid prototyping" of complicated geometrical parts. Compared with conventional techniques, this process dramatically reduces the time from product idea to an actual product. Since time is one of the most important factors of future competition, rapid prototyping will affect whole industries.

Simplification of processes has always been the aim of manufacturing engineers. Figure 1.12 shows how boring, tapping, and deburring can be integrated into one single process. In this example the motion of the tool has to be controlled by computerized numerical control (CNC). Increase of flexibility and reduction of required tools leads to significant economical advantages in small and medium-sized lot production. Mass production, on the other side, now as before will be dominated by specialized processes.

Traditionally, workpieces produced by casting or forging need further op-

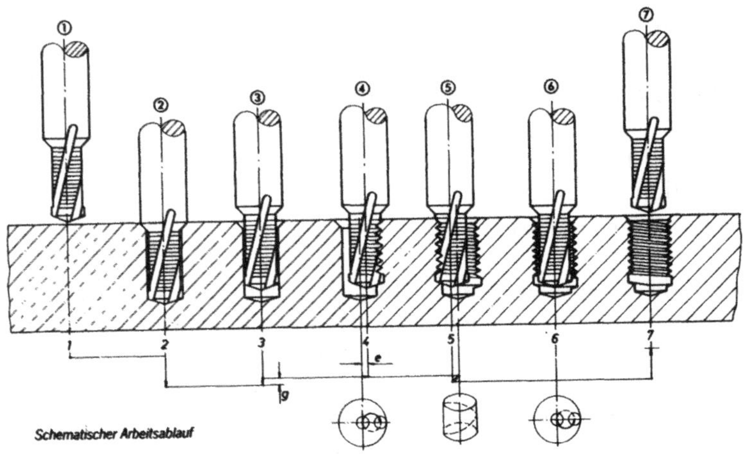

Fig. 1.12. Boring, tapping, and deburring with one tool (J&E Link GmbH, 1990).

erations to provide accuracy and surface quality. Progress in process design and process control make it possible to eliminate this final operation for a remarkable number of applications.

New construction materials are favorable in terms of their specific performance under load, temperature, and so forth. For example, the number of single parts can significantly be reduced by application of composite materials. Fewer parts means also less assembly effort and, last not least, less possible sources of error. An overall evaluation of lifetime costs further advances *integral part design* as a design paradigm (Fig. 1.13).

1.8 MACHINE TOOLS

Machine tool design has shown a significant change within the last decade. Integration of different manufacturing processes into one machine tool allows for a reduction of lead times. Perhaps of more importance, however, is the reduction of organizational effort: Fewer manufacturing devices also means less effort to control the whole production process.

Some years ago the first lathes were equipped with integrated boring and milling devices. Today this additional equipment has become state of the art. It is offered by nearly every producer. For special applications (in terms of required power and range of axes) special milling devices for lathes have been developed.

Meanwhile the idea of combined processes has been extended to machining centers. They allow for complex turning, milling, and boring operations without setup changes. The machine shown in Fig. 1.14 has a movable round table. Automatic change of workpieces is carried out by movable pallets. The

Fig. 1.13. Integral part design (Hoechst AG, 1990).

spindle can be changed from vertical to horizontal position, and therefore allows for five-sided machining. Turning and milling tools are equipped with identical adapters. Therefore only one magazine is needed to store all tools. (See also Fig. 1.15.)

The concept of *bringing processes to the workpiece* can be expected to

Fig. 1.14. Machining center for combined operations (SHW GmbH, 1985).

Fig. 1.15. Four-axes CNC lathe (Gildemeister AG, 1991).

influence the future development of machine tools. More and more processes are integrated into a single machine, like grinding, laser treatment, or hardening (Fig. 1.16).

The application of new materials creates a need for suitable manufacturing processes. Intensive research is underway to determine optimum parameters, for example, for chip-forming operations. Causes of tool wear significantly

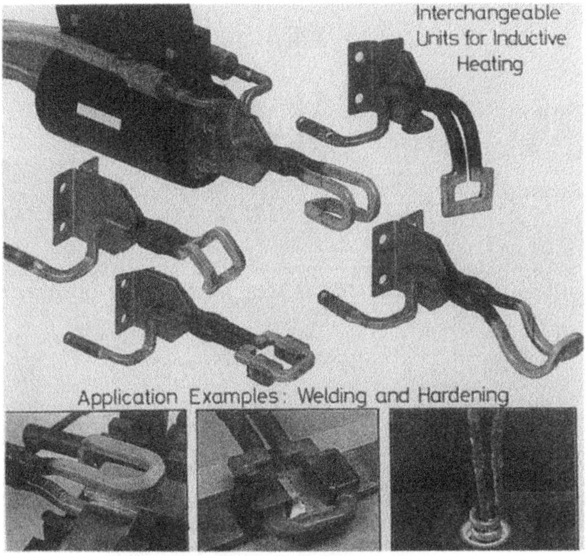

Fig. 1.16. Inductive heating for welding and hardening (ELVA, 1989).

Fig. 1.17. Workpiece handling.

change when the tools are applied to different materials. The results influence the design of machine tools: A demand for high-speed spindle drives will increase. Accuracy of motion will be put forward by direct drives with advanced dynamic behavior.

1.9 MACHINE PERIPHERY

The reduction of unproductive periods is the main reason for investments in advanced machine periphery. Automatic handling devices for workpieces are mainly derived from solutions for turned parts (Fig. 1.17). Additional benefits (time and accuracy) for prismatic parts are expected from pallet changers.

Regarding the whole production process, automatic tool change is another prerequisite for extended utilization of machine tools. Analyses on milling parts indicated that seldom are more than 40 different tools necessary for the whole spectrum of machining processes. These 40 tools can be stored in common tool magazines. However, frequently used tools have to be changed within short periods of time. This change must not reduce the productive time. Therefore, the external change of tools must be completed within the time interval of the actual process. However, this time often is very short. An analysis of Fraunhofer-Institute for Manufacturing Engineering and Automation (Stuttgart, Germany), carried out together with a machine tool manufacturer on alloy parts, indicated the following:

- Of all single processes 90 percent are finished within 2 minutes.
- Of these 75 percent are finished within 0.5 minutes.

Fig. 1.18. AGV with cassette magazine (Schenck AG, 1989).

For steel and cast iron these times are slightly higher.

Consequently, many manufacturers have developed modular tool storage devices. Figure 1.18 shows an example with cassette magazines, which can be transported by automated guided vehicles (AGV) from a central tool storing facility to the machine tool.

Tool management has turned out to be a functional area with a high rationalization potential. For an effective use of tool resources, data-based management systems have been developed to coordinate all activities related to tools:

- Storage
- Reconditioning
- Preparation
- Commissioning
- Transport
- Replacement

Modular tools enable a minimization of tool investment without lacking availability. Most tool manufacturers offer such systems (Fig. 1.19). Adapters within these systems, however, are not standardized. Therefore, a decision for the wrong supplier might be very cost-intensive in the long run.

1.10 FLEXIBLE MANUFACTURING

FMS stands for flexible manufacturing system or flexible machining system (Warnecke, 1988). It is used to describe a wide range of flexible automation

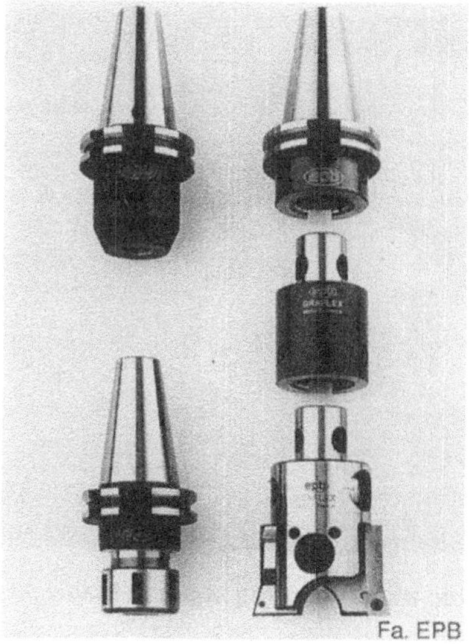

Fig. 1.19. Modular tools (EPB GmbH, 1991).

technologies stretching from simple cells to complex systems. Flexible manufacturing systems are not the only answer to market demands. They are the third stage in the development toward more flexibility. The first stage is NC machine tools, and the second stage is flexible manufacturing cells (FMCs).

An FMC consists of a CNC machine tool, a workpiece storage, devices for handling and changing workpieces and tools, and automatic control and supervision subsystems (Fig. 1.20). Hence, the machine tool is capable of performing more than one operation automatically on more than two work-

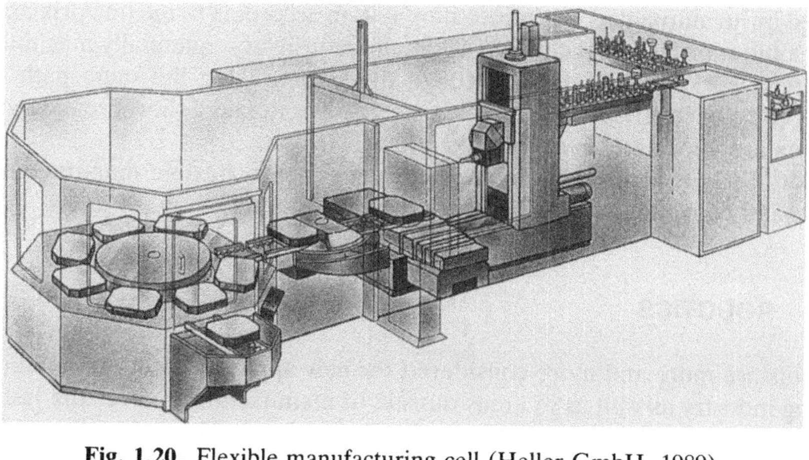

Fig. 1.20. Flexible manufacturing cell (Heller GmbH, 1989).

Fig. 1.21. Flexible manufacturing system (Maho AG, 1991).

pieces. Further devices may be integrated. Such manufacturing cells have experienced a rapid growth in both number of suppliers and users. Their appearance and frequent application closed an important gap between the individual machine tool and interlinked systems.

An FMS contains several automated machine tools of the universal or special type as well as flexible manufacturing cells. Further manual or automated workstations may be included. These elements are interlinked by an automated workpiece flow system in a way that enables the simultaneous machining of different routes. Thus, multistep and multiproduct manufacturing is possible in an FMS (Fig. 1.21).

A flexible transfer line contains several automated universal or special-purpose machine tools and further automated workstations. They are interlinked by an automated workpiece flow system according to the line principle. A flexible transfer line is capable of simultaneously or sequentially machining different workpieces that run through the system along the same path. In order to balance differences in cycle time, setup times, and short breakdowns, buffers are allocated between the stations.

Borders between flexible manufacturing systems, flexible manufacturing lines, and conventional transfer lines overlap smoothly in many cases.

1.11 ROBOTICS

Robots are more and more considered for new applications in the manufacturing industry as well as in areas outside of manufacturing. Over the last 10 years the average annual growth rate has been almost 35 percent (Fig. 1.22). These figures are impressive. However, the installations are generally, with

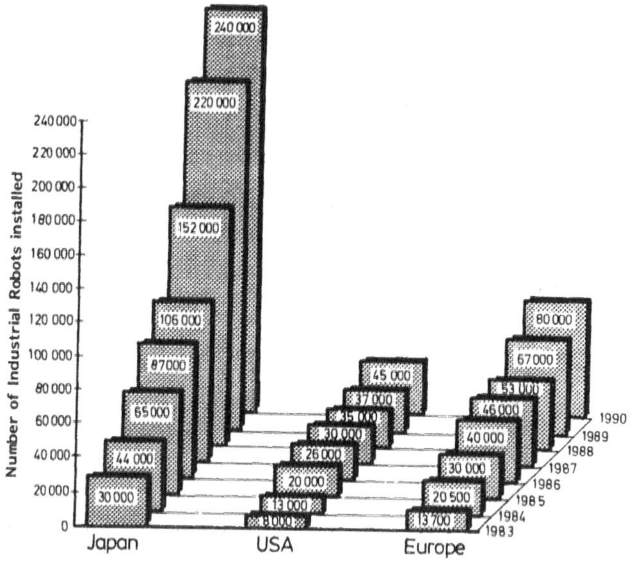

Fig. 1.22. Application of industrial robots.

few exceptions, low-valued repetitive tasks such as spotwelding, spray painting, and machine loading and unloading. For more sophisticated applications, for example, flexible assembly, industrial robots still fall short, requiring redesign of products and expensive equipment.

One of the largest fields in which robots are used is welding. Industry applies robots for spot, stud, stick, metal-inert gas (MIG), and tungsten-inert gas (TIG) welding. Spot and stud welding are resistance welding processes. Stick, MIG, and TIG are arc welding processes.

Spot welding is the most common technique used in welding because it is the easiest and most often demanded. At the beginning of assembly operations on automobiles, spot welding is used to interconnect the pieces of sheet metal. Hundreds of spot welds are made to keep the pieces of metal together, providing a tight body without rattles and squeaks.

The most common type of arc welding is MIG. The welding station consists of a wire feeder, an inert gas supply, a gas flow meter, and a welding gun. The welding operation itself is a continuous process once it is started. An inert gas shields the liquid metal from the oxygen in the air.

Workpiece handling must be fast, accurate, smooth, and reliable if productivity improvements are to be achieved in modern industry. Manufacturing efficiency increases in direct relationship to the controlled work flow rates and precise part positioning. These essential factors allow plant processing equipment to be operated at full capacity for prolonged periods, with downtime limited to scheduled tool changes and maintenance periods.

Loading and unloading is the second main field of application for robots.

These machines do jobs humans do not care to do since they may be in locations less than favorable for human habitation. The use of robots reduces the personnel injury rate and increases productivity. Excessive heat and noise, lack of light, dirt, and pollution do not bother the robots as they would humans. Using robots can lower production costs since safety equipment does not have to be purchased and maintained.

The robot's wrist is the usual location for a spray painting gun. A human operator takes the robot arm through the motions of spray painting. Usually a continuous-path robot with fast motion control is used for this task. Spray painting by robots has several advantages. One advantage is the withdrawal of the human operator from the fumes, making it possible to use different types of paints as many of today's coatings are toxic. Another advantage is that the robots can spray paint in very thin coats with high accuracy. Areas that could not be reached by human painters can be reached by robot arms with up to nine axes.

1.12 ASSEMBLY

Industrially produced, finished products consist mainly of several individual parts, which have mostly been manufactured at different times and in separate locations. Assembly thereby presents a cross section of the problems within the whole of production engineering (Schraft, 1991).

Robots have been used in assembling calculators, watches, printed circuit boards, electric motors, and alternators for automobiles. The future will see robots assembling everything that can be broken down into simple step-by-step operations, and that includes almost everything.

More and more attention is being paid to the possibilities of using robots for the tedious jobs of assembling small and large parts. One of the problems with converting a product to automated assembly is its design. Some products are not designed to be assembled by machines. Designers must take a close look at each product and consider the ability of the machines that do the assembly work. By studying both the design requirements of a product and the ability of the robot or other machines, it is possible to design a device of better quality and have it assembled by a machine. Robot assembly relieves workers from performing boring, repetitive tasks day after day.

One of the most outstanding assembly automation projects in Germany during the 1980s was the automated assembly of the VW Golf plant, where 25 percent of the final assembly work is carried out automatically, not only by robots but also by a number of special assembly machines designed for a specific purpose. The tasks completed by robots include the automated insertion, tensioning, and tightening of the V-belt, the insertion of the spare wheel, and insertion and bolting of the battery (Fig. 1.23).

The entire vehicle concept had to be modified to enable these assembly operations: Fitting of the front is the last assembly task, with the front end

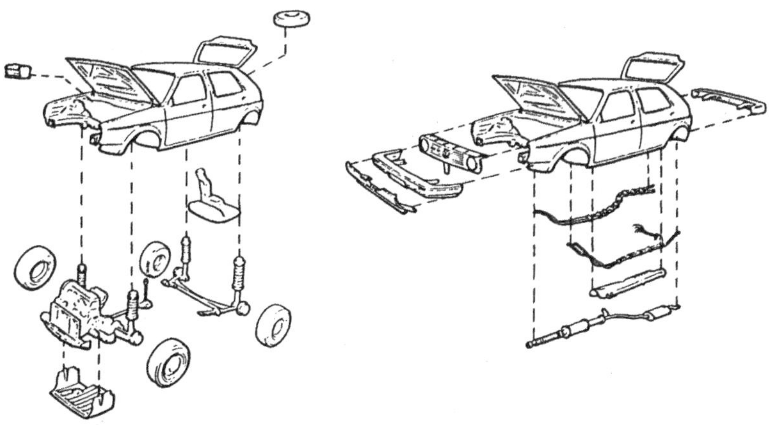

Fig. 1.23. Automated car assembly (Courtesy of Volkswagen).

staying open until all parts in the engine compartment have been mounted. This last operation is carried out by a linear joining movement, the entire front section being preassembled, including adjustment of the headlamps. This assembly system has been running for almost 10 years. It is state of the art and has proven its reliability. Today, new assembly systems with industrial robots are in use for windscreen assembly, wheel assembly, dashboard assembly, and many of the subassembly tasks in the automotive industry.

In the last few years a trend toward flexible systems capable of guaranteeing economical production has developed in assembly. However, assembly has not yet reached the level of automation achieved in flexible production (especially chip-forming processes).

1.13 MATERIAL FLOW

Material flow systems are the natural links between the stations a workpiece has to pass through in manufacturing. Therefore, they are a central element within a factory. Potential for improvements is very high, since logistic costs reach a significant share of a product's value.

Actual development focuses on the integration aspect. This is of major importance for all material flow systems, for they often have to cope with interface problems. Manufacturers of equipment have become system suppliers. They offer solutions that include all hardware and software elements. Naturally, standardization reduces the expenses for development. However, there will be individual solutions adapted to the various tasks of transport.

In general, material flow systems are a complex subject. In small-batch production their tasks change continuously, whereas in mass production the interaction of system elements can hardly be predicted. Therefore, simulation

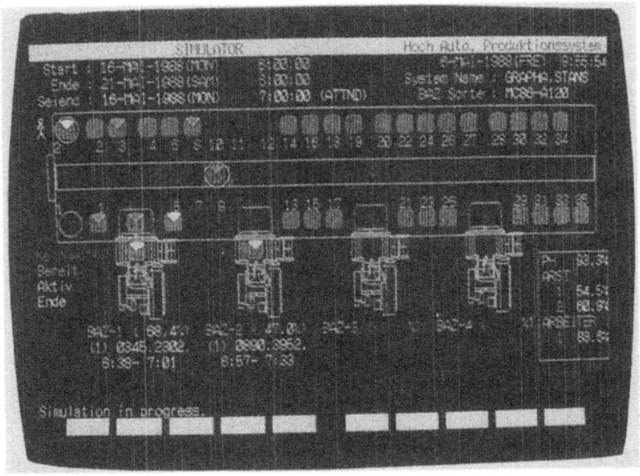

Fig. 1.24. Simulation of material flow within an FMS.

has become a powerful planning tool. The dynamic behavior of the whole system can then be tested under various conditions. In this way the best configuration can be found (Fig. 1.24). This approach avoids cost-intensive adjustments after installation.

Advanced transport on the shop floor is dominated by AGVs. Most of the installations are guided by wires in the floor. Flexibility of these applications is limited. To enhance the variety of tasks, latest developments have aimed at new guiding technologies. Some of the vehicles are nowadays able to leave the wire path temporarily, for example, for docking at a pallet station. In this way installation expenses are reduced. Basic research goes far beyond, however. Autonomously guided vehicles navigate within the factory on their own. They are equipped with sensors and a special control software to find the ideal route. Changes of factory layout have only to be recorded to the software. No physical installations are necessary any more. In addition, the vehicle is capable of reacting to temporary obstacles that are detected by the sensors. In this case another route is chosen by the AGV in order to arrive at its destination.

Communication with superior control systems is carried out wireless: The AGV in Fig. 1.25 is equipped with an infrared communications interface.

When the frequency of transport exceeds a certain level (depending on the individual conditions), discrete transport systems are no longer applicable. Continuously working transport systems usually demand for a high investment. They are designed for one special task, and changes are very cost-intensive. Mass production usually justifies this investment.

From this fact we learn that future manufacturing in general will be characterized by a huge variety of solutions. The main task in factory management

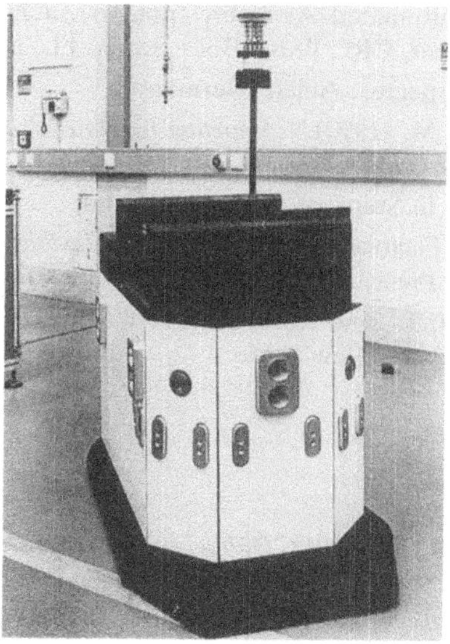

Fig. 1.25. Sensor-guided AGV.

is to set up structures in which homogeneous manufacturing systems can be realized.

REFERENCES

ELVA (1989). Prospectus, Aulnay (France).

EPB GmbH (1991). Prospectus, Achern (Germany).

Ford Motor Company and McKinsey & Co., Inc. (1988). Case study (not published).

Gildemeister AG (1991). Prospectus, Bielefeld (Germany).

Heller GmbH (1989). Prospectus, Nürtingen (Germany).

Hoechst AG (1990). Prospectus, Frankfurt (Germany).

Jaikumar, R. (1986). Postindustrial manufacturing, *Harvard Business Review*, **6**, 69–76.

J+E Link GmbH (1990). Prospectus, Stuttgart (Germany).

Kaempf, R. (1985). Concept of Manufacturing Control for a Highly Automated Factory. Proceedings of the 8th Conference on Production Research, Stuttgart.

Kersten, G. (1991). Steuerung und Unterstützung von Produkt- und Prozeßentwicklung durch Methoden per präventiven Qualitätssicherung, VDI-Z 133 (1991) Special IV, pp. 20–33.

Maho AG (1991). Prospectus, Pfronten (Germany).

Porter, M. E. (1985). *Competitive Advantage*, Free Press, New York.

Schenck AG (1989). Prospectus, Darmstadt (Germany).

Schraft, R.-D (1991). Automated Assembly Systems, in *Factory Automation and Information Management*, CRC Press, Boca Raton, FL, pp. 20–32.

SHW GmbH (1985). Prospectus, Aalen (Germany).

Stalk, G., and Hout, Th. M. (1990). *Competing in Time*, Free Press, New York.

Warnecke, H.J., Ed. (1988). *Flexible Manufacturing Systems*, 20th IPA Working Conference 13/14.9.88 in Stuttgart, Springer, Berlin.

Warnecke, H.J. (1991). Factory of the Future, in *Flexible Manufacturing Systems Handbook*, Academic Press, San Diego.

Womack, J. P., Jones, D. T., and Roos, D. (1990). *The Machine That Changed the World*, Rawson Associates, New York.

BIBLIOGRAPHY

Beitz, W., and Küttner, K.-H. (1991). *Dubbel—Taschenbuch für den Maschinenbau*, Springer, Berlin.

Eidenmüller, B. (1989). *Die Produktion als Wettbewerbsfaktor*, Köln, TÜV Rheinland.

Krallmann, H. (1990). *CIM—Expertenwissen für die Praxis*, Oldenbourg-Verlag, München.

Schraft, R.-D., and Warnecke, H. J. (1989). *Industrieroboter-Katalog '89*, Krausskopf, Mainz.

Ulich, E. (1989). *Arbeitsform mit Zukunft: ganzheitlich flexibel statt arbeitsteilig*, Lang, Bern.

Warnecke, H.J. (1988). Grundlegende Gesetzmäßigkeiten in der Produktion, in: *Vorträge Fertigungstechnisches Kolloquium Stuttgart FTK'88*, Springer, Berlin.

Warnecke, H.J. (1993). *The Fractal Company—A Revolution in Corporate Culture*, Springer, Berlin.

Wiendahl, H.-P. (1987). *Belastungsorientierte Fertigungssteuerung*, Hanser, München.

CHAPTER 2

EVOLUTION AND DIFFUSION OF ADVANCED MANUFACTURING SYSTEMS

J. RANTA
Technical Research Centre of Finland
Laboratory of Electrical and Automation Engineering
Espoo, Finland

2.1 INTRODUCTION AND BACKGROUND

The extensive use of information technologies in manufacturing has been one of the driving forces in the manufacturing change. This change has brought a new category of advance manufacturing systems, in which the systems aspects of the manufacturing processes and the integration functions are the essential features. Many new acronyms used to describe the new aspects reflect this systems aspect: JIT (just-in-time), CIM (computer-integrated manufacturing), FMS (flexible manufacturing systems). This development has resulted in a quantum leap from a functional workshop to a flow-based, cellular manufacturing.

It has not been only a technological change, but to a great extent also an operational and organizational change, where human factors, skills, and managing the change play as important a role as the technology itself. We can

Organization and Management of Advanced Manufacturing. Edited by Waldemar Karwowski and Gavriel Salvendy.
ISBN 0-471-55508-8

put this more directly: In the key role is not the technology per se (it can be bought off the selves) but the organizational knowledge and intelligence, which defines how the systems are used and how the manufacturing systems and processes are and will be developed.

At the same time, although there has been a technology push, the recent development has been also a reply to a new demand and to a new competitive environment. The customizing of products, the search for flexibility, and the competition in time have also demanded to search for new ways to operate and new technological and organizational solutions.

Therefore, the evolution and diffusion of advanced manufacturing systems is also a multidimensional process: We have to understand also the institutional framework in which the systems are used and to understand also the gained relative benefits in different application areas. Also the technological constraints are of different importance in different industrial sectors, and this partially explains different diffusion patterns. We can also claim that a specific institutional setting may be more suitable for advanced manufacturing systems than some others, and this may explain the relative diffusion pattern in different countries: see Krafcik (1988) and Maly (1988, 1989).

In many discussions the industrial robot is usually regarded as a symbol of the automated production, advanced manufacturing systems, and the structural change taking place in manufacturing, especially in the metal product industries. This is understandable because the robot has been a visible part of the manufacturing change, and there are a lot of mature, well-established applications, such as welding, painting, pick-and-place, and surface finishing. Moreover, these more or less stand-alone applications were used first in labor-intensive sectors and were replacing human labor, which raised a public worry about the impact of robot applications. Robots were a starting point toward more automated production.

Today, however, it is misleading to say that the robot density represents the degree of automation and the mature stand-alone solutions would be a cornerstone of modern automation. Robots are now used extensively as part of automated manufacturing systems or in special applications utilizing information processing and intelligent signal processing. These new type of applications are expanding the possibilities of the classical stand-alone applications. At the same time the impact assessment is becoming more difficult because the benefits and costs are no longer direct; the critical part of the benefits and costs are indirect and qualitative.

Therefore, when assessing the economic and other impacts of advanced manufacturing systems, we have to look beyond the stand-alone applications of robots and focus more on the systems applications, such as the FMS or flexible assembly systems (FAS). It is also obvious that we must look at the broader system concepts: the integration of several functions, such as design, planning, and manufacturing.

It is quite often argued that the major source of benefits is from the labor savings. Also, labor reduction is quite often the key goal in applying production automation. This is true in the case of the simple stand-alone applications of industrial robots. In the case of manufacturing systems and intelligent robots this is not the case. There are surprising elements in the cost factors and the benefits can form complex and indirect dependences, making the assessment difficult. Also there are usually many unintended impacts. This leads us to argue that the implementation issues and practices are critical factors behind success. In fact there is much practical evidence resulting from the many case studies (see Jaikumar, 1986; Meredith, 1987a,b,c; Ranta 1989b, 1990) that proves that the most critical issues of the application of advanced production automation technologies are managerial and organizational aspects of the systems implementation.

However, the situation as a whole is more difficult because the basic technologies are not yet invariant and not at a mature stage of development. Thus it is reasonable to expect that technological factors can still considerably contribute to the performance and economic efficiency of different automation and production technologies. We can also expect that many systems applications are simply dependent on the economic efficiency and capital intensity of different manufacturing technologies. Thus many of the stated goals and targets of production automation, integration systems, and especially FMS technologies—such as flexibility, accuracy, processing speed, complexity of products and parts—are mainly dependent on the trade-off between the technological capabilities and the costs of technological solutions.

One of the main technological driving forces, as already mentioned, has been the application of information technologies, that is, mainly basic electronics, computer technologies, and communication technologies, in *both* the products and the production of manufacturing industries. We can expect that this process will still be intensified when the possibilities of information technologies are fully utilized. This also means that the key carrier branches of technological change will be the manufacturing equipment industries, which utilize the dynamics of technological change both in their products and in their production and which have *forward linkages* with other manufacturing industries providing them with systems and advanced tools. The motive branches of change will be electronics, the computer, communications, and electric machinery industries, which provide tools and means for the manufacturing equipment industry. Thus the manufacturing equipment industry has *backward linkages* with information technologies and related industries. Therefore we can conclude that the evolution of advanced manufacturing systems depends very much on information technologies. Also the diffusion of the systems depend indirectly on information technology through the cost efficiency ratio of the new solutions.

This chapter analyzes possible forms of the future applications of production automation and makes an assessment of different carrier technologies, that is, electronics and information technologies and their impacts on future manufacturing technologies. Empirical data on the costs, benefits, and success factors of systems applications are then presented and some implementation strategies are suggested. However, first, let us discuss what are the driving forces, why advanced production automation is applied, and what are the means to realize the stated goal in practice. Throughout our discussion we will use the acronym CIM (computer-integrated manufacturing) as a symbol for the trends toward the integration of different manufacturing functions.

The key part of the empirical data is based on the findings and results of the IIASA CIM project, which has a 4-year impact assessment project of the CIM technologies during the years 1986–1990. Among other things the project produced cost benefit analyses of 900 FMSs worldwide. The key results are reported elsewhere in more detail (Ayres et al., 1991; Maly, 1988, 1989; Ranta et al., 1989; Ranta and Tchijov, 1990; Ranta et al., 1992; Tani, 1987a,b,c,; Tchijov et al., 1987–1989a,b,c).

2.2 THE CHANGING ENVIRONMENT AND A NEW WAY TO COMPETE

It has been already stated that applying advanced information technologies in production has been and will be the major technical driving force in manufacturing change. The information technologies give new possibilities to produce a variety of products in an efficient way. Modern manufacturing is achieving the flexibility of classical production, but on the productivity level, which is far ahead of the classical production and even of the mass production principles. Also it is evident that information technologies aid in producing superior quality compared to the old organizational principles, and the new production systems are much smaller in scale in terms of machines and personnel than the mass production (Clark et al., 1990; Jaikumar, 1988).

Why is the flexibility needed now? In general, we can state that there has always been a need for flexibility and customizing of products and to have special niche products. However, this was not possible previously in a cost-efficient way because of technological constraints. So we can say the production was stiff and the customers were flexible. Because of the development of information technologies, combining efficiency and flexibility is becoming possible. The customers can start to be inflexible and the production has to be flexible and adaptive.

Thus we can state that the industrial change is now mainly technology driven. The information technologies provide new options to look for competitive advantages and to take off from the classical mass production in a competitive and beneficial way. This means that in order to benefit from information technologies, companies have actively to look for new options. Previously it was necessary to provide a single product by specialized machinery and skills to many global markets. Now it is necessary to make many products by flexible machinery and multiskilled personnel to specific, segmented markets (see Krasa-Sethi et al., 1988; Yamashina et al., 1991).

What all this indicates, however, is that parallel to the technology push there are changes in the business environment. These changes then necessitate a shift to more flexible operation principles. To some extent, we can say there is a social need for flexibility. Thus it is not only a question of the technology push, but the new production technologies and operation principles are reacting with the markets in an interactive way responding to the new needs and at the same time creating new needs. We can describe the changing environment in the following way (see National Research Council, 1991).

There is the *time dimension* of the new environment (see Johnston and Lawrence, 1988; Stalk, 1988). First of all, because of rapid technological development, the product life cycles have become shorter. This is especially true in consumer electronics, the automobile industry, and consumer durables. Also in the fashion industries there is a need for continuous and rapid changes of styles and products. Thus, in general, the product renewal rate has become higher. To be efficient and competitive in this environment companies need flexible design and planning systems and flexible production to introduce rapidly new products and designs into production and to the markets without expensive technical changes in the production.

Parallel to the higher renewal rates and shorter product life cycles the total delivery time has become an important competitive factor. This is closely related to the tendency of increased customizing of products and to just-in-time production principles. The competition with total delivery times calls for short offering time, short throughput time in production, and quick logistics and distribution. To cope with these demands the information technologies are needed to provide flexibility in customizing and designing of products, in production planning and control, in manufacturing, and also in the supervising of logistics activities. Usually the reduction of time is also connected to capital savings in the form of inventories and work in progress.

The second dimension in the new environment is *product variety*. The tendency of the increased product variety is related to the customizing of products and producing of special products to special market segments. This demand exists not only because in capital goods production providing application-specific machines and devices can be more cost efficient to the final user but also because customizing consumer goods production fulfills individual needs better. Also the increased income level in the Western industrialized countries is creating more demand for customized products because consumers are ready to pay for customized products (Gupta and Buzacott, 1990; Lei et al., 1990).

Furthermore, the customizing of products and the increased product variety are creating more design actions, more complex production planning and control, thus increasing part variety in manufacturing and more complex and difficult logistics. To cope with these problems information technologies are needed in planning and production to provide both efficiency and flexibility. Note, however, that customizing products is not always a reaction to the changed competitive environment. The flexibility can be used also in an active

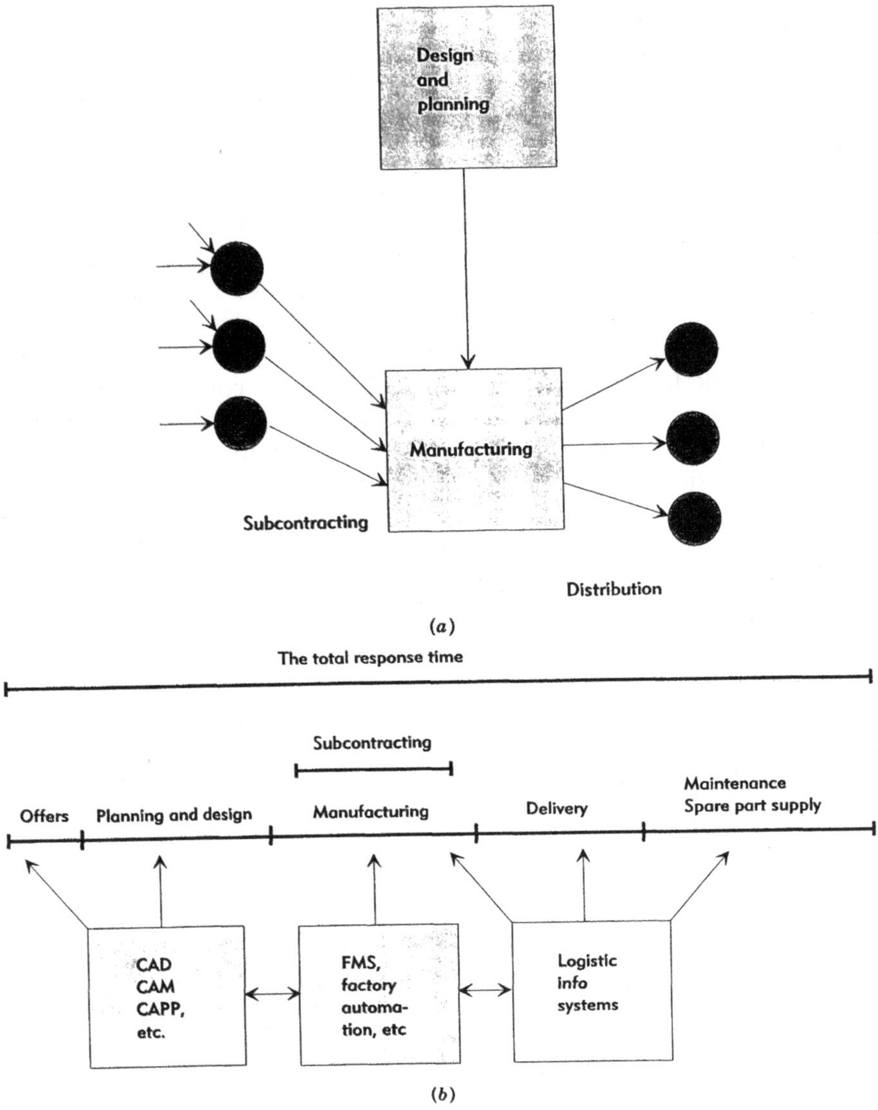

Fig. 2.1. Technical tools of flexibility.

way to create new market options and new competitive advantages. Figure 2.1 demonstrates this.

The cornerstone of operational flexibility in manufacturing is the flexible manufacturing system (FMS). It is usually a necessary step in achieving the basic forms of flexibility. However, it not the only step. In terms of time, the design, planning, and logistic information systems are more critical and usually more benefits can be created from the planning and logistic systems. Even in the case of operational flexibility, the planning (production planning and control) systems are as important as FMS. Also the subcontracting system has to be as flexible as the manufacturing. Therefore the logistic information systems play a critical role.

In the case of tactical flexibility, for example, coping with the high renewal rate of products, the design, planning, and logistic systems are usually the starting point along with critical technical elements. However, to manage the product changes in an efficient way in the production, flexible manufacturing is needed.

To manage strategic flexibility or to cope with the long-term market and product changes, it is necessary to build up a flexible subcontracting and collaborative network. This usually can react rapidly to completely new demands. Then the logistics systems play a critical role.

Apart from technological innovations new organizational solutions are also needed. These are needed to cope with the increasing integration of the production systems, to guarantee high utilization rate of the systems, to cope with quality requests, and to cope with continuous changes.

Now we can see what CIM means in practice. We can define it as a method or an approach to integrate different operations and functions, like planning, product design, production planning and control, manufacturing control, and so forth, by the means of information technologies. Already from our discussion it is clear that in practice there will be, and needs to be, different approaches toward the implementation of CIM technologies depending on the environment. Also the goals and intended benefits as well as realized benefits are different in different business environments.

Figure 2.2 explains this situation. Instead of the classical volume–product variety axes, the product renewal rate–product variety axes are used as key variables. The production volume or capacity is expressed as isocapacity curves, isoquants in the figure. For instance, it means that there are different areas and environments for an economic use of flexible manufacturing systems. It also means that the strategy toward a CIM implementation should be different in different environments. The four corners present four different extremes.

The high-capacity corner presents the area where an efficient and cost-effective manufacturing system is an essential starting point. In practice, this still means specialized and fixed systems and can be a typical area of the stand-alone robot applications.

In the left upper corner the high product renewal rate is a characteristic feature of the environment. We can say that this environment is still technology driven. To cope with the changing products, design and planning systems are needed as well as flexible manufacturing. It is also in this environment that high-capacity FMSs and wide applications of logistic systems are found. The industrial robots are typically used in assembly systems.

The right lower corner shows customized products and special market niches. An essential starting point is design, planning, and logistic systems.

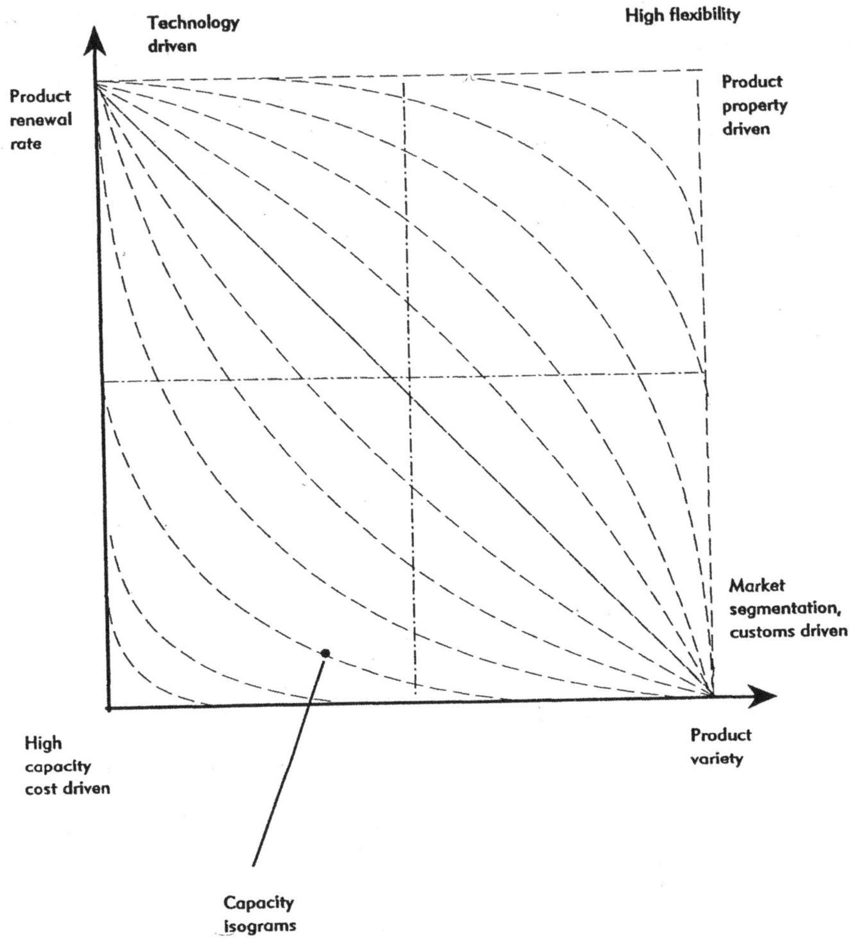

Fig. 2.2. Interaction of product variety and product renewal rate.

Flexible manufacturing systems are used to cope with an increasing demand and with a need to expand capacity.

The right upper corner is the area of a semiproduct or subassembly vendor or supplier. There can be many and rapidly changing products in production. Also here design, planning, and logistic systems are essential, and FMSs are used to cope with the need to increase capacity. The robots are typically applied as part of the systems or in specific stand-alone applications.

What has been said shows that the goals for production automation are not simple, but in practice today manufacturing automation is applied in a multicriteria environment. Also the automation systems are not only used to reduce costs but to cope with changing environment and to create new com-

	Technology base	Essential knowledge base
Planning methods	Applications in different industries	Customized needs, application know-how, project management methods, organization design, impact analysis
Production planning and control CAD/CAM CAPP	Planning and engineering Project deliveries	
	Systems Flexible manufacturing units (FMU), flexible manufacturing cells (FMC) and systems (FMS) Factory automation Software systems Production control systems	System engineering Manufacturing engineering Software and computer technology Information technology in general (especially communication)
	Machine automation NC, robots, automatic storages, automatic vehicles, etc. Special production machines	Mechanical engineering Electronics and software technology "Mechatronics" Control engineering Physics
	Production interfaces Sensors, transmitters, servo-mechanisms, switching devices, etc. Special devices	Mechanical engineering Electronics Special methods: signal processing, pattern recognition, image processing

Fig. 2.3. The systems view of production automation.

petitive advantages and options. This analysis also underlines that robots have to be studied as part of systems applications.

2.3 TOWARD INTEGRATED MANUFACTURING SYSTEMS: COMPONENTS, SYSTEMS, AND TRENDS

Production Automation—as a Whole and in Part

No attempts will be made to give a complete definition of computer-integrated manufacturing and all aspects of modern production automation. In the previous section CIM was defined as a process toward the integration of different systems. An illustrative definition is presented in Fig. 2.3. Accordingly, CIM or integrated production automation systems are integrating different technologies as well as different fields of engineering and knowledge.

We can differentiate four basic levels. The heart of production is, of course, the machine level. The machine automation, especially the numerically controlled (NC) machine tool has been one of the starting points of manufacturing automation and one of the first application of electronics and information technologies in manufacturing. The most typical examples of the machine automation are the applications of information technology in computer numerically controlled (CNC) machines, robots, automatic storage, transportation devices, and special machines in different industrial sectors, such as cutting machines in the clothing industry. The automatic machines have two kinds of interfaces: one downward interface toward production with sensors, transmitting devices, servomechanisms, and switching devices and one upward interface toward systems level and toward the planning process with different communication systems. For example, a modern NC tool or robot can have its own microprocessor control for each axis of the motion, custom-designed semiconductors in servomechanisms, as well as sensing and information processing devices to take care of the downward and upward interfaces and the all-over control of the machines. These are real mechatronic products, combining mechanical engineering, electrical engineering, and electronics.

The systems applications are the second level, which integrate single machines into autonomous manufacturing or production systems with the help of computers and communication technology and material handling and transportation devices. There are flexible manufacturing cells (FMC), their combination in the form of flexible manufacturing systems (FMS) or even an integration of several FMSs in the form of a factory-scale system. For such an integration, systems or single machines have to be able exchange data and information. Therefore a combination of communication technology with computer technology is important—not only for the integration but also for the upward interfaces to different planning systems. Another crucial issue is the mechanical integration with the help of transportation devices and storage systems.

The systems concept is at its emerging stage because there are no ready-made or standard concepts for the software and the communication parts of the system. Therefore software engineering is one of the key issues on the systems level. Software engineering is at the moment the most important cost driver and also a technical constraint in the evolution and diffusion process of advanced manufacturing systems. It is not wrong to say that the future development of the systems—both the economics and technical performance—depend first on the software and second on the mechanical integration, like pallets, transportation, grippers, and fixtures. These two technical factors broadly define the technological evolution of the systems in the future.

The third level consists of different planning and design systems, including computer-aided design (CAD), computer-aided manufacturing (CAM), and computer-aided process planning (CAPP), as well as production planning and control. The planning level has been the second starting point for information technology applications in manufacturing, and it has been far ahead of the systems level. For example, different CAD and production control systems are well established and intensive users of computers. However, there have been problems with the integration of different planning technologies, such as CAD, CAM, and CAPP, with the manufacturing systems in order to produce directly from the design databases the required tooling steps, routing instructions, and NC programs. The same is true for the downward interfaces with the manufacturing systems level. Thus, considering the whole CIM concept, one of the key questions is again software engineering and communication technology and the integration of design and planning with manufacturing. On top of the planning systems we have to add logistic systems, which will have an important role in integrating distribution and subcontracting networks with planning and manufacturing.

The production control systems are also faced with new problems when the control of flexible tooling and assembly systems are considered. The flexible manufacturing systems themselves can be regarded analogous to the time-sharing systems: The router has to decide how to route a workpiece (or in fact parallel several workpieces) through the system; in other words this means that we allocate resources (cutting tools, machining time) to the tasks (workpieces). As a real-time task, taking into account that we might have 100 different parts, 5 to 15 machining centres, 100 to 200 cutting tools to be routed or allocated, the scheduling and routing can be extremely difficult. Also managing disturbances and re-routing work lines and the task allocation should be an elementary part of the production control, and this increases the complexity of the real-time control.

This control task and structure is repeated on the factory level. Instead of cells and workstations the production control systems have to coordinate and schedule several flexible manufacturing systems to form task priorities and to allocate the resources. The problem is the same as for flexible system control. We can expand this concept to supplier network control, which is the uppermost level of the flexible production control.

Again it is easy to understand why software plays such a critical role: It is really a cost factor, availability factor, and even one of the critical constraints for the development of the systems. We can speculate that the trend is toward two kinds of total control structures. On one hand the system can be hierarchically organized, autonomous, and self-configurable, where the upper level sees the lower level only as a set of resources that have to be allocated to the tasks and the lower level performs the tasks by the optimal structure (see also National Research Council, 1991; Ranta 1989b).

On the other hand it is also possible to have a flexible network structure in which the control is performed in the form of capacity auction or the capacity itself will be a commodity. Thus within the system there will be cooperation and competition at the same time.

Carriers of the Evolution: Development Factors and Linkages

In order to forecast the future applications of technologies, it is essential to know the routes of diffusion and the ways to utilize modern technologies. We call these routes forward and backward linkages. The forward linkages present practical needs, or the market pull, and the, of course, the backward linkages can be regarded as the technological push.

The first-order carrier branches, or the key applier of the motive technologies, consist of the machine tool industry, the robot industry, and related industries. The development of these branches has been much impacted by information technologies; they also provide components for systems integration, which presents the second part of the first-order carrier branches. The systems integration, as stated, is mainly based on extensive applications of computers and communication technologies. Important special issues are distributed processing and databases as well as local area networks. Therefore, software engineering also plays a crucial role in this context. On the machine level the driving forces are, apart form computer technologies, the integration of mechanical engineering and electronics, and even of the basic component technologies, into the machine design. Also electrical engineering plays a special role because of its servos and drives.

The main application areas are called the second-order carrier branches because they are, apart from the machine tool industry itself, the main users and appliers of the systems and the basic machine technologies. So far the main branches have been general machinery, the automobile and car industry, and other transportation equipment industry, as well as the electronic and electric product industry or the component suppliers to these industries. The majority of the present advanced manufacturing systems can be found within these industries. These branches will also be the main appliers in the future and they will also demonstrate the future solutions and give the references for the other industries.

However, depending on technological capabilities and economic efficiency, we can expect the diffusion of compact systems also to extend to small and medium-scale enterprises as well as to the so-called marginal branches, like wood product industries. There are furthermore some special branches, like the clothing industry, which are extensive users of information technologies in the form of modern production and planning technologies, but they are somewhat outside the mainstream of the development because of special requirements and special machinery. The reasons might be difficulties in material handling, possibly necessitating special types of machines, or other technology-related reasons.

With regard to the backward linkages it is safe to expect that the manufacturing industries will not be the key industries influencing the future trends in the semiconductor and computer industries. This influence does, to some extent, exist. But in general we can regard the semiconductor and computer industries from the manufacturing viewpoint as an autonomous force, which will proceed in the near future according to its own laws or which will partly be driven by the space, aircraft, telecommunication, and military industries. Thus we can analyze the impacts resulting from the backward linkages to be a real technological push.

As mentioned before, the key element of change is the machine tool and the related manufacturing systems industries. It has so far been the technological driving force, as it has provided the basic components for systems integration and as it has been the first user of FMS in its own production. In that way it has been the key element of the systems integration and has been moving from a component supplier and a systems applier to a systems vendor.

Another important path to systems integration has been through electrical components and special controls. Usually this industry has provided special components and controls to the machine tool industry or has even produced robots. It has also been an early user of the systems in its own production. It is quite a logical step to move from systems control to integration and related software products. The third main path has been from the computer and software industry to manufacturing software integration. This is, presently, the growing group, and there is a special need or a special market niche for that kind of activity because the systems and related software products are still mainly customized products due to a lack of common standards, and so far there is no common control structure for the systems and for their architectures.

It has been and still is common that systems integration is carried out as an in-house activity directly by the final user of the system in the application area. This is to a great extent due the customizing of the systems. At the present time there is also a common phenomenon that systems are supplied by a consortium of companies, which can be even international. Such forms of cooperation usually consist of a systems integrator and several component suppliers, where the systems integrator has the main responsibility. Such consortia seem to be quite sustainable forms of collaboration.

Motive Technologies: Electronics and Information Technology

As pointed out several times, one of the basic technological driving forces in flexible manufacturing systems and in production automation has been the possibilities created by electronics and computers. It is safe to expect that this trend will also continue in the near future. Already today electronics and software systems are a critical part of the systems on all levels, that is, from sensors and servos to systems control and architecture. There are many technical performance indexes such as speed of tooling, accuracy of the motion, accuracy of measurements, and communication possibilities of system parts that can still be improved by the application of basic technologies and will be impacted by the fast development of the motive technologies.

It is reasonable to expect that this fast development and the existing basic trends of information technology will continue, so that we will have even more efficient, more dense, and cheaper circuits and processors in the future. This development has several advantages with regard to manufacturing automation. The small size improves the possibilities of combining mechanical and electronic design—or to make mechatronic devices; the higher density decreases mechanical connections, making systems more reliable and suitable for the so-called embedded systems; and the decreased power consumption also makes systems more reliable and makes it easier to develop embedded systems, which are important issues in industrial applications. Finally, the decreasing price, of course, makes new cost-effective applications available.

It is, of course, not only these quantitative numbers that matter but the development will also include a lot of qualitative aspects, which are more essential for the industrial automation. These aspects will be emphasized below. It seems to be safe to forecast that the following trends will take place (Gelsinger et al., 1989; Perry, 1989; Ranta, 1989a,b; Ranta, 1991):

- The current trends of miniaturization will continue in the next 10 to 15 years, which will lead to an improvement of the component density by a factor of 20 to 40.
- This is made possible both by the decreasing line thickness and the lower operating voltage of circuits.
- The main transistor technology in use will be the so-called field-effect transistors, most probably the CMOS technology, providing a gate delay on the order of 0.2 ns and a chip density on the order of 16 to 20 million devices per chip, offering 256-Mbit memory chips and providing an operating speed of 30 to 60 MIPS in microprocessors or a fully single-chip computer in the speed range of 1 to 3 MIPS.
- Optical communication and optoelectronics circuits in general will be on the order of 10 times cheaper than today and most likely in wide use.
- Advanced architectures will proceed, and it is most likely that there will be commercially feasible parallel processing computers as well as other special-type architectures for signal processing, graphics, logic programming, and so forth.
- Optoelectronics and optical communication will be in general use in the future.

It is obvious that all these trend will have a development push also on advanced manufacturing systems. Figures 2.4 and 2.5 present the application possibilities of different motive technologies.

Development Tendencies of the Components of Integrated and Automated Manufacturing Systems

NC Machines and Machine Automation Numerically controlled machines are key components of FMS and CIM for the part manufacturing of metal and wood products and a cornerstone for automated production. There have been a lot of changes in the market balance and the product segmentation in the NC markets.

Behind these changes there have also been changes in technological factors. Perhaps the two important factors are the incorporation of electronics and improved cost efficiency as well as the development of new types of compact machines for small-batch production.

Electronics and software engineering can be applied on many levels (see Fig. 2.6). Usually the control of axis motion and the basic tooling process are now based on electronically controlled servo systems. Most advanced systems

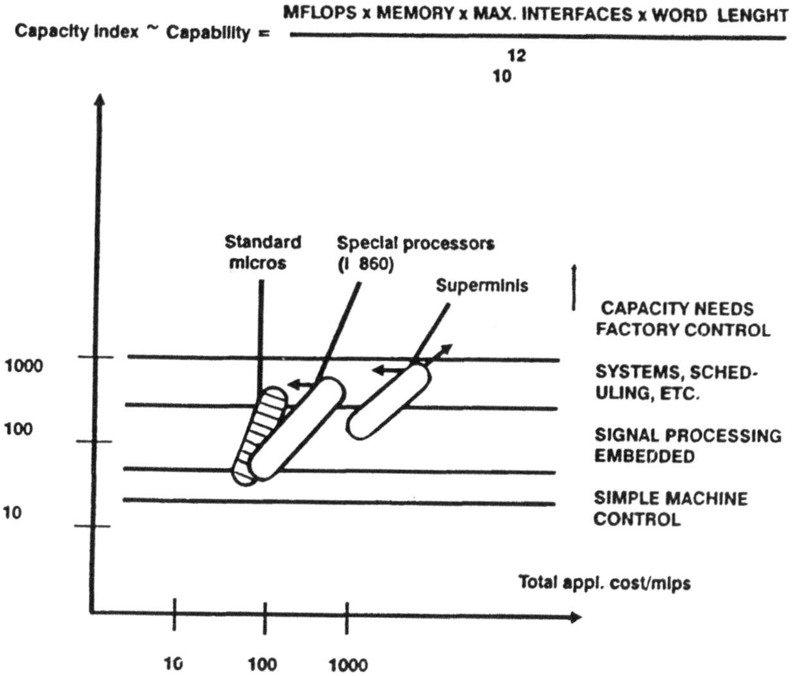

Fig. 2.4. Application of different computer categories.

already utilize custom-designed circuits, or so-called ASIC. The direct measurement of the tooling process has so far been a difficult task. The control is basically an open control or based on indirect measurement, such as power, torque, or stress. For diagnostic purposes advanced signal processing, like acoustic noise analysis or other stochastic analyses, can be utilized. The impact of electronics has been an increased efficiency without any major cost increase—or an improved cost–performance ratio (see also Wright and Bourne, 1988; Wright, 1991).

The NC machine controllers are nowadays usually realized by programmable controls or by a free programmable microprocessor system. So fair it has been a common habit to obtain the basic machine and numerical control from separate sources. In this case a separate interface is needed to connect the numerical controls and the servos. However, the trend is toward more integrated systems and, again, an improved cost–performance ratio. It is reasonable to expect that new software engineering tools and the foreseen development of processors will furthermore improve the cost–performance ratio.

There are many special tools and high-level languages to produce basic NC programs. However, usually a separate micro- or minicomputer system

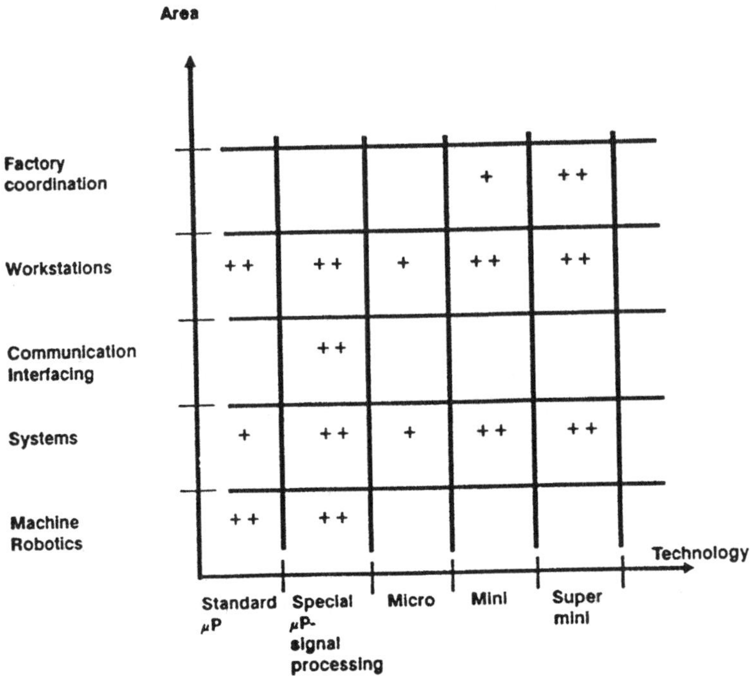

Fig. 2.5. Application areas of different computing technologies.

is used for this purpose. Then a separate compiling system (post-processor) is needed to interface the programming system and the numerical controller. The recent trends indicate, however, that integrated systems will be increasingly available, which consist of the programming environment, numerical controls, and all required interfaces. It is reasonable to expect this trend to become even stronger because of the advances in processing devices, particularly of special processors and workstations. This again will improve the cost–performance ratio and might also provide tools to facilitate flexibility (e.g., to produce new part programs).

All these changes have caused NC machines to become more diversified and markets to become segmented, according to capacity and basic tooling functions. Therefore there are also compact systems available for small-batch production.

However, at the same time there are other interesting trends and signs: Incorporation of electronics, software functions, and the flexibility of production processes *open a new potential* for special-purpose machines, which, with their high performance, satisfy the needs of a special application. After many years there is now a growing class of small manufacturers who have built up competitive power based on special segments and customized ma-

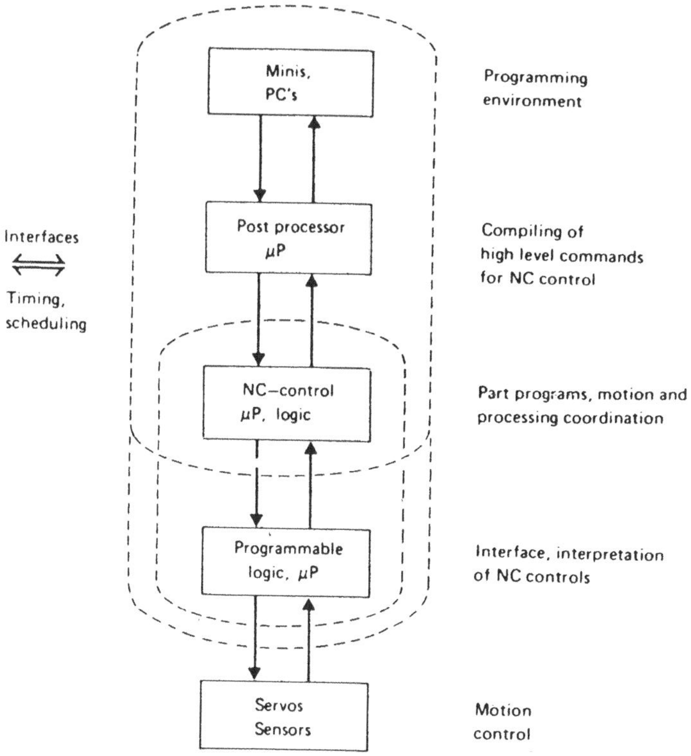

Fig. 2.6. The control hierarchy of NC machines.

chines. Moreover, we can expect that the previously described trends in software engineering will facilitate the customizing process and open new possibilities for product development.

In any case, the main problem still is to integrate basic tooling functions, turning, drilling and milling, into a universal machining center. There are weak signals of a slow progress in this respect. Some machining centers can already do elementary turning functions or have a horizontal axis, and, similarly, some turning centers have a vertical axis to make drilling operations. Electronic and software development will yield completely new prospects in this respect—it might help create real flexibility in a cost-efficient manner.

On the systems level important aspects of NC machines are new measurement technologies, fixtures and pallets, and tool and workpiece changers. The development in all these aspects means an increasing level of flexibility and increasing performance measures of FMSs. However, these mechanical auxiliary devices may be the key technological obstacles to creating operational flexibility, and they will remain to be key cost factors. There are no rapid changes foreseen in this area.

Note that the machine level provides the basic framework for flexibility. Software functions cannot help to provide any extra flexibility if the machines are hard and expensive to change and to set up. The development in this respect seems to be rather slow. The IIASA questionnaire study on the cost trends and major development problems in production automation indicates that many machine-level problems are key factors also on the systems level.

On the other hand the cost estimates can be regarded rather optimistically: We can expect multipurpose fixtures to exist in 10 years by the cost factor, which is only one tenth of the present costs, and also the machining speed will be 59 percent higher by the same cost than today. But in the trends and problems many machine-level problems like fixtures, measurements, and grippers are emphasized. This is an indication that mechanical integration and engineering problems are as important as software problems. It is also worth noting that the key problems and the cost trends indicated are relevant not only for integration problems but they are also constraints for intelligent stand-alone applications.

In terms of general trends we can conclude (see also Wright and Bourne, 1988):

- There will be a slow trend toward universal machining centers or multipurpose CNC machines capable of parallel tooling and of several standard tooling functions.
- Special-purpose, multifunctional machines will become cost effective, providing an efficiency increase in the areas where one special machine can replace many standard, single-purpose machines.
- Incorporating electronics and software engineering with the mechanical design will bring flexible and general-purpose fixtures into commercial use in a 10-year perspective, increasing the machine flexibility and decreasing the costs of flexibility.
- New programming environments and methods make it easier to implement new NC programs and improve the cost efficiency ratio of NC machines.
- All these trends mean that the efficiency and flexibility of the basic machinery will increase without a considerable cost increase.

One radical innovation that could change the whole picture is laser processing. If the technical reliability of lasers increases, they could become an effective means of increasing flexibility (milling, drilling, and turning by the same tool; no tool maintenance and drift; flexibility of software; applicability to different materials). This could be a real qualitative change from the viewpoints of both production and the NC machine business. In general we can claim that optics could, in a wide sense, be the technology that necessitates a new technological form in material processing and possibly also information processing.

Robotics Robot manufacturing shows some of the same tendencies as the NC machine industry: There are indications of a mature industry. The so-called standard robots are an area in which competition is highly cost-oriented. There have been considerable changes in the market balance: Many companies have given up and there are only a few strong manufacturers left. At the same time new possibilities and potentials have been created in specialized robotics for very narrow applications by adding specific technical properties (speed, accuracy, interfaces, signal processing, image processing) with the help of electronics and software engineering. Specialized robots have also opened possibilities for small manufacturers.

Robotics is also analogous to machine automation with control hierarchy. Figure 2.6 is valid only with appropriate terminological changes. Basically all that has been said regarding the control technology is also valid (see, e.g., Knasel, 1987; Wright and Bourne, 1988).

The effects of robots seem to have two major trends. First, there are many stand-alone applications of standard robots, such as point-and-arch welding, painting, and other surface finishing tasks. The diffusion of robots seems to depend mainly on the costs of standard robots compared to replaced labor (Tani, 1987a,b,c). Since their costs are decreasing and their efficiency is increasing, we can expect a steady diffusion of the standard applications.

The second main application trend is the use of robots as a part of manufacturing systems (flexible manufacturing unit [FMU], FMC, FMS, assembly systems, etc.). The technical features and the performance of robots are essential for these applications. The capabilities of the robots can have *a considerable influence on the flexibility* and the technoeconomic performance measures of FMU, FMC, and FMS and also of assembly cells and systems. In particular, reliability, accuracy, speed, flexible grippers, and intelligent interfaces (tactile sensors, vision, other signal processing) play an important role. New achievements in these areas always mean new possibilities on the systems level and also increase flexibility.

The advancement of sensors and grippers is especially critical for assembly systems. The flexibility of robots is critical both with respect to technological performance and cost efficiency of the systems. The second critical factor is speed and accuracy, which present slightly conflicting goals. On the systems level it is no longer the simple labor replacement that matters, but there are more complicated performance requirements. However, it can be expected that the development will be rather slow and real generic and flexible grippers and reliable and cheap tactile sensors will be commercially available only after 1995 at a reasonable price (10 percent of the robot price) (see Ranta 1991; Wright, 1988, 1991).

It has been commonly thought that the so-called intelligent robots will gradually substitute for the standard robots, such as playback and fixed-sequence robots. *Basically, this is not true*. The intelligent robots are more likely to expand robot applications to completely new fields and create new markets. But because this process has been slower than commonly forecasted—due to technological and economic reasons—there were too many optimistic expectations for the market growth. This is one of the origins of the problems of robot manufacturers.

The substitution process is more likely to happen between different generations of playback and fixed-sequence robots because their cost–performance ratio has also improved and will improve even further. This presents basically the same fact as stated earlier concerning stand-alone robots. From the economic and technical point of view there are no reasons to buy an expensive visual system or flexible grippers if a simple fixed-sequence robot with simple grippers can perform the task reliably enough; and vice versa, there are tasks that simple robots cannot perform. There has been a lot of disbelief in this respect. These tendencies also indicate that as a product robots are and will differentiate and markets are and will be segmented.

The following trends might materialize in a 10-year perspective:

- Grippers will no longer be technical and economic obstacles in assembly systems, and, because of the increased speed and accuracy of robots, the general efficiency of assembly systems will be improved.
- Robots will no longer be a practical limitation in tooling systems.
- A completely new set of applications, such as mobile robots, will be available and make autonomous and self-configurable systems available and possible.

Integration Technologies Software engineering or software systems play a key role in system integration. This has many performance consequences; in particular it has a strong impact on the reliability and availability of a system. On the other hand, the software systems can also improve the flexibility.

On the software side standards will play an important role. Today, customizing and integration of the systems are carried out with the help of software. Each system requires, in principle, basic software and communication software modules *that must be developed separately* for each system. This is nowadays a significant cost factor and is, in fact, also a major entry barrier for newcomers to the system integration business (FMS and CIM systems). It also increases the costs of the specific applications, and customizing is also a major source of software errors and software quality problems an thus the key factor of the system availability and reliability. The standard modules, both for communication and basic functions, which could be used in many applications, can considerably decrease the development costs of a system.

In general it is difficult to identify a life-cycle model for software products. But we can say that the *standardization* of the basic software modules corresponds to *the maturity stage* of the product life cycle. Standardization usually means a cost decrease and thus opens possibilities for newcomers and specialized systems.

Thus, if, for example, Manufacturing Automation Protocol (MAP) development is to succeed, we can expect that in addition to the standardization it will *open* possibilities or potentials for cost-effective means to realize specialized systems architectures, or in other words, to increase *the flexibility* in the systems development process itself (as will usually happen at the mature stage). This means that:

- The economic barriers for new entries will decrease.
- The special-purpose systems or the subsystems can be economically realized, which means the creation of new business segments and increased differentiation of the system products.
- The technology-oriented, specialized subsystems or systems (signal processing, image processing) can be economically realized as a part of the standard system; this again means new growth in market segments.
- In a longer perspective all these trends reflect on the applications of FMS: Economic barriers will slowly decrease, application areas will extend, and the concept of flexibility will broaden with respect to part families (shapes and sizes).
- In longer term we can expect a totally new concept for the timing and routing, that is, the scheduling of the flexible production and flexible part supplier network. This opens completely new aspects like a network-type system or autonomous and self-configurable systems.

The basic elements of the integration are manufacturing systems, such as FMC and FMS. As already indicated, software cannot help very much on the FMS level. The flexibility as well as the cost–performance ratio depend also on the fixtures and tool changers, apart from the software as already indicated. However, the advancement of the factory level is completely dependent on software and communication issues. Even more critical is the software for the integration of CAD/CAM with the manufacturing level. There is still a long way to go for a complete integration.

An optimistic forecast could be the following for the 10-year perspective:

- There exists a common standard for the systems architecture (communication, interfacing, etc.) that makes it possible to develop modular-type automation systems for manufacturing (FMS) control and a new type of production organization and control principles.
- This modular approach makes it possible to decrease software costs of manufacturing systems, to extend systems, to develop systems, and to decrease the risk of software errors and software quality.
- Design databases are integrated into manufacturing systems and the first, low-cost commercial transformation programs are available to generate NC programs from the design databases.

On the manufacturing level mechanical integration is also critical. Transportation and storage systems are significant cost factors, and they are, together with related interfaces, important performance factors, too. Now the changes of the systems architecture tend to cause discontinuities in the cost–performance ratio. If a modular-type system, which is easy to expand as system requirements change, will be commercially available, it might have a great impact on the systems level. Now many systems are based on compact-type, standard architectures—especially at the low-cost end of application—which are difficult to change and to expand. This is a special risk factor of investments.

Because the manufacturing systems play a crucial role in the future a little bit more attention will be paid on systems integration and on the implementation issues.

2.4 SPECIAL ASPECTS OF SYSTEMS TECHNOLOGIES

In order to understand the diffusion factors of FMS and how conventional systems are substituted by FMS, the IIASA CIM project collected empirical data on costs, benefits, and relative advantages as well as on implementation and design practice of FMSs (Tchijov, 1989b, 1989c). Using the worldwide empirical data, we want to demonstrate and understand which are typical systems and user characteristics, how different advantages benefits and advantages are associated with each other, and how different are dependent on the application and the business environment. Moreover, these basic analyses are then used to understand the factors behind successful applications and implementation practices, and they are also used to project future applications.

Moreover, some of the problems described, typical applications and the main users found, can be used to assess the evolution and the diffusion of the systems. We can indicate already that many of the forecasts given are relevant taking into account the present problems and needs.

These analyses are based on data, which IIASA has on more than 900 FMSs in the world and 60 thorough case studies in different countries (see Ranta et al., 1990, 1992; Tchijov et al., 1989a, 1989b, 1989c, 1990). These data are used to analyze the characteristics of relative advantages and the interconnections among the advantages. Technological characteristics and factors behind costs and benefits are then studied, and their role with regard to the benefits and the type of substitution of conventional systems is explained.

In the analysis we use three categories for the FMS systems: compact systems, which cost less than $4 million, and high-efficiency, large-scale system systems, which cost more than $15 million, and the mid-range systems, which are in between. As will be shown, these system types show rather different cots–benefit characteristics.

The diffusion of FMSs and the FMS density are presented in the Table 2.1 (based on the IIASA database). The numbers as such reflect the fact that

TABLE 2.1 Distribution of FMSs and the FMS Density[a]

N		*N*/*I*		*N*/*E*	
Japan	167	CSSR	135	Sweden	89
United States	137	Finland	105	Finland	71
United Kingdom	93	Sweden	87	Bulgaria	58
FRG	74	United Kingdom	65	United Kingdom	37
France	67	GDR	63	France	35
USSR	56	France	49	Japan	33
Italy	37	Hungary	44	Austria	32
Sweden	36	Bulgaria	38	GDR	25
GDR	28	Holland	38	Holland	24
CSSR	23	Austria	37	CSSR	24
Bulgaria	15	United States	36	Belgium	23
Finland	12	Japan	29	FRG	21
Holland	8	Belgium	27	Italy	20
Hungary	7	FRG	19	Switzerland	19
Austria	6	Italy	17	Norway	18
Switzerland	6	Poland	12	Ireland	17
Belgium	6	USSR	9	Hungary	17
Poland	5	Switzerland	9	United States	16
Taiwan	5	Taiwan	8	Canada	9
Canada	3	Israel	6	Poland	4
Spain	2	Canada	3	Romania	0.8
South Kore	2	Spain	3	Spain	
Norway	1	South-Kore	2	South Korea	
Israel	1	Romania	2	USSR	
Ireland	1	Ireland		Israel	
Romania	1	Norway		Taiwan	
Total	799				

[a] *N*, number of systems; *I*, total machine tool investments, 1988; *E*, labor force in metal product and engineering industries.

Source: Tchijov, 1989b.

there might exist significant institutional factors, which are different by countries and which explain the uneven distribution of the systems.

The relative benefits of FMS applications are presented in the Table 2.2. As such, the average figures tell only the potential benefits associated with FMSs. There are remarkable country-specific differences, and there are remarkable systems-specific and application-dependent differences.

A very surprising fact was found (see Ranta and Tchijov, 1990). The relative benefits seem to depend on the size of the system and the total amount of investments for the system in a V-shaped manner. This is more interesting phenomena than the average figures can tell us. Therefore the FMSs are divided into three classes according to investment costs: compact systems,

TABLE 2.2 Comparison of Average Figures for Basic and Compressed Versions of the Data

Operation rate (OPR)	2.7	2.3	2.4	2.6
Number of unmanned shifts (UNM)	1.0	0.8	1.6	1.0
Number of product variants (PV)	163	1138	88	216
Batch size (BS)	207	71	324	188
Investments (INV)	5.7	3.1	5.9	5.6
Payback time (PBT)	3.8	3.1	3.6	3.8
Lead time reduction (LTR)	5.1	9.5	10.1	5.4
In-process time reduction (IPT)	7.3	4.0	4.1	5.9
Inventories reduction (INR)	3.9	***	***	4.2
Work-in-progress reduction (WIP)	3.7	***	***	4.0
Personnel reduction (PER)	4.2	1.8	4.1	3.9
Number of machines reduction (NOM)	4.0	***	***	4.1
Floor space reduction (FLS)	3.1	1.7	2.2	2.8
Capacity utilization increase (CAP)	1.8	***	***	1.8
Unit cost reduction (UCR)	1.7	2.4	2.4	1.8

mid-range systems, and high-capacity systems, which are characterized separately.

A typical compact, "cheap" system (see Ranta et al., 1989b; Shah, 1987) consists of two to four CNC machine tools or machining centers, a conveyor, and/or automatic storage and retrieval system, sometimes, two robots for material handling, and a programmable controller for systems control. Usually the costs of the system are less than $3 million and the cost breakdown is approximately the following: CNC machines (50–55 percent), material handling and robots (15–20 percent), control, communication, and other systems-level software (20–25 percent), planning and training (10 percent).

It is also typical that the system architecture of a compact system is closed so that is is hard to extend and to add new features without major new investments. This can also be a major economic risk of investment. The technologies used for the system realization, for example, programmable controllers or small-capacity computers with a simple communication solution, as well as limited capacity for transportation and storage, make the system difficult to expand economically. Usually a major revision is needed as well as a new solution for the systems control. Therefore the system can be called closed—it has a limited expendability and adaptability. On the other hand, the system has relatively low initial investments.

A typical large-scale, "expensive" system consists (see Bose, 1988; Shah, 1987) of 15 to 30 CNC tools, automated guided vehicles (AGVs), and an automated storage and retrieval system for material handling, a local area network and distributed microcomputer-based cell, machine control systems, and usually two supermini-type computers for coordination, scheduling, and database management. It usually has a backup computer system to secure the availability of the system and advanced algorithms and a software system for the coordination of the system. The average costs are $10 to $15 million and the cost breakdown is approximately the following: CNC machines (35–40 percent), transportation and material handling (15 percent), control and communication and other system software (25–30 percent), and planning and training (15–20 percent).

It is also typical that the systems architecture is open and systems can be extended in a stepwise manner. New features can be added without a major new design effort. Because of the technology used, there is usually some reserve capacity to expand the system; some new features, like machining centers or more AGVs can also be added. The communication and computer system provides support for such additions, as common interfacing is used. Of course, this open-system structure and expandability has been achieved by relatively high initial investments.

Thus we can see that the expanding systems structure will typically lead to an increased software complexity and a more complex control architecture.

We can call those systems, which are between these two basic categories, mid-range systems. Typically the investment costs are between $4 and $8 million and the system consists of 5 to 10 CNC machines, and most likely also of AGVs for transportation. The all-over control is based on superminicomputers, and the system may even have a local area network for coordination and communication. The systems are relatively expensive, the total cost per CNC tool is quite high, but their efficiency lies only in the mid-range.

All of these three basic categories also show different cost-efficiency figures, which will be explained in more detail in the following.

Table 2.3 and Fig. 2.7 summarize the characteristics of the systems as well as of different cost–benefit figures.

As we can see cost–benefit data demonstrate the typical V-shaped form of the relative benefit measures. Thus it seems that:

- The compact and small-scale systems have the best cost–benefit ratio.
- The medium-range systems are inefficient from the viewpoint of their economic justification.
- Only relatively large investments and relatively complex systems provide the same efficiency and roughly the same cost–benefit ratio as the compact systems.

Figure 2.7 can be interpreted so that the increased capacity of systems and the increased complexity will increase the systems costs/machining unit in a stepwise manner. This is because of the need for more efficient machinery when a certain level of complexity is reached. In small size systems, it is enough to have a compact type of material handling system, such as a conveyor, and simple systems control based on programmable logic. When the complexity increases, a more sophisticated material handling system is needed, such as automated guided vehicles, and the systems control has to be based

TABLE 2.3 Characteristics of Different FMS Categories

System	Total Costs (million dollars)	Share of CNC (%)	Costs, Mat. Hand. (%)	System Software (%)	Planning, Training (%)	Number of CNC	Control Architecture	Algorithms	Relative Benefits[a]	Payback Time (years)
Compact	3	50–55	15–20	20–25	10	2–4	Closed, centralized 1–2 levels	"Simple" scheduling	3–7	1.5–3.5
Midrange	4–9	40–50	15	25	15	5–15	Semiopen, distributed 2–3 levels	Complex scheduling and tool management	1–3	5
High capacity	10–15	35–40	15	25–30	15–20	15–20	Open, distributed LAN-based 3–4 levels	Complex scheduling, tool management, timing, and diagnostics	3–7	1.5–4

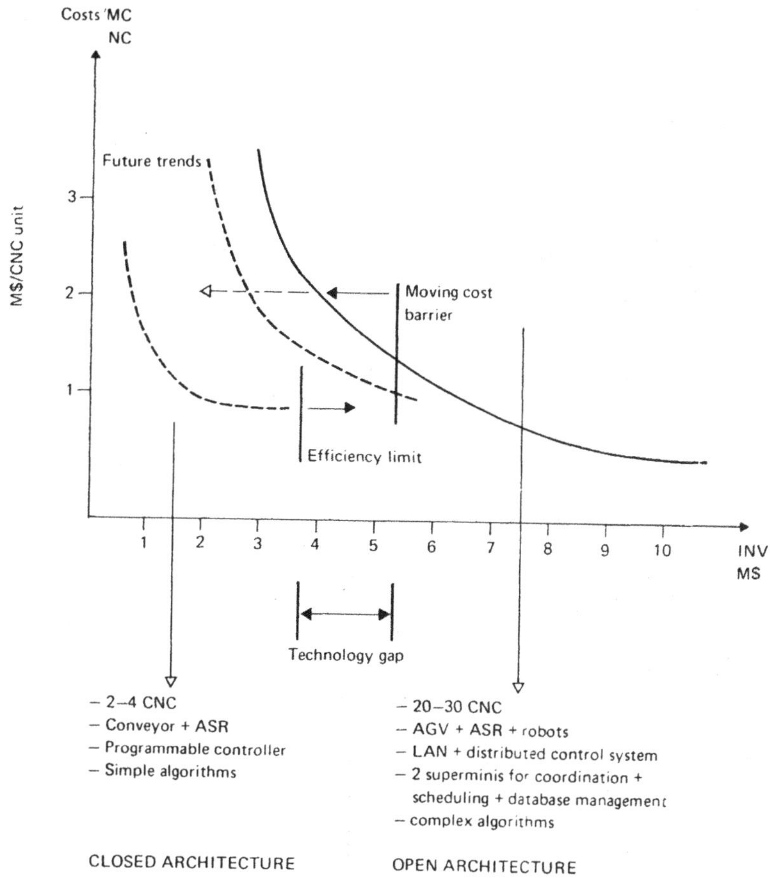

Fig. 2.7. The relative costs of different FMS.

on computers, distributed databases and integrating communication systems. These changes in system complexity tend to occur in a stepwise manner.

On the lower end of the applications, modest benefits can be achieved by a compact system and by low investment costs. There are also many standard-type solutions for the compact systems. This brings the software costs down and improves the software quality as well. If the basic software is not longer customized but can be reproduced for many applications, the costs of single software modules will drastically decrease, or "scale" down. Also the quality and reliability of the single modules can be "learned" through many applications, which make the system design and realization shorter and cheaper when the standard structure is reapplied.

On the upper end of the applications, there are possibilities for substantial savings and benefits, although the investment costs as well as the complexity of the system are high. The potential benefits usually justify the higher investments. The second factor, which generally conforms with the use of complex systems, consists of real learning curve effect or economies of scale in software production. This is the same "scale" and "learning" effect as already explained in the case of the compact systems, but for different reasons.

When the level that necessitates the changes in the basic systems architecture has been reached, there are many possibilities to repeat (or simply copy) the basic software modules and use these same modules in different interfaces and in systems coordination and timing. The larger the scale of the system under design, the more immediate are the benefits of software repetitiveness. This helps again with regard to the high costs of customizing. The repetitiveness of the basic modules of the software scale down the unit cost and also help with learning to improve the quality and to make the realization cheaper and shorter. However, the increasing overall complexity of the systems control and timing usually conflicts with these trends.

The medium-scale systems are critical from the economic point of view. It might happen that a sophisticated systems architecture based on distributed databases and communication is needed, but the potential benefits are not high enough to justify the system investments. The system complexity is not high enough to draw the benefits from the economies of scale effects. However, the software is fully customized. This remark is also consistent with the empirical data, which show that compact, small-scale systems and very complicated, large-scale systems clearly have a shorter payback time than medium-scale systems. This also leads to the following conclusion: A critical technical issue for future applications is the possibility of a module-type control structure and a transportation device, which allows for a soft extendability of the system without drastic architectural changes.

This problem can also be called a complexity dilemma. The higher number of CNC machines combined with a large variety of part type (part variety or part family) usually results in such a complexity of systems coordination (e.g., routing, scheduling, tool management) that high software and planning costs cannot be avoided. The only way to reduce relative costs is to have a modular systems structure and standardized software modules. The benefits of standard systems structures are already clearly visible in compact systems. It is common practice that the same basic system layout and architecture is usable in many applications, providing learning curve advantages for different applications, as explained before.

This conclusion also emphasizes the problems pointed out earlier on production control, scheduling, and routing. These two successful system categories demonstrate that a new approach is needed in order to have a successful systems control and scheduling. If the number of both the parts and machine is high, the present systems architectures, software engineering tools, algorithms, and methods are not able to manage the systems control and scheduling.

Thus, if there is no real technical breakthroughs in realizing systems controls and all-around architectures, which guarantee module-based design of systems and an easy extendability, it is reasonable to expect that the basic diffusion paths of the flexible manufacturing systems will be of the following two types: highly efficient, high-capacity, complex systems replacing rigid transfer lines, and, on the other hand, compact, small-scale systems replacing conventional semimanual, NC-tool-based production. The economy and applicability of the middle-range system will be highly dependent on systems control and communication software as well as on flexible transportation devices.

Thus it is obvious from the previous discussion that there are two basic implementation strategies of flexible manufacturing systems: highly efficient systems replacing transfer lines and fixed automation or highly flexible, compact systems replacing semimanual small-batch production.

The starting point for the first implementation strategy is usually a fixed automation or transfer line in a mass production of a big company. The main goal is to increase flexibility, save capital, and decrease the lead times as well as to cope with the changing environment and demand in the future. As the FMS replaces the highly automated lines, labor savings play only a minor role in this case. The major benefits are really coming from the potential flexibility: The product can be modified and quickly redesigned without expensive changes in the part manufacturing. The typical industrial sectors are automobile industry, consumer electronics, and the personal computer industries.

The starting point of the second implementation strategy is usually semiautomatic or even manual production in small or medium-scale companies. The strategy is a simple capacity increase and quality improvement strategy, while sustaining the already existing flexibility. The capacity can be increased without major investments in new buildings and also without having more labor. Then the productivity increase can be quite remarkable (see Ranta and Tchijov, 1990). The typical industry using these is small-batch manufacturing, which is usually producing customized and specialized products.

REFERENCES

Ayres, R. U., Merchant, M. E., Ranta, J., and Warnecke, H-J. (1991). *Computer Integrated Manufacturing, Systems and Technology: The Past, the Present and the Future*, Chapman and Hall, London.

Bose, P. P. (1988). FMS automates F-16 production, *American Machinist*, April, 43–47.

Clark, K. B., and Fujimoto, T. (1990). The power of product integrity, *Harvard Business Review*, November–December, 107–118.

Gelsinger, P. P., Gargini, P., Parker, G. H., and Yu, A. Y. C. (1989). Microprocessors circa 2000, *IEEE Spectrum*, **26**(10), 43–47.

Gupta, D., and Buzacott, J. A. (1990). A framework for understanding flexibility of manufacturing systems, *Journal of Manufacturing Systems*, **8**(2), 89–97.

Jaikumar, R. (1986). Postindustrial manufacturing, *Harvard Business Review*, November–December, 69–79.

Jaikumar, R. (1988). From Filling and Fitting to Flexible Manufacturing: A Study in the Evolution of Process Control, Harvard Business School, Working Paper 88-045, February.

Johnston, R., and Lawrence, P. R. (1988). Beyond vertical integration—the rise of the value-added partnership, *Harvard Business Review*, July–August, 94–101.

Krafcik, J. F. (1988). Triumph of the Lean Production System, *Sloan Management Review*, **30**(1), 41–52.

Krasa-Sethi, A., and Sethi, S. P. (1988). Flexibility in Manufacturing: A Survey, TIMS/ORSA Meeting, Washington, D.C., April.

Lei, D., and Goldhar, J. (1990). Multiple niche competition: The strategic use of CIM technology, *Manufacturing Review*, **3**(3), 195–206.

Maly, M. (1988). Strategic, Organizational and Social Issues of CIM: International Comparative Analysis (Part I), WP-88-077, IIASA, Laxenburg, September.

Maly, M. (1989). Strategic, Organisational, Economic and Social Issues of CIM: International Comparative Analysis (Part II), WP-89-90, IIASA, Laxenburg, November.

Meredith, J. (1987a). Implementing the automated factory, *Journal of Manufacturing Systems*, **6**(1), 1–13.

Meredith, J. (1987b). Managing factory automation projects, *Journal of Manufacturing Systems,* **6**(2), 75–91.

Meredith, J. (1987c). Implementation of new manufacturing technologies: Managerial lessons over the FMS life cycle, *Interfaces*, **17**(6), 51–62.

National Research Council, (1991). *The Competitive Edge—Research Priorities for U.S. Manufacturing*, National Academy Press, Washington, D.C.

Perry, T. S. (1989). Intel's secret is out, *IEEE Spectrum*, **26**(4), 22–28.

Ranta, J. (1989a). The impact of electronics and information technology on the future trends and applications of CIM technologies, *Technological Forecasting and Social Change*, **35**(2–3), 231–260.

Ranta, J. (1989b). Tomorrow's industries, structural changes, the future of work organizations—options, possibilities and new way to compete, in *European Competitiveness in the 21st Century—Integration of Work Technology and Culture*, M. Cooley, Ed., FAST Programme Publications, CEC, Brussels.

Ranta, J. (1991). The technology push: Electronics, computers and communication, in *Computer Integrated Manufacturing, Systems and Technology: The Past, the Present and the Future*, Ayres et al., Eds., pp. 43–59.

Ranta, J., and Tchijov, I. (1990). Economics and success factors of flexible manufacturing systems—the conventional explanation revisited, *Int. J. Flexible Manufacturing Systems*, **2**(2), 169–190.

Ranta, J., Ettlie, J., and Jaikumar, R. (1993). *Fewer and Faster: A Story of Technological, Organizational and Managerial Innovation in the Manufacturing Enterprice*, Harvard Business Press, Cambridge, MA (to appear 1993).

Shah, R. (1987). Flexible Fertigungssysteme in Europa—Erfahrungen der Anwender, *VDI-Z*, **129**(10), 13–21.

Stalk, G. (1988). Time—The next source of competitive advantage, *Harvard Business Review, July–August, 41–51.*

Tani, A. (1987a). Future Penetration of Advanced Industrial Robots in the Japanese Manufacturing Industry, IIASA Working Paper WP-87-55.

Tani, A. (1987b). Enterprise Size and Its Impact on Penetration of Industrial Robots—an Application of Econometric Analysis, IIASA Working Paper WP-87-108.

Tani, A. (1987c). An International Comparison of Industrial Robot Penetration, IIASA Working Paper WP-87-125.

Tchijov, I., and Alabian, A. (1988). Flexible Manufacturing Systems (FMS): Main Economic Advantages, WP-88-83, IIASA, Laxenburg, September.

Tchijov, I. (1989a). CIM introduction: Some socioeconomic aspects, *Technological Forecasting and Social Change*, **35**(2–3), 261–275.

Tchijov, I. (1989b). FMS World Data Bank, WP-89-33, IIASA, Laxenburg, May.

Tchijov, I. (1989c). FMS in Use: An International Comparative Study, WP-89-45, IIASA, Laxenburg, August.

Tchijov, I., and Sheinin, R. (1989). Flexible Manufacturing Systems (FMS): Current diffusion and main advantages, *Technological Forecasting and Social Change*, **35** (2–3), 277–293.

Wright, P. K. (1991). Manipulation and robotics, in *Computer Integrated Manufacturing, Systems and Technology: The Past, the Present and the Future*, Ayres et al., Eds., pp. 113–132.

Wright, P. K., and Bourne, D. A. (1988). *Manufacturing Intelligence*, Addison-Wesley, New York.

Yamashina, H., Matsumoto, K., and Inoue, I. (1991). Prerequisites for implementing CIM—Moving towards CIM in Japan, in *CIM—Computer Integrated Manufacturing*, ECE-Unido, UN Publications, Geneve, Vienna, pp. 141–174.

CHAPTER 3

HUMAN ROLES IN ADVANCED MANUFACTURING TECHNOLOGY

W. WOBBE and T. CHARLES
Commission of the European Communities
MONITOR/FAST Programme
Brussels, Belgium

3.1 WHAT IS ADVANCED MANUFACTURING?

Only a few years ago the world was quite simple for the engineering and research community regarding advanced manufacturing (AM). The indicator for progress in manufacturing matters was the *degree of exclusion of human beings* in all stages of manufacturing, from design, through planning, scheduling, control to direct intervention. The information technology based revolution of technical instruments seemed to penetrate all production functions and displace human intervention, with all its inherent advantages over the unpredictability of human beings in areas of wage negotiations, boycotts, strikes, hazards, accidents, and work time restrictions.

The question as to whether a factory based on group and teamwork principles could be considered to operate in terms of advanced manufacturing would have been answered by an overwhelming majority as not being pertinent. Today, full discussions are taking place on the *nature* of advanced manufacturing. The scope has been broadened and technologies have there-

Organization and Management of Advanced Manufacturing. Edited by Waldemar Karwowski and Gavriel Salvendy.
ISBN 0-471-55508-8

fore to be placed in a wider context. First, the question to be posed is what main factors have to be considered in manufacturing, and what role is played by technology in connection with the other factors? Evidently, the human role has to be reconsidered. Social organization also plays an eminent part. Therefore, discussions about advanced manufacturing today center around the integration of three factors: people, organization, and technology, and the dynamics among them.

The present juncture, both in theory and practice, is marked by uncertainty concerning the new forms of advanced manufacturing following the crisis of mass production, of Fordism, and of Taylorism. Discussions on advanced manufacturing concern the development of whole production systems in their socioeconomic context—variously described as an emerging *lean production model* (Womack et al., 1990), *flexible specialization* (Piore and Sabel, 1984), new *technoeconomic paradigms* (Perez, 1983), or, more hesitantly, simply as *the new model of production* (Boyer, 1989). Theoretical debates concerning the contours of new production systems are paralleled by widespread managerial awareness and experiments with new forms of work organization, new supplier relationships, and radical changes to management in the wake of the Japanese ascendency in world manufacturing.

To answer the question what constitutes advanced manufacturing, an exploratory research study composed of several research networks (Brandt, 1991; Kidd, 1990; Lehner, 1991; Rauner and Ruth, 1991) of the MONITOR/FAST Programme[1] of the Commission of the European Communities decided to look outside the framework in which technological progress is the measuring stick for the degree of progress (Wobbe, 1991). We are of the opinion that the assessment criteria for AM will be long-term economic success as well as immediate competitiveness.

Economic performance criteria in the 1990s go far beyond the traditional manufacturing concern for efficiency in terms of cost per unit of output (Charles et al., 1990). Product innovation must focus increasingly on quality of life issues and energy conservation. Process innovation can no longer be assessed by the level of automation but by the effective integration of advanced technology with new forms of work organization to meet the new parameters of production in world markets.

Social compatibility refers to the need for manufacturing industry to be responsive to increasing social demands for more rewarding and meaningful work in manufacturing. Skill shortages in certain sectors and a declining proportion of school leavers in the demographic profile of advanced industrial societies will increase the demand for improvement in conditions of work.

Advanced manufacturing has therefore to offer the potential to cope with the latest market developments and thus take on a structure that allows for adaptability, flexibility, and innovativeness. Finally, assumptions about eco-

[1]This is a program for forecasting and assessment set up in the late 1970s. Its purpose is to advise on European Community policies

nomic developments, markets, and products are decisive criteria for defining AM.

The MONITOR-FAST research has called the manufacturing systems capable of coping with these developments "anthropocentric production systems," in short, APS. Briefly, they can be defined as forms of advanced, computer-aided manufacturing that depend on human skills, collaborative work organization, and adapted technologies. They are based on the new principles of work and factory organization, reduced division of labor, the unity of conception and execution of work, decentralized decision making, skill-enhancing job design, as well as human-centered technology design (Brödner, 1990; Charles et al., 1990; Warner et al., 1990).

Having reviewed the research body on flexible manufacturing via flexible manufacturing technology looking at flexible manufacturing systems (FMS), robots, and materials requirements planning (MRP) systems (Cooley et al., 1989, Fix-Sterz et al., 1987; Wobbe, 1987), the study came to the conclusion that continuous improvement and innovation has to be ensured by the organization itself: It has to be *inbuilt* (Kidd, 1990). Therefore a plant has to be organized as a learning organization, as a steady upgrading process. This cannot be provided by technology, nor by flexible technology, nor by so-called artificial intelligence. Continuous improvement can only be assured by the mobilization of all the knowledge and skills of the work force and an organization that assures collaboration and motivation of the employees. Therefore the human role is at the very forefront in advanced manufacturing.

3.2 THE SHORTCOMINGS OF TECHNOLOGICAL SOLUTIONS IN AM

One of the big problems advanced technologies had to face in manufacturing was always return on investment, and managers have struggled to achieve this in reality (not just on paper). Finally, at the end of the 1980s, we witnessed a productivity paradox as well as a lot of cases of mislead investments. To illustrate what is meant by this, the following might be taken as examples:

- The market for *industrial robots* lags far behind all predictions made in the late 1970s. In the United States no indigenous manufacturer has survived.
- Misinvestment in material resource planning was estimated in 1990 at 100 million DM in Germany because a proper functioning of the systems was not achieved.
- An established manufacturer of FMS in Germany has refused to deliver systems to clients who do not provide proper training and work organization because he does not wish to gain a bad reputation should the systems fail to operate properly through lack of this provision.
- Also, with regard to the United Kingdom, indications of high failure

rates with FMS and production planning systems have been reported (Haywood and Bessant, 1990; Lawrence, 1986).

- Volkswagen—the leading European car manufacturer and also leader in car assembly automation—has indicated that it will not increase levels of factory automation for future projects. It will experiment with group work organization.
- The International Labor Organization (ILO) estimated in 1987 that 30 percent of all CAP/CAM systems were abandoned within the first year of installation.
- A Danish study (Gjerding et al., 1990) confirmed that in the mid-1980s Denmark witnessed a strong growth in production, investments, and employment, but productivity in manufacturing was falling, contrary to the forecast of economists. The reason for this was judged to be the "technological fascination" of management. Management put so much emphasis on the power of new technology for solving problems, and far less on the labor force, management, and training, to the detriment of productivity.

3.3 FROM WHERE DOES THE PRODUCTIVITY PARADOX STEM?

Besides the fascination that sophisticated technologies hold not only for engineers and managers, the technology-dominated view of advanced manufacturing has been heavily influenced by the industrial military complex. Numerical control (NC) machines, FMS, computer-integrated manufacturing (CIM), microelectronics itself, as well as new materials have been developed by military-sponsored research, first and foremost in the United States. In the same way in the former Soviet Union, advanced production has only existed in the military sector, which, according to economists at the Academy of Sciences, made up to 60 percent of the whole Soviet economy toward the end of the 1980s. Military and state-funded projects do not have to concern themselves with market economies.

This was the central advantage in the rise of Japanese manufacturing. They took a closer look at market economies when they developed their manufacturing concepts and were therefore always concerned with productivity and cost reduction, and at the same time increasing quality.

Also, advanced manufacturing has always been viewed in a framework of mass production, particularly in the Western context. Automated facilities, if they are not in a military complex, have always aimed at mass consumption goods, which justify the heavy investments in, or at least they aim at, volume production. Mass markets will not disappear in the future, particularly in view of the potential demand from Eastern European countries and the developing world. Nonetheless, new market economies have come to the forefront and are in the course of dominating the most developed industrial countries in

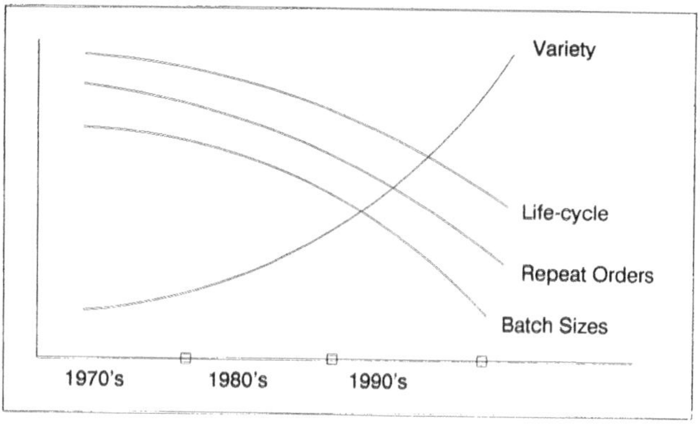

Fig. 3.1. Development of economic parameters for manufacturing.

Western Europe, the United States, and Japan, which offer sufficient purchasing power for sophisticated products.

3.4 THE INCREASE IN CUSTOMIZED QUALITY PRODUCTS

The Western economy had been transformed in the last decade into more and more segmentation based on increasing customization, high-quality demands, and more service inputs. This has allowed a firm to increase the variety of a given product and to reduce its life cycles—a requirement contrary to the manufacturing practices based on a high division of labor and automated technology.

These economic requirements are demonstrated in the graph in Fig. 3.1. They are taken empirically from a large electrical supplier in the United Kingdom.

The demand for upgrading product quality and the customization of products has therefore pushed new manufacturing concepts to the fore. In order to gain responsiveness to new market requirements, manufacturing organization is challenged at all levels. Innovation in flexible manufacturing and in product quality is not longer the domain of production and design engineers: It involves all collaborators in the plant, from design to the shop floor and vice versa. To produce one-of-a-kind products (Browne, 1990) in an economic way with short time delays has therefore become the ultimate aim of real advanced manufacturing.

3.5 THE HUMAN ROLE IN MANUFACTURING

All these developments demand viable economic manufacturing patterns that place human interaction, innovativeness, and skills at the forefront.

- The more complex products become, and this is particularly relevant to the engineering industry sector, the more quality is dependent on permanent upgrading of all stages of manufacturing and demands the full dedication of employees at all levels. Quality control, improvement proposals, individual responsibility are all factors that demand a high level of motivation and excellent coordination at all stages of product manufacture.
- The more sophisticated manufacturing technology becomes, the more it is vulnerable and dependent on human skills for control and maintenance.
- The more customized products are, the more human intervention is necessary with regard to innovation, for the product to be treated in a holistic way. Change over, setting up machines, adaptation, adjustment, and control demand considerable human intervention.
- If products demand a high service input and after-sales service and maintenance, skilled people are required to deal with this.
- The shorter the life cycle of products becomes, the more innovativeness comes into play; take-off phases occur more frequently, and their mastery is dependent on experienced personnel with formal and tacit knowledge to overcome new challenges connected with the start of a new product.

All this challenges the individual workplace and the responsibility and motivation of the individual employee. Moreover, it demands vertical cooperation patterns, teamwork, and opposes a deep and hierarchic organization that is only able to cope with continuous and repetitive tasks. Therefore collaboration, flat hierarchies, and decentralized decisions are of crucial importance of an organic working and manufacturing environment.

Although these principles are essential for small-batch production, they are also viable in mass production environments. An MIT study (Womack et al., 1990), which compares automotive production around the world, had a bombshell effect on the understanding of advanced manufacturing in mass production environments. The term *lean production* was utilized to describe the employment of new collaboration concepts in manufacturing and the upgrading of the human role. This concept of manufacturing organization is considered doubly as productive as traditional concepts over all its production chains. Technological factors are not considered decisive in generating this result, more so organization and human factors.

3.6 IT IS NECESSARY TO CHANGE THE ENGINEERING PARADIGM

In today's discussions about advanced manufacturing, completely new themes are now relevant, which would previously have been rejected in the mid-

1980s, for example the human role in groups or teamwork, factory organization, interdepartmental collaboration, the tacit dimension of knowledge, national cultural factors in influencing working attitudes, and so forth.

What do we mean by an engineering paradigm: Managers and engineers, like other professionals, hold a set of core beliefs and assumptions that are specific and relevant to their profession and that are shared in common with other members of the profession. A paradigm is taken for granted and not seen as problematic by those who adhere to it. Paradigms are difficult to change and to challenge because they are more than intellectual exercises, they are fundamental, nonexplicit beliefs. Paradigms might be changed in fundamental ways, not by external experts but by members of the same professional community and by constructive criticism (Kidd, 1990).

Paradigms are fundamental perceptions of how to organize and run a business. However, the new economic challenges require a fundamental shift in manufacturing design paradigms, and engineers should no longer use design methods aimed at manufacturing practices that have been developed for market circumstances that no longer exist.

Therefore, the traditional engineering paradigm has to change in order to adapt to the new advanced manufacturing concepts, for example, lean production, or like the European version for batch production called anthropocentric production systems, or APS.

In the following, we shall briefly characterize the important elements of anthropocentric production systems, which could be called the ensemble of advanced manufacturing in the 1990s.

3.7 ANTHROPOCENTRIC PRODUCTION SYSTEMS COPE WITH THE NEW REQUIREMENTS

The principles of APS start with the assumption that human beings have a central role in manufacturing, and they appeal to the work organization of an enterprise, its management structure, and its industrial culture. The principles also provide guidelines for technology development. In order to achieve these principles, technology has to allow for these collaborative production structures from the workshop to the design and planning departments. Software principles are challenged and particular requirements are needed to adopt the technology.

The collaborative production structure is laid down in organizational design at different plant levels. These levels stretch from workplace organization to factory organization in total. There are particular measures taken to establish APS and to organize the different levels of the factory toward APS principles. These measures and principles are spelled out in Fig. 3.2.

Anthropocentric production systems can be perceived as a *complete model of a factory*, integrating all aspects of organization, people, and technology. However, the system is composed of different elements and encompasses

LEVELS	*MEASURES & PRINCIPLES*
FACTORY	* *small decentralised production units* * *product shops* * *companies within a company* * *delegation of responsibility to lower levels*
INTERDEPARTMENTAL RELATIONS	* *cooperation between design and manufacturing* * *interactivity between workshop and engineering departments concerning programming of machine tools* * *integration of business department, technical department and shop floor concerning planning and scheduling*
GROUP	* *installation of production islands* * *group work in Flexible Manufacturing Systems (FMS)* * *semi-autonomous assembly groups*
WORKPLACE	* *workshop programming* * *integration of intellectual and manual functions* * *in highly automated areas: integration of programming, planning, maintenance and processing tasks* * *in low automated assembly areas: work enrichment with decision space on execution sequence and performance*

Fig. 3.2. APS organization design on different plant levels.

various levels of the factory. These elements stretch from the individual human–machine level to the collective level; from the workplace to the working group. We can also identify the technical tools, as well as the "organizational tools," in the sense of how to develop human collaboration, to be deployed in anthropocentric production systems.

The Workplace

The individual workplace has to allow the optimal use of the skills of the worker. It also has to guarantee development and learning possibilities, so that motivation is challenged, as well as innovation and ingenuity, and that tacit knowledge and experiences can be put into operation. This means that the work has to be holistic with a selection of tasks. For a machine operator in the engineering industry this involves planning, programming, sequential decisions of executive work, maintenance, quality responsibility, communication, and collaboration tasks. In order to guarantee this structure, factory management has to ensure that both, this and the technology, is suitable.

Anthropocentric production system requirements for the workplace can be applied, for example, in discrete parts production, for machine tools, in the assembly areas, in the process industries, and on the shop floor. They also have implications for the technical offices and apply to administrative workplaces as well.

Group Work

It is advantageous to embed work places in a team structure to increase cooperation. The group structure is built systematically around collaborative synergies. Group work allows decentralized decisions. In particular, the work scheduling and capacity utilization plays a strong part in group concepts allowing flexible adaptation. Group work already has a long tradition in assembly areas of car production, in production islands for concrete parts, particularly with product families, as well as in flexible manufacturing systems, in design islands, or administrative functions.

Technical Tools

Although the central part of the agenda of APS involves changes in the work organization, the technology has to be adapted or has to aid the form of organization chosen. This concerns the workplace, when, for example, the programming has to be carried out on the shop floor. Then an adequate system has to be provided, or, if a group has to make autonomous decisions in planning and scheduling, then the system has to support these decisions and provide interfaces for the general planning system. In the same way, communication between the design offices and the shop floor has to be guaranteed so that not only does the designer download his or her data to the shop floor but also the other way round. Software ergonomics has to provide procedures so that the user of information systems can work with traditional logic and not vice versa, which would compel users to learn the logic procedures of the system. The technological system has to be adapted to the experience and knowledge of the user of the system.

Organization Tools

The traditional division of work in its hierarchical structure and high specialization of jobs has led to rigidity, quality problems, and low productivity. The traditional forms of organization are nowadays deeply rooted in industry and the heads of industrial management. It has become a weak industrial culture.

The steps for change start with the redesign of production facilities. The new organization aims at holistic, cooperative work structures with flat hierarchies. These structures require new management procedures and participative structures to unfold its productive strength. The introduction of APS has therefore to be understood as a process that is based on cooperative industrial relations and personnel participation. The employees have to be educated and trained in a continuous process and the management has to understand and provide these processes.

Education for Workers and Managers

General education, further education, and training play a crucial role in advanced manufacturing and have to be made available to those in the workplace and those working in groups. They are of importance not only on the shop floor level but also for white-collar employees. In the end, APS is based on well-educated and competent personnel with a basic education, which is receptive to further education and training insofar as the new production processes require it. In order to succeed in the restructuring of plants, further education on APS principles and practices for middle and upper management is needed. Insofar as the company culture is concerned, engineers and technicians have to be educated within the framework of a new industrial culture, if we are to move toward a collaborative social environment without the blue/white-collar barrier, which is the particular strength of APS.

3.8 NEW PRODUCTION CONCEPTS

These, in brief, are the core elements of the new production systems we have termed anthropocentric (APS). They constitute a challenge to more traditional tayloristic and technology-led manufacturing strategies, and they provoke a complete reappraisal of the human role in production. The principles of APS, which we have briefly outlined, have their origins in European manufacturing practices and the debate about "new production concepts" (Brödner, 1990; Hyman and Streeck, 1988; Kern and Schumann, 1984). The reawakening of interest in the human aspects of advanced manufacturing in the late 1980s reflects both the changing conditions of production and the increased recognition of the importance of human and organizational factors for successful innovation and implementation of new technology (den Hertog, 1991; OECD, 1992). However, there is a strong continuity and elective affinity between current conceptions of human-centered technology and earlier research in the human relations and sociotechnical approaches of the Tavistock Institute in the 1950s and earlier in the British traditions of human factor industrial psychology led by C. S. Myers in the 1920s (Rose, 1975). Despite this strong continuity with earlier British, North American, and Scandinavian approaches to sociotechnical systems design, anthropocentric, or human-centered, technology comprises a deeper and critical analysis of the "black box" of technology and systems architecture (Corbett et al. 1991; Latour, 1987) and a wider appraisal of the human role in manufacturing beyond the task level to encompass enterprise organization and authority structures, interfirm relations, and organizational culture (Badham and Naschold, 1991; Gordon and Krieger, 1990; McLoughlin and Clarke, 1988; Wood, 1989).

Current experiments with new production concepts are marked by a diversity of manufacturing strategies that cannot be encompassed by a single model of the transition from Fordism to post or neo-Fordism. In the same way that fordist mass production and Taylorism were never implemented in their entirety but adapted and modified in a European context (Littler, 1982), new production concepts are adapted to local employment conditions, national traditions of education and training, product markets, and type of process technology.

In northern Italy, Piore and Sabel (1984) have identified *flexible specialization* as a new production system based on craft work, advanced technology, and networks of cooperation between small firms in localized agglomerations. This model is associated with a general crisis of fordist mass production and the ascendancy of post-Fordism based on flexible manufacturing. There are two main problems with this model as an attempt to capture the key features of advanced manufacturing. First, rather than a master trend of development in manufacturing, the features of flexible specialization are limited to a specific social and economic spatial context, one which, in the case of Emilia-Romagna, is undergoing further evolution from localized to global networks, and, is some cases, to the dominance of large, transnational corporations. Second, the model is premised upon a romanticized craft ideal, but the regeneration of artisanal labor with new technology constitutes only one route for flexible manufacturing rather than the end of large-scale production (Williams, 1987; Wood, 1989).

3.9 LEAN PRODUCTION

In recent years discussion of new production concepts has been dominated by analysis of Japanese success in manufacturing—by superior productivity in both indigenous Japanese firms and so-called transplants in Western Europe and the United States. We will not add to the already voluminous literature attempting to explain the reasons for Japanese manufacturing competitivity, but we will focus briefly on the MIT concept of *lean production* (Womack et al., 1990) since it attempts to define the key elements of a new production paradigm replacing traditional mass production; a paradigm that has its origins in Japanese car production, but which, the authors suggest, is slowly becoming a new, universal "best practice" across nations and sectors. Contrary to the flexible specialization thesis, lean production is postulated to combine the advantages of both craft and mass production—high volume but diversified production using flexible automation, multiskilled teamwork, just-in-time supply relationships, and an overall corporate culture of continuous improvement and high quality. Lean production relies less upon full-scale automation and more on human commitment and involvement in production and changes to work organization.

The success of lean production systems has generated a widespread managerial concern with new forms of work organization and "human resource" as opposed to personnel management (Streek, 1987). In both Europe and the United States the car industry is restructuring and reorganizing production in an attempt to match Japanese performance. We have cast doubt on the universal applicability of flexible specialization as a new production system, similarly we can question whether lean production is potentially universal or culturally specific. The relative success of Japanese transplants in the United States and Europe would appear to confirm the view that lean principles can be applied outside the Japanese social structure. On the other hand, studies of the UK transplants (Oliver and Wilkinson, 1988) reveal that Japanese practices are adapted to local conditions and applied selectively—lifetime employment, for example, is not applied in the UK context. It is also an open question as to the potential applicability of lean production to sectors other than automobiles where very different market and production conditions exist. Nor is the Japanese production system static—there is union pressure to reduce working hours, there are labor shortages, and plans by some manufacturers to increase the level of automation in assembly in the future.

The transplant phenomenon and spread of new production concepts occurs through a process of learning and trial and error, hastened by the increasingly global nature of advanced manufacturing and R&D in the "triad" (Petrella, 1991), but the limits to this process and the diversity of production concepts reveal that transfer of manufacturing knowledge and practice is not simply a matter of rational choice in response to market pressures. National infrastructures of education and training, industrial relations, institutions, and power relationships in the labor market play a critical role in the development of advanced manufacturing.

3.10 NEW PRODUCTION CONCEPTS AND THE HUMAN ROLE IN ADVANCED MANUFACTURING

What are the similarities and differences between our description of anthropocentric production systems and these new production concepts? Each marks a departure from the central organizational principles of traditional mass production: increased division of labor, semi or unskilled work, and tall, bureaucratic hierarchies. The new production concepts each accord direct labor on the shop floor a central place in improving production, ensuring the flexibility of the production process, and they underline the increased importance of human skills, even with increased capital intensity. Proponents of lean production argue that the human role in manufacturing is more rewarding, involves more control over work, allows for problem solving and learning in the workplace, and leads to higher employee morale and motivation (Womack et al., 1990). Critics (see, e.g., Thompson and McHugh, 1990) note variously—greater work intensification, management by stress,

	FORDISM	LEAN PRODUCTION	ANTHROPOCENTRIC PRODUCTION
Labour Market	numerical flexibility core/periphery	core of high skill, segmented core/ periphery	core of high skill and skill up-grading for the less skilled
Employment Relationship	high-substitutability of labour low trust	high-dependency on skilled labour high trust	high dependency but labour mobility high trust
Industrial Relations	adversarial	corporatist company unions	corporatist or pluralist participative social contract
Qualifications	low skill minimal training	high - training horizontal/multi-skilling company based	high education & training, multi-skilling transferable skills (craft origins) vertical/horizontal skilling
Design Paradigm	Taylorist mechanistic metaphor	organic system metaphor	organic tool metaphor
Organisation Structure	hierarchic divisionalised/ functional departments bureaucratic	leadership less hierarchy collaborative networks and team work post-bureaucratic	flat hierarchies decentralised semi-autonomous work groups post-bureaucratic
Organisational Culture	'one best way' scientific management	'one best way' teamwork collectivist harmony consensus	group based plus occupational identity, pluralist, partnership

Fig. 3.3. Comparison of manufacturing paradigms.

longer working hours, and, in the case of certain transplants, careful recruitment policies on greenfield sites and weakened union representation.

Lean production and APS differ in certain important respects—reflecting their respective institutional and cultural origins. Figure 3.3 compares important features of the ideal types of lean production and APS, contrasted with the fordist paradigm.

Both ideal lean production and anthropocentric production types involve a significant increase in shop floor responsibility for production and quality combined with various forms of group technology. They also imply fundamental changes in interdepartmental relations (frequently from functional to product or project organization), which affect the whole enterprise. They differ particularly with respect to the nature of group work, industrial relations practices, and the structure of authority in the enterprise. In APS, group work is based on the relative autonomy of work groups in production scheduling and the distribution of work tasks in the context of a partnership or social contract for enterprise participation. Lean production, at least in case studies in the automobile industry (Womack et al., 1990), entails higher levels of supervision and less trade union involvement in decision making. In part, these differences reflect the type of process technology, since cases of APS (Brandt, 1991) are predominantly from volume or small-batch production such as the machine tool industry.

In the next section we move beyond comparison of ideal type constructions of new production concepts to examine the extent to which APS principles are realized in practice, drawing upon the empirical results of the FAST survey of the prospects for APS in Europe (Wobbe, 1991).

3.11 PROSPECTS FOR APS IN EUROPE

It is clear from our discussion so far that our conception of advanced manufacturing technology goes far beyond innovation in hardware and software. So-called human factors considerations cannot follow after technology design since they form part of the whole process of technological change from design to implementation. Advanced manufacturing technology therefore comprises a system of production that is socially organized and shaped by the wider culture and socioeconomic context. Consequently, research teams from the European Community (EC) countries carried out an assessment of those factors that shape the design and use of advanced manufacturing technology, including:

- Technology design and technical infrastructure
- Production regimes and work organization
- Skills, education, and training
- Industrial relations
- Public policy on science and technology

The surveys provided the basis for comparisons between EC countries (Lehner, 1991) as well as the United States and Japan (Rauner and Ruth, 1991); they also allow, as far as possible, sectoral comparison, with examples from the machine tool industry, automobiles, textiles, clothing, and electronics. An APS is a multidimensional concept, and we would expect considerable variation in the ways in which APS principles might be realized in practice, reflecting the characteristic diversity of European manufacturing. Indicators of APS include shop floor programming and decentralized control and information systems, interactive and transparent computer systems that allow operator intervention, group production concepts, multiskilling, and new organizational forms that reduce hierarchical levels and increase interdepartmental collaboration; namely, developments in advanced manufacturing that mark a significant departure from tayloristic principles.

Technology design for advanced manufacturing is not predetermined—programmable technology allows the possibility of alternative types of work design and user control with existing hardware and software (Kidd, 1990). Germany and Denmark exhibit wider experimentation with decentralized systems and workshop programming, whereas UK firms are more likely to assign programming predominantly to white-collar staff and specialists away from the shop floor—a reflection of differences in the skill profile and preexisting division of labor prior to the introduction of new technology. Experiments with human-centered technology design such as interactive and object-oriented software are in their early stages. These observations confirm our earlier argument that the continued predominance of the traditional "technocratic" engineering paradigm is an obstacle to the further diffusion of APS and results in a failure to exploit the full potential of new programmable technology in terms of flexibility, product variety, innovation, and improved quality. The critical factor here is the organizational dimension of manufacturing, rather than technology narrowly defined as hardware and software.

3.12 WORK ORGANIZATION AND MANAGEMENT STRUCTURES

Organizational innovation and changes in corporate culture have, until recently, been considered as the "soft" aspects of the enterprise compared to the "hard" areas of technical change and accounting. This situation has changed with extensive debate in management circles about organizational development, managing change, and corporate culture, but despite this increased awareness many European enterprises are characterized by organizational conservatism and rigidity. Changes to organizational structure tend to be confined to work organization on the shop floor (Lehner, 1991). Among the reasons for organizational rigidity, the FAST studies identified a lack of managerial expertise in strategic aspects of human resources, resistance due to existing status differences and power hierarchies in management, and a lack of participation by both employee representatives and personnel managers in the early stages of technology design. Most importantly, organizational changes conducive to APS are limited by an underlying culture of "low trust" in several EC countries.

However, there are an increasing number of European enterprises experimenting with new organization forms, such as group technology, semiautonomous work groups, flatter hierarchies, quality circles, and integrated product structures that reduce the horizontal division of labor. Examples are more in evidence in Germany, Denmark, and the Netherlands and in firms using leading edge technology (Lehner, 1991). France and the United Kingdom exhibit greater rigidity in terms of the vertical or horizontal division of labor, respectively (Lane, 1989).

3.13 SKILLS, EDUCATION, AND TRAINING

National variation in the patterns of general and vocational education provide different points of entry as well as limitations toward the realization of APS. Countries such as Germany with a well-developed system of vocational training, complemented by high-quality professional and general education, provide a supportive infrastructure for APS—especially at enterprise level, where a high skill profile of employees provides a platform for decentralized decision making and continuous learning for innovation. In addition to problems of skill shortages in some sectors, there is also a need for new types of skill for group working that include communication and interpersonal skills as well as the ability to solve problems and understand the production process holistically. Despite many European initiatives developing new training programs, educational systems in the United Kingdom, France, and Italy still tend to produce narrow vocational skills and a polarized structure between vocational skills and academic knowledge. The challenge for education and training systems is to develop opportunities for the less skilled, who are often excluded from technological change, to reduce the gap between blue- and white-collar employees and provide an environment for continuous learning.

Industrial relations systems exhibit considerable diversity in Europe between the "corporatist" and legally codified approaches of Germany and the Nordic countries and patterns of segmented trade unionism and low-trust relations that characterize the United Kingdom, France, Greece, and Portugal. The former provides a more conducive environment for APS; nevertheless, there is a lack of participation in decision making for technology design and work organization throughout Europe (Fröhlich et al., 1988).

This brief outline of the factors that foster or inhibit the development of the human role in manufacturing in Europe could be extended to a comparison of production cultures in the United States, Europe, and Japan (Rauner and Ruth, 1991). We will limit our discussion here to the major conclusion of the FAST international comparisons: Global trends in technology innovation and manufacturing indicate some convergence toward cooperative and group-based production. However, this process conceals the existence of different trajectories for production cultures and competing production concepts. The United States, the heartland of fordist mass production, is identified as the weakest example of APS developments (Gordon and Krieger, 1990). European manufacturing, we have seen, is characterized by diversity—practical examples of the implementation of elements of APS exist among many enterprises and the body of case study evidence is increasing, aided by various European and national research programs (Brandt, 1991; den Hertog, 1991). Results indicate significant improvements in productive performance deriving from group technology, enhancement of shop floor skills, and organizational integration. Many obstacles remain to be overcome, however, to facilitate the broader diffusion of APS—of which the most intransigent is a deep-rooted combination of neo-taylorist management strategies and a technocratic design paradigm for new technology. Anthropocentric production system does not constitute a new "one best way" but reflects the diversity of the European manufacturing base comprising small and medium enterprises (SMEs) (65 percent of EC GNP) as well as large multinational corporations and the diversity of local and regional bases of manufacturing innovation (Bruno et al., 1991).

3.14 CONCLUSION

Advanced manufacturing today has gone far beyond the pure concept of production functions, fulfilled preferably by sophisticated machines, or information technology led automated facilities. Nonetheless the human role in these concepts of manufacturing is seen in planning and control. Bridging automation gaps has also gained fresh attention. The main reason might be due to today's improved understanding of the shortcomings of automation. In particular, world market trends toward steadily growing segmentation, demands for better quality, product innovation, and more customization have fostered a reappraisal of the human role and of human capability. The contribution of skilled workers toward innovation, technology implementation, improving product quality, and flexibility of the whole production process can no longer be neglected nor simply substituted by increased automation.

All advanced new production concepts such as lean production, flexible specialization and anthropocentric production systems aim at reducing the division of labor and levels of hierarchy in order to gain the collaboration of people and departments and to avoid frictions in the production chain. In group and team work the development of human ability aims at raising productivity and improving the manufacturing process. Therefore what was previously defined as the soft part of manufacturing systems has become the leading edge in competitivity of both firms and nations. The social and cultural dimension of manufacturing, meaning the different forms of industrial relations, educational systems, authority, and participation in organizations has to be given more attention because it shapes the human and organizational choices to be taken for advanced manufacturing systems.

Today, anthropocentric production systems in particular seem to fulfill the new market requirements toward customized quality production. They contain the three elements of advanced manufacturing: a balance among skilled people, collaborative organization, and adapted technologies. These elements can be taken as a hybrid manufacturing system because all the elements have to be tackled in an integrated way. Contrary to a decade or more ago, today the human being has returned to the center of these systems in order to render them more productive, more flexible, and more innovative. Time will tell whether these advanced manufacturing systems can successfully combine improved performance and competitiveness with social compatibility in terms of improved conditions of work and environmentally sensitive products and processes.

REFERENCES

Badham, R. and Naschold, F. (1991). New technology policy concepts, in *Technology Policy: Towards an Integration of social and Ecological Concerns*, G. Aichholsr and G. Schienstock, Eds., De Gruyter, Berlin.

Boyer, R. (1989). New Directions in Management Practices and Work Organisation, Mimeograph CEPREMAP, Helsinki.

Brandt, D. (1991). *Advanced Experiences with APS*, FOP 246, EC-FAST, Brussels.

Brödner, P. (1990). *The Shape of Future Technology*, Springer, London.

Browne, J. (1990). Future integrated manufacturing system: A business driven approach, in *Organisation, People and Technology in European Manufacturing*, P. Kidd, Ed., FOP 247, EC-FAST, Brussels.

Bruno, S., Cohendet, P., Desmartin, F., Llerena, D., and Sorge, A. (1991). *Modes of Usage and Diffusion of New Technology and New Knowledge*, FOP 227, EC-FAST, Brussels.

Charles, T., Charles, R., and Roulstone, A. (1990). *Prospects for Anthropocentric Production Systems in Britain*, FOP 252, EC-FAST, Brussels.

Cooley, M., d'Irbarne, A., Martin, T., Ranta, J., and Wobbe, W. (1989). *European Competitiveness in the 21st Century—Integration of Work, Culture and Technology*, EC-FAST, Brussels.

Corbett, J., Rasmussen, L., and Rauner, F. (1991). *Crossing the Border*, Springer Verlag, London.

den Hertog, F. (1991). *Options for Anthropocentric R & D*, FOP 267, EC-FAST, Brussels.

Fix-Sterz, J., Lay, G., Schultz-Wild, R., and Wengel, J. (1987). *Flexible Manufacturing Systems and Cells in the Scope of New Production Systems in Germany*, FOP 135, EC-FAST, Brussels.

Fröhlich, D., Fuchs, D., and Krieger, H. (1988). *New Information Technology and Participation in Europe*, European Foundation for the Improvement of Living and Working Conditions, Dublin.

Gjerding, A., Johnson, B., Kallehauge, L., Lundvall, B. A., and Madsen, T. (1990). *Den Forsvundene Produktivitet*. Charlottenlund, Okonomiforbundets Forlag.

Gordon, R., and Krieger, J. (1990). *Prospects for Anthropocentric Systems in the United States*, FOP 262, EC-FAST, Brussels.

Haywood, B., and Bessant, J. (1990). Organisation and integration of production systems, in *New Technology and Manufacturing Management*, Warner et al., Eds., Wiley, Chichester.

Hyman, R., and Streeck, W., Eds. (1988). *New Technology and Industrial Relations*, Basil Blackwell, Oxford.

Kern, H., and Schumann, M. (1984). *Das Ende der Arbeitsteilung*, Beck, Munich.

Kidd, P., Ed. (1990). *Organisation, People and Technology in European Manufacturing*, FOP 247, EC-FAST, Brussels.

Lane, C. (1989). *Management and Labour in Europe*, Gower, Aldershot.

Latour, B. (1987). *Science in Action*. Harvard University Press, Cambridge.

Lawrence, A. (1986). Are CAPM systems too complex? *Industrial Computing* (September vol.).

Lehner, F. (1991). *Anthropocentric Production Systems. The European Response to Advanced Manufacturing and Globalisation*, FOP 248, EC-FAST, Brussels.

Littler, C. (1982). *The Development of the Labour Process in Capitalist Societies*. Heinemann, London.

McLoughlin, I., and Clarke, J. (1988). *Technological Change at Work*, Open University Press, Milton Keynes.

OECD (1992). *Technology and Economy. The Key Relationships*, OECD, Paris.

Oliver, N., and Wilkinson, B. (1988). *The Japanisation of British Industry*, Basil Blackwell, Oxford.

Perez, C. (1983). Structural change and assimilation of new technologies in the economic and social system. *Futures*, **15**(4),

Petrella, R. (1991). *Four Analyses of Globalisation of Technology and Economy*, EC-FAST, Brussels.

Piore, R., and Sabel, C. (1984). *The Second Industrial Divide*, Basic Books, New York.

Rauner, F., and Ruth, K. (1991). *The Prospects of Anthropocentric Production Systems: A World Comparison of Production Models*, FOP 249, EC-FAST, Brussels.

Rose, M. (1975). *Industrial Behaviour*, Penguin, London.

Streeck, W. (1987). The uncertainties of management in the management of uncertainty: Employers, labour relations and industrial adjustment in the 1980's. *Work, Employment and Society*, **1**(3), 281–308.

Thompson, P., and McHugh, D. (1990). *Work Organisations*. Macmillan, London.

Warner, M., Wobbe, W., and Brödner, P., Eds. (1990). *New Technology and Manufacturing Management*, Wiley, Chichester.

Williams, K. (1987). The end of mass production? *Economy and Society*, **16**(3), 405–439.

Wobbe, W., Ed. (1987). *Flexible Manufacturing in Europe—State of the Art of Approaches and Diffusion Patterns*, FOP 155, EC-FAST, Brussels.

Wobbe, W., (1991). *Anthropocentric Production Systems: A Strategic Issue for Europe*, FOP 245, EC-FAST, Brussels.

Womack, J., Jones, D., and Roos, D. (1990). *The Machine that Changed the World*. Rawson Associates, New York.

Wood, S., Ed. (1989). *The Transformation of Work?* Hutchinson, London.

CHAPTER 4

EMPLOYEE COMPENSATION

B. GERHART and R. D. BRETZ, JR.
Center for Advanced Human Resource Studies
Cornell University
Ithaca, New York

4.1 INTRODUCTION

The globalization of product markets has intensified competition in an increasingly wide array of industries, including automobiles, consumer electronics, steel, and computer chips to name just a few. In manufacturing as a whole during the last 30 years, productivity growth in the United States has lagged significantly behind that of Japan, Germany, Sweden, and many other industrialized countries. For example, between 1960 and 1985, the annual growth in manufacturing productivity (output per hour) was 2.7 percent in the United States compared with 8.0 percent in Japan. Unless this trend can be turned around, U.S. companies will find it increasingly difficult to compete in the world market.[1]

Automation is one important factor in productivity and productivity growth. In particular, organizations are increasingly turning to advanced manufacturing technology (AMT) to enhance their ability to compete. Advanced manufacturing technology typically refers to flexible manufacturing systems

[1]On a brighter note, U.S. productivity growth was more competitive during the late 1980s (Gerhart and Milkovich, 1992).

Organization and Management of Advanced Manufacturing. Edited by Waldemar Karwowski and Gavriel Salvendy.
ISBN 0-471-55508-8

(FMS), computer numerical control (CNC), computer-aided design and computer-aided manufacturing (CAD/CAM), and computer-integrated manufacturing (CIM). It is sometimes also used to include total quality management (TQM) and just-in-time (JIT) strategies (Harrison, 1990).

As the chapters in this volume indicate, however, the payoff to AMT depends on developing management structures and practices that take full advantage of employee skills and abilities. Hayes and Jaikumar (1988) have argued that U.S. managers have too often "acquired new equipment much in the way a family buys a new car. Drive out the old, drive in the new, enjoy the faster, smoother, more economical ride—and go on with life as before. With the new technology, however, 'as before' can mean disaster" (p. 77). A company cannot "exploit the full potential of advanced equipment unless it organized to do so" (p. 83).

In the chapter our focus is on the specific role that compensation and related activities like performance appraisal play in the effective management of human resources in AMT settings. As Gerhart and Milkovich (1992) have argued, compensation decisions are of key strategic importance, with some empirical evidence suggesting that such decisions may have significant consequences for organizations' financial performance (Gerhart and Milkovich, 1990). This is not surprising because compensation has at least two important attributes. First, employee pay and benefits typically account for a substantial portion of total operating costs. Although this share is ordinarily smaller in AMT settings because of the automation of many labor-intensive activities, it is still significant, and thus requires careful management. Second, beyond cost, compensation decisions can have important consequences for key outcomes like job satisfaction, attraction, retention, performance, flexibility, initiative, commitment, cooperation, skill acquisition, individual performance, and, as indicated above, ultimately organization performance. Putting the cost (investment) and behavioral/performance (return) aspects together makes it clear that compensation can have an important influence on productivity and, therefore, the ability of an organization to compete in the global marketplace.

Prior to our discussion of the link between compensation and productivity, it may be useful to provide an example of the magnitude of productivity differences that can be found across organizations within a particular industry. A comprehensive study of the automobile industry (Womack et al., 1990) is very helpful in this regard, particularly given the important role of automation in this sector of the economy. Womack et al. note that the U.S. and Canadian share of world motor vehicle production has fallen from over 70 percent in 1955 to approximately 25 percent in the late 1980s. The share of the U.S. domestic motor vehicle market held by U.S. manufacturers has fallen from nearly 100 percent in 1955 to just over 60 percent by the late 1980s. What explains this decline in the market share of U.S. companies?

Comparisons with Japanese producers shed some light on the issue. Womack et al. (1990) report that the average Japanese plant requires fewer assembly hours per vehicle (17 versus 25), one-third less physical space, and carries a dramatically lower inventory (0.2 days versus 2.9 days). Moreover, the Japanese plants demonstrate that such efficiency does not have to come at the expense of quality. In fact, their quality (e.g., defects per 100 vehicles) is 25 percent better than that of U.S. manufacturers (Womack et al., 1990, p. 92).[2] Finally, Japanese producers achieve all of this despite having lower labor costs (Capdevielle, 1989).[3]

How is the average Japanese plant able to produce motor vehicles with greater efficiency and quality, as well as lower labor costs? One explanation revolves around automation. The average Japanese motor vehicle plant is more heavily automated than its U.S. counterpart, and Womack et al. (1990) show that higher automation significantly reduces the number of assembly hours per vehicle ($r = -0.67$). Further, this high investment in automation has not come at the price of reduced flexibility. As Adler (1988, p. 40) has noted, AMT strategies such as flexible automation "undermine a deeply and widely felt intuition . . . that there is a fundamental tension between innovation and efficiency" (see also Hayes and Wheelwright, 1984).

However, consistent with the Hayes and Jaikumar (1988) argument, Womack et al. (1990) also concluded that the Japanese advantage in automation was only one (albeit an important) part of the story, explaining perhaps one-third of total productivity differences between plants. As they put it, "automation and manufacturability are important, but gaining the full potential of either requires superior plant management" (p. 98). Womack et al.'s (1990) focus here is on *lean production*, which has two key features: "It transfers the maximum number of tasks and responsibilities to those workers actually adding value to the car on the line, and it has in place a system for detecting defects that quickly traces every problem, once discovered, to its ultimate cause" (p. 99).

4.2 COMPENSATION AND HUMAN RESOURCE IMPLICATIONS OF AMT

What do AMT and lean production require of employees and employment relations? According to Womack et al. (1990), it is the following: teamwork, open communication and information-sharing, and skill breadth (e.g., to fill in for each other and be able to quickly solve any problems that arise).

[2]Deming's work also questions the widespread belief that there is a trade-off between quantity and quality, suggesting that they are positively related (as in the automotive data example in the text).

[3]Fluctuations in currency exchange rates make such comparisons difficult. However, over the years, hourly compensation costs (including benefits) have been lower in Japan.

Employees also "need encouragement to think actively, indeed proactively, so they can devise solutions before problems become too serious." Further, they argue that "workers respond only when there exists some sense of reciprocal obligation, a sense that management actually values skilled workers, will make sacrifices to retain them, and is willing to delegate responsibility to the team" (p. 99). Again, comparisons between Japanese and U.S. motor vehicle producers suggest significant differences on these dimensions. The Japanese have a greater percentage of the work force in teams (69 versus 17 percent), more suggestions per employee (62 versus 0.4), fewer job classes (12 versus 67), more training for new production workers (380 versus 46 hours), and significantly greater use of job rotation (3 versus 0.9 on a scale ranging from 0 = none to 4 = frequent). Employment stability is also significantly greater in both the automotive industry (Womack et al. 1990) and economy wide (see Gerhart, 1991, for a partial review).

Consistent with the above, a multi-industry study of practices in 24 AMT settings (Walton and Susman, 1987) concluded that the following factors characterize the AMT workplace: (a) greater interdependence among employees and their activities, with smaller organizational units that make an entire product or part; (b) different (often higher) skill requirements and fewer job classifications; (c) increased cost of errors or malfunctions; (d) skill, knowledge and mental effort are more important than physical effort; (e) more monitoring, maintenance, and other long-cycle activities that can make it easy for employees to become inattentive, although it "requires workers to be alert and take initiatives"; (f) continuous change and the need for adaptability and the ability to learn; (g) higher investment per employee and leaner staffing levels; and (h) more decentralized decision making.

These consequences of AMT have implications for each of the basic areas of compensation (e.g, Gerhart and Milkovich, 1992): structure, level, individual differences (e.g., merit pay), and benefits. We will examine each in turn (and also briefly touch on the topic of top management pay). As a general point, we note that the influence of compensation practices on important objectives can be very direct. Skill-based pay, for example, directly pays employees to acquire skills. Just as important, however, are the "indirect" effects of compensation. For example, the compensation system can either facilitate or constrain the success of other human resource management activities in areas such as recruiting, selection, training, development, participative decision making, and team approaches. At times, in fact, a compensation program can be particularly successful for what it does not do. Deming (1986), for example, argues that most organizations do more harm than good when they reward managers for achieving narrow (e.g., individual, departmental), numerical, short-term goals. The implication is that removal of rewards for such behaviors would benefit the organization.

We now turn to a discussion of the components of compensation, beginning with pay structures.

4.3 STRUCTURE

Paying the Job

Pay structures are typically defined in terms of jobs. Tasks and responsibilities for jobs are clearly specified in job descriptions. Job evaluation is used to measure the worth of jobs to the organization and to allow comparisons with jobs in other organizations for the purpose of establishing market-based rates of pay (Schwab, 1980).

However, Lawler (1990) has argued that job-based pay structures can be characterized by one or more of the following potential disadvantages. First, paying the job instead of the individual may encourage a bureaucratic orientation (see also Kanter, 1989). The job description sets out specific tasks and activities for which the incumbent is responsible and, by implication, those for which the incumbent is not responsible. Although this facilitates performance evaluation and control by the manager, it can also encourage a lack of flexibility and a lack of initiative on the part of employees as in "Why should I do that?—It's not my job description." Second, the hierarchical nature of this system entails mostly top-down decision making and information flow as well as status differentials. This is not consistent with the increasing emphasis on taking advantage of the skills and knowledge of those closest to production or with reciprocal feedback and information sharing. Third, the bureaucracy required to generate and update job descriptions and job evaluations can become a barrier to change because wholesale changes to job descriptions can involve a tremendous amount of time and cost. Fourth, the job-based pay system may not reward desired behaviors, particularly in a rapidly changing environment where the knowledge, skills, and abilities needed yesterday may not be very helpful today and tomorrow. Lawler suggests that promotion-seeking behavior is typically rewarded, which may not be consistent with goals such as skill acquisition and continuous improvement that are being increasingly embraced by AMT and other organizations.

One response by some organizations has been to reduce the number of job classifications and levels, sometimes dramatically. For example, General Electric replaced its 29 pay grades with 5 grades. Pay grades 8 through 11, for instance, were combined into one "band" ranging from $33,000 to $74,000 (*Business Week*, December 10, 1990), giving GE considerably more flexibility in job assignments and in assigning merit increases. In the automobile industry, Saturn, the new division of General Motors, has only a few job classifications, in sharp contrast to most other U.S. automobile manufacturers where the average is 67, but very similar to Japanese manufacturers where the average is 12 (Womack et al., 1990). One possible consequence, however, of a reduction in job classifications is a reduced opportunity for promotion. Therefore, organizations need to consider what they will offer employees instead. Continuous training, greater responsibility, and pay tied to seniority or performance are some of the approaches being used (*Business Week*, 1990).

Finally, reducing job classifications as a means of simply getting more work out of employees and without decentralizing decision making may not be a recipe for success (Keefe and Katz, 1990).

Paying the Person: Pay for Skill and Knowledge

As another response to the drawbacks of job-based systems, some organizations have turned to an alternative type of structure based on individual characteristics such as skill or knowledge (Lawler, 1990; Ledford, 1991). The idea is very simple. If you want employees to learn more skills and become more flexible in the jobs they perform, pay them to do it. It is, however, "a fundamental departure" because employees are now "paid for the skills they are capable of using, not for the job they are performing at a particular point in time" (Ledford, 1991, p. 199).

Conceptually, skill-based systems seen to fit well with the increased breadth and depth of skill that many authors believe is necessary to maximize AMT effectiveness (Adler, 1988; Cummings and Blumberg, 1987; Gupta, 1989); Walton and Susman, 1987; Womack et al., 1990). For example, instead of being limited to loading, monitoring, and unloading parts from a machine, operators might be expected to take responsibility for maintenance and troubleshooting, reducing variance in the quality of materials, adjusting machines to within needed tolerances, or modifying computer programs (Wall et al., 1990). Cummings and Blumberg (1987, p. 48, cited in Wall et al., 1990) recommended that production workers should be "given the necessary skills, information and freedom to respond to unforseen circumstances." In similar fashion, Toyota concluded years ago that "none of the specialists [e.g., quality inspectors, many managers, and foremen] beyond the assembly worker was actually adding any value to the car. What's more . . . assembly workers could probably do most of the functions of specialists much better because of their direct acquaintance with conditions on the line" (Womack et al., 1990, p. 56).

Thus, the key advantage of skill-based pay in an AMT environment is its contribution to increased flexibility and the ability to decentralize decision making to those who are most knowledgeable. This flexibility can help in covering employee turnover or absenteeism (Lawler, 1990), thus permitting learner staffing levels. In addition, multiskilled employees are important in cases where different products are manufactured requiring different processes, or where supply shortages or other problems call for adaptive or flexible responses (Lawler)—characteristics typical of many AMT environments (e.g., flexible manufacturing, just-in-time systems). So, if there is an equipment failure or a need to shift resources on short notice to a particular part of production, multiskilled employees should be better prepared. Lawler also observes that in addition to the content of the particular skills acquired, skill-based plans may carry a more general benefit in that they foster a climate of learning and adaptability. Finally, employees with a broader view of the production process and organization should be in a better position to participate in decision making and make constructive suggestions for continuous improvement.

On the other hand, skill-based approaches also have a number of potential drawbacks (Gupta et al., 1986; Gerhart and Milkovich, 1990; Lawler, 1990). First, such plans can become very costly as employees acquire more skills that are not being utilized but must, nevertheless, be compensated. Thus, it is essential that the work process be structured so as to take full advantage of this costly investment. Obviously, creating multiskilled employees who continue to work on one narrow job will not offer a good return on investment. Second, as with job-based systems, "topping out" can be a problem if employees reach a point where they have acquired all the skills and are therefore not eligible for further pay increases. Third, to effectively administer the system, a large bureaucracy may arise, ironically, very similar to that which occurs with job-based systems. Training programs need to be developed. Skills (e.g., skill blocks) must be described, measured, and assigned monetary values. Certification tests of some sort must be constructed to determine whether an employee has in fact acquired a certain skill. As Ledford (1991) points out, this can be a very contentious process, and has no real counterpart in job-based systems. Finally, it may be difficult to determine market rates under skill-based systems, partly because relatively few companies are available for comparison.

4.4 PAY LEVEL AND LABOR COST

Initially, it will be helpful to think of pay level as cost per employee. As discussed, skill level and breadth requirements often increase with the introduction of AMT. For several (related) reasons, this may result in organizations choosing (or being pressured) to raise pay levels. Higher levels and breadth of skill typically bring higher pay in the labor market (e.g., Campion and Berger, 1990). Not surprisingly then, a substantial body of evidence indicates that employees expect higher pay in return for increasing their skill level and breadth, and without what they see as fair (higher) pay, they may strongly resist the introduction of AMT technology (Cummings and Blumberg, 1987; Wall et al., 1990). Some organizations act proactively by using skill-based pay programs, which as discussed earlier, can also result in higher pay levels.

Second, employee movement can be very costly in an AMT environment. As the organization's investment in its employees grows, so too does the cost of employee movement. Therefore, a higher pay level policy may be chosen to enhance retention and, thus, the probability of realizing an adequate return on the investment in training. Employee retention is also important in avoiding possible disruptions in production while replacement employees are found. The large investment in capital equipment in AMT settings can make such disruptions very costly if it results in equipment being left idle.

Third, if as Walton and Susman (1987) suggest, the cost of errors increases under AMT, a high-quality work force becomes more important. Consequently, a high pay level may be chosen to achieve greater selectivity in hiring decisions.

A final factor relevant to pay level in AMT is the ratio of total labor costs to either total costs or revenues. These ratios are typically much lower under AMT because of a combination of large investments in technology and low staffing levels. The fact that labor costs are a smaller percentage of total costs does not necessarily mean that pay level will be higher, but it may permit greater flexibility on the part of the organization to choose a higher pay level.

4.5 INDIVIDUAL DIFFERENCES IN PAY

Pay structures include definitions of pay averages and ranges. Within ranges, especially those that are job based, individual differences in pay are often determined through one or more "pay for performance" programs such as merit pay, individual incentives, group incentives, gain sharing, profit sharing, and ownership.[4] In describing specific programs, it may be useful to keep in mind a classification scheme based on two dimensions (Milkovich and Wigdor, 1991): (a) whether pay changes are rolled into the base (yes for merit pay, but not for most others), and (b) the criterion (e.g., organization profits, group output, individual performance appraisal) on which pay is based. The first dimension has implications for the long-term accumulation of labor costs, and the second dimension has consequences for the types of behaviors encouraged (and by implication, discouraged) under the compensation system.

Traditionally, individual differences in pay (and to some extent, pay structures) have been viewed as most important in influencing the behaviors and attitudes of current employees, whereas pay level and benefits have been seen as more critical to achieving attraction and retention objectives. However, as Gerhart and Milkovich (1992) have noted, this view leaves out the potentially important influence of individual pay programs on the composition of the work force (see also Milkovich and Wigdor, 1991). It may be, for instance, that high performers will be most attracted to organizations that have a strong pay for (individual) performance relationship. For example, evidence indicates that such organizations are more attractive among those that have a high need for achievement (Bretz et al., 1989) or strongly value individual achievement (Judge and Bretz, 1992). By the same logic, it is also possible that people who prefer to work in teams may be more attracted to organizations that link pay to team performance. If so, this reinforces the

[4]In some cases rate ranges are not used. Instead all employees in a particular job are paid a single rate. In other cases a range may exist, but pay increases are based on seniority. These provisions are most common in the union sector (Freeman and Medoff, 1984).

need to make sure that compensation design is consistent with other human resource policies and objectives.

As a final general observation, we suggest that the following pay programs should not be viewed as competing alternatives in making design decisions. Instead, because organizations have multiple, sometimes conflicting objectives, multiple compensation programs are likely to be necessary (Gerhart and Milkovich, 1992; Lawler, 1990; Majchrzak, 1988). So, profit sharing might be well-suited to helping control fixed labor costs and promoting a broad view, but at the same time, there are individual and group level objectives that may be balanced through the simultaneous use of merit pay and team awards.

Merit Pay

Merit pay ties increases in base pay to individual performance. As a report by Bretz et al. (1992) indicates, merit pay continues to be widely used in U.S. companies. As such, we give it and related activities such as performance appraisal considerable attention here.

The traditional approach to performance assessment is often described as having the following features. First, there is a focus on identifying individual differences in performance. These are assumed to reflect differences in ability or motivation. By implication, system constraints on performance are not seen as significant. Second, the majority of information on individual performance is collected from the immediate supervisor. Peer and subordinate ratings are rare and where they do exist they tend to receive less weight than supervisory ratings (Bretz et al., 1992). Third, there is a policy of linking pay increases to performance appraisal results (Bretz et al., 1992). Fourth, the feedback under such systems tends to occur infrequently, often once per year at the formal performance review session. Fifth, the flow of feedback tends to be largely unidirectional, from supervisor to subordinate.

Despite its widespread use, merit pay as typically practiced is being increasingly questioned, especially in the context of AMT, where teams, interdependence, flexibility, and decentralization are often key considerations. Best known perhaps are several criticisms raised by Deming (1986). First, he argues that it is unfair to rate individual performance because "apparent differences between people arise almost entirely from the system that they work in, not from the people themselves. A man not promoted is unable to understand why his performance is lower than someone else's. No wonder; his rating was the result of a lottery" (1986, p. 110). As examples of system factors, Deming cites a person's co-workers, the job, materials, equipment, customers, management, supervision, and environmental conditions. He argues that these are, for the most part, outside of the worker's control, instead falling under management's responsibility.

Second, Deming (1986) argues that merit pay discourages teamwork. As he puts it: "Everyone propels himself forward, or tries to, for his own good, on his own life preserver. The organization is the loser" (p. 102). As one example, he notes that if people in the purchasing department are evaluated on something like the number of contracts negotiated per worker-year, they may not take much interest in improving the quality of materials because that might help people in manufacturing, but not necessarily them (i.e., their pay). So, they do not seek such improvement even though the organization suffers as a result. (See also Hayes and Jaikumar, 1988, on this point.)

Deming's suggested solution is to eliminate the link between individual performance and pay. What is used instead? Schuler and Harris's (1991) case study of a small electronic instrument manufacturer provides some insight. The company has attempted to follow the Deming philosophy. It pays hourly workers on the basis of seniority (not performance) and a profit-sharing plan. (It is not clear whether seniority also determines promotions.) Performance appraisals are used for developmental purposes and where necessary, to identify employees who are "out of the system."[5]

Salaried employees are also on the profit-sharing plan, but in other respects, the determination of their pay is "a little different" (Schuler and Harris, 1991, p. 201). This appears to be something of an understatement. They note that "greater competition for skilled employees forces [the company] to adjust to the market" (p. 201), and that significant adjustments take place "when the employee and management become aware of an imbalance between the employee's skills and the public market valuation of these skills" (p. 202).

This approach sounds very much like pay for performance to us and illustrates the limitations of arguments that call for completely ignoring individual differences in performance, especially in key jobs. As discussed earlier, high performers may be more attracted to organizations that do pay for performance. Organizations that do not pay individual performance face the risk of losing higher performers (Gerhart, 1990). At some point, this disadvantage will negate any advantage derived from an egalitarian approach to pay. Therefore, although we agree with Deming that individual differences in performance may be given too much emphasis in many cases, and certain steps can be taken to reduce this emphasis (e.g., Scholtes, 1987), the answer is not to completely eliminate merit pay. This would only replace one set of problems with another. Rather, as discussed earlier, the goal is to design a mix of pay programs that will be effective in accomplishing a mix of objectives.

Another issue concerns the heavy reliance on the supervisor as the source of performance information and feedback. In a team environment, for example, there may not be a supervisor or if there is, he or she may delegate many of the traditional roles to the team. In such cases peer ratings may be considered as an alternative to supervisory ratings. What is the evidence on peer ratings?

On some dimensions, peer ratings have received high praise. For example,

[5]A typical rule of thumb is three or more standard deviations from the average.

McEvoy and Buller's (1987) analysis of the literature suggested that peer ratings are more stable than supervisory ratings and more likely to focus on task-relevant abilities and competencies. Wexley and Klimoski (1984, cited in McEvoy and Buller, 1987, p. 785), in fact, concluded that peer ratings were "potentially the most accurate judgments of employee behavior" (p. 60). This may be due, in large part, to the ample opportunity that peers have to observe other's performance.

Nevertheless, peer ratings are very rare in practice, with one study finding that "less than 3% of the rating information used in the final appraisal decision comes from [peer and other nontraditional] sources" (Bretz et al., 1992, p. 9). Why is there this discrepancy between research and practice?

The most common explanation in the academic literature is that employees simply do not like peer ratings (see empirical findings by Cederblom and Lounsbury, 1980; Downey et al., 1976; Love, 1981) for a variety of reasons (DeNisi and Mitchell, 1978; Kane and Lawler, 1978; Smith, 1976; see McEvoy and Buller, 1987, for a summary), including: (a) manager's reluctance to turn over evaluation power to others, (b) the belief that peer ratings are a popularity contest, (c) reliance of peers on stereotypes, and (d) the possibility of retaliation if one provides a peer with a low rating.

However, there are some potential problems with both the empirical evidence and the logic employed in the peer ratings literature. Without exception, the empirical literature has focused on the following question: How much do employees like peer ratings in an absolute sense? As discussed, there is typically a lack of enthusiasm. But, we would argue, the more relevant question concerns how employees view peer ratings *relative* to supervisory ratings. This question has not been asked, but evidence from other sources suggests that supervisory ratings are not terribly popular among either supervisors (Rice, 1985) or the employees they rate.[6] Thus, it may be a mistake to make too much of the lack of user acceptance of peer ratings found in the literature. Moreover, McEvoy and Buller (1987) argue that the low levels of user acceptance found in some previous studies of peer ratings (e.g., Cederblom and Lounsbury, 1980) may be due, in large part, to the fact that there was not confidentiality of raters.

Closer examination of the suggested drawbacks of peer ratings [see points (a)–(d)above] suggests similar doubts. In particular, problems (b), (c), and (d) are not unique to peer ratings—they apply equally well to supervisory ratings. Again, the more relevant (but unstudied) question is whether peer ratings are more susceptible to such problems than are supervisory ratings.

---

[6]Unions, for example, have typically opposed using supervisory ratings as a basis for pay. "Because the value of worker's contribution to a firm is extremely difficult to measure and different supervisors may read the same facts in different ways, the union will seek to protect the membership from the uncertainty of arbitrary supervisory decisions by pressing for a one rate-per-*job pay policy*" (Rice, 1985, p. 80). Note the similarity between this view and Deming's (1986) criticisms of merit pay.

There are, in fact, several reasons to believe that peer ratings can be superior to supervisory ratings under certain conditions. Performance rating accuracy can be thought of as a function of ability and motivation. Key factors on the ability side are the opportunity to observe performance and knowledge of what constitutes good performance. In a team setting, peers are likely to be superior to supervisors on both counts. The motivation side is less straightforward, with key considerations being the reward system and the degree of decentralization of actual decision making. For example, where pay is linked to individual performance and is relative (e.g., ranking), the motivation to rate accurately may not be high. On the other hand, in cases where pay is based on team performance and the team has the formal authority to take remedial (or even disciplinary) action, motivation to rate accurately can be quite high.

Whatever the source of performance rating information, acceptance of the process is likely to be greater if employees believe it is procedurally fair. This may be more likely in cases where employees (a) believe feedback is well-grounded in evidence, (b) feel that consistent performance standards are used, and (c) have an opportunity to provide input prior to a recommendation (Folger and Konovsky, 1989).

Individual Incentives

Like merit pay, individual incentives reward individual performance. But, in this case, payments are not rolled into the base and the performance measure is usually some type of physical output rather than a subjective assessment. Individual incentives have the potential to significantly impact performance, but are relatively rare for a variety of reasons (Gerhart and Milkovich, 1992). First, they simply are not feasible for many jobs because there is no physical output measure. Second, there are a host of potential administrative problems (e.g., in setting and maintaining acceptable standards) that often prove to be intractable. Third, individual incentives may do too good of a job of motivating employees in the following sense: They do whatever they get paid for and nothing else.

Individual incentives can be particularly ill-suited to many AMT environments. For instance, individual incentives typically do not fit well with the team approach. The payment system does not encourage cooperation or any sort of organizational citizenship behavior, which has been defined as discretionary behaviors "that are not part of employees' formal role requirements, but nevertheless promote the effective functioning of the organization" (Organ, 1988, p. 4). Individual incentives may also be inconsistent with multi-skilling and proactive problem solving. Learning new skills can entail a large opportunity cost because of lost income. In addition, income can be maximized by specializing in one task (e.g., a particular machine). Proactive problem solving is likely to have a very narrow focus, limited to maximizing output (and thus income) on a particular task. Therefore, although individual incentives carry potential advantages, these are not likely to include contributing to a flexible, proactive, problem-solving work force.

Finally, in cases where individual incentives have been in place for some time, they may be particularly difficult to remove. If, as suggested above, different compensation systems attract different types of employees, it may be the case that individual incentives attract the most individualistic employees. An easy transition to cooperation and teamwork may be especially difficult in this case. Somewhat related, as suggested in the case of merit pay, high performers earning high pay may be more prone to question what they stand to gain from a more egalitarian approach to pay.

Profit Sharing and Employee Ownership

At the other end of the continuum from individual plans are profit-sharing and stock ownership plans. Under profit sharing, payments are based on an organization level performance measure (profits) and the payments do not become part of the base salary. Profit sharing has two potential advantages. First, it may encourage employees to think more like owners, taking a broad view of what needs to be done to make the organization more effective. Thus, the narrow self-interest encouraged by individual incentive plans (and perhaps also by merit pay) would presumably be less of an issue. Rather, greater levels of cooperation and citizenship behaviors would be expected. Second, because payments do not become part of base pay, it automatically reduces labor costs during difficult economic times and shares the wealth during good times. As one consequence, organizations may not need to rely on layoffs as much to reduce costs during tough times (Weitzman, 1984).

Although there is some evidence that employment stability is greater under profit sharing (Chelius and Smith, 1990; Gerhart, 1990; Kruse, 1991), the evidence on performance effects is less clear. Weitzman and Kruse (1990) conclude that profit sharing does increase organization performance, but Gerhart and Milkovich (1992) question the nature of much of this evidence. Conceptually, Deming's point about the relative influence of the individual versus the system comes to mind here. It is not clear that any single employee (except for top executives and other key employees) is likely to see much of a link between his or her performance and profits, because the latter are influenced by so many factors beyond his or her control and outside the scope of the job. Consequently, profit sharing may not only fail to increase performance motivation, but employees may also view as unfair the fact that they are expected to earn less money during business downturns if they feel that they are not to blame.[7] In addition, at lower pay levels, income reductions

[7]The much publicized profit-sharing plan for the Dupont Fibers Division (Santora, 1991) was eliminated when division profits were down and employees were about to actually experience "downside risk." Saturn has also recently scaled back its reliance on profit sharing because of lower than expected profits.

due to low profits can be very painful. Therefore, although profit sharing may be useful as one component of a compensation system (e.g., to enhance identification with broad organizational goals), it may need to be complemented with other pay programs that more closely link pay to outcomes that individuals or teams can control (or "own").

Employee ownership vehicles such as stock options are similar to profit sharing in that employees are encouraged to identify with broad organization goals. As owners, this identification may be stronger. In addition, under employee stock ownership plans (ESOPs), employees must be permitted to vote their securities if they are registered on a national exchange (Conte and Svejnar, 1990). In this sense participation in decision making may be given a boost. Finally, ESOPs can carry important tax and financing advantages and are sometimes used as a takeover defense (under the assumption that employee owners will be "friendly" to management). Given the similarity with profit sharing, stock ownership also has the same potential disadvantages—a questionable impact on individual motivation and significant income risk.

Gain Sharing, Group Incentives, and Team Awards

These plans are similar in that they (a) do not roll payments into the base, and (b) they attempt to blend the advantages of individual and organizational level pay plans. As Milkovich and Wigdor (1991) have suggested, these plans "may provide a way to accommodate the complexity and interdependence of jobs, the need for work group cooperation, and the existence of work group performance norms and still offer the motivational potential of clear goals, clear pay-to-performance links, and relatively large pay increases" (p. 86).

One way to distinguish the plans is in terms of coverage breadth. Gain-sharing plans may encompass an entire business unit (e.g, division, department, plant), whereas group incentives and team awards would typically cover one of the groups or teams within the business unit (Welbourne and Gomez-Mejia, 1991). Perhaps the main distinction between group incentives and team awards is that the former typically refers to plans where performance is measured as physical output, whereas team awards can be tied to a variety of performance measures (e.g., cost savings, successful completion of product design, meeting deadlines; see Welbourne and Gomez-Mejia, 1991). Gain-sharing plans can also include a similar variety of performance measures.

The similarity of group incentive plans to individual incentive plans suggests some of the same potential drawbacks such as narrowness of focus, as well as lack of cooperation (in this case, between groups rather than individuals). Team awards also raise the question of how interteam relationships can be effectively handled (Welbourne and Gomez-Mejia, 1991).

Because of its wider coverage, gain sharing may offer something of a solution to competition between groups or teams. This can be a particularly important consideration, especially in AMT environments, where coordination and cooperation between groups and individuals in different functions and departments is often critical. Like team awards, gain-sharing plans can make use of a wide variety of performance measures, including productivity ratios such as labor cost/sales, actual hours/standard hours, or even return on investment at the plant level if the accounting system permits.

Gain sharing plans are often confused with profit sharing plans, but there are important differences. In contrast to profit sharing, (a) payment is based on a measure of productivity (often plant level) not profits (usually organization level); (b) payments are distributed more frequently (e.g., monthly or quarterly); and (c) closely related, the payments are not deferred. (They are under most profit sharing plans.)

More generally, in practice, gain sharing plans often include nonpay components such as periodic meetings, problem-solving teams, newsletters, steering committees, and suggestion systems (Conference Board, 1990). As several authors (Hammer, 1988; Milkovich and Wigdor, 1991; Mitchell et al., 1990) have noted, these nonpay aspects may well explain some of the success associated with gain-sharing plans. In one study, for example, goal setting and feedback interventions appeared to be much more important than monetary group incentives in generating performance improvements (Pritchard et al., 1988). However, Pritchard et al. suggested that the monetary incentives may have been important in sustaining such performance improvements over the longer run. In addition, evidence from other studies indicates that compensation can have a significant independent effect on performance (Schuster, 1990), as illustrated, for example, by the positive experiences with Improshare (Fein, 1981) and other pay-oriented plans (Kaufman, 1992; Wagner et al., 1988).

Ross and Ross (1991) argue that the following contextual factors are conducive to success of gain-sharing programs:

a. Management commitment
b. Either a need to change or a strong commitment to continuous improvement
c. Management acceptance and encouragement of employee input
d. High levels of cooperation and interaction
e. Employment security
f. Information sharing on productivity and costs
g. Goal setting
h. Commitment of all involved parties to the process of change and improvement
i. Agreement on a performance standard and calculation that is understandable, seen as fair, and closely related to managerial objectives

Finally, one administrative challenge that can arise with any of these three plans concerns the question: Who's on the team (or in the group)?[8] A project development team, for example, might include people from marketing, research, engineering, and so forth. How long is each individual involved in the project and how much time does each allocate while involved? Each person may be on multiple "teams" at any point in time, which may be difficult to define because they are fluid. The result can be a significant administrative challenge.

4.6 BENEFITS

Several features of AMT have important consequences for the design of benefits. First, given the large capital investment in automation, production downtime is expensive and needs to be minimized. Variation in production can also cause difficulties under many circumstances, such as under a just-in-time system. Underutilization of equipment can stem from several sources, including breakdowns, tool availability, setup time, rejected products, and lack of operator availability (Lenz, 1989). Benefits programs can have a significant influence on the latter through their influence on both employee attendance and retention (Gerhart and Milkovich, 1992). Thus, basic decisions about sick leave policies (amount paid, disciplinary actions for excessive sick leave) take on added importance because of the increased concern with avoiding staffing problems.

Second, as discussed earlier, AMT often entails higher skill requirements (Walton and Susman, 1987). Similar to its investment in physical capital, the organization needs to maximize the return on its investment in employees' human capital (Hayes and Jaikumar, 1988). Therefore, employee retention takes on added importance. In addition to pay level decisions, benefits programs such as pensions and medical insurance need to be designed with this in mind to reduce voluntary quits. More broadly, benefits such as family leave and wellness programs can also be viewed as playing an important role in making sure that human capital is retained (and maintained in good condition).

Third, the organization must create a climate of trust and mutual obligation if it expects employees to accept new technology, learn new skills, and be more proactive. A common theme is that employment security may be necessary to achieve this objective (Hayes and Jaikumar, 1988; Majchrzak, 1988; Walton and Susman, 1987; Womack et al., 1990). As Schuler and Harris (1991) note, although such a practice "strikes many as out-dated and impractical . . . companies interested in quality are rethinking this . . . [because] they recognize that repeated downsizing to reduce costs destroys credibility in HR [human resource] practices that claim to view employees as important"

[8]Thanks to Robert Mulkey for sharing his experiences with this issue.

(p. 196). They also point out that the Malcom Baldridge Award places significant weight on employment security.

4.7 TOP MANAGEMENT PAY

Top managers have great potential to influence organizational success. Consequently, their pay may warrant some addition discussion. One might argue that a long-term perspective in decision making takes on added importance in an AMT environment. Automation typically entails a large investment with returns often not realized until several years down the road. Unfortunately, U.S. managers are often described as having a short-run orientation (Hayes and Abernathy, 1980), with a significant amount of the blame being attributed to the design of their compensation (e.g., linking bonuses to short-term, e.g., quarterly financial performance; e.g., Rappaport, 1978). Deming's (1986) criticism of the use of numerical goals is in the same general vein. As with individual incentive systems, precise, difficult goals can often be achieved quite successfully. One problem, however, is that trade-offs may be made with other important (but not as clearly specified) goals, such as quality and cooperation.

To encourage long-term (and sometimes risky) investments, greater emphasis on vehicles such as restricted stock, stock options, and so forth can be considered. Some evidence, for example, suggests that their usage is linked to increased capital investment (Larcker, 1983). In addition, research suggests that an emphasis on long-term goals in executive compensation packages can decrease year-to-year fluctuations in employment levels (Gerhart, 1991), which may be desired to preserve investments in human capital and to show employees that there is a sense of mutual obligation.

Finally, we should note the continued attention in the popular press to the level of top management pay in the United States and how it compares with pay in other countries (e.g., *Business Week*, 1992). In 1989, Towers, Perrin, Forster, and Crosby (TPF&C) reported that the average total remuneration (base, bonus, long-term incentives, benefits, perquisites) of U.S. chief executive and chief financial officers in companies with at least $250 million in annual sales was $543,000. By comparison, the average was $352,000 in Japan, $287,000 in the former West Germany, and $130,000 in South Korea (*CompFlash*, 1990).[9] In addition, the ratio of chief executive pay to that of hourly production workers was 35 to 1 in the United States, compared with 15 to 1 in Japan and 20 to 1 in Europe (*Industry Week*, 1990).

There is no consensus on what these numbers tell us (Gerhart and Milkovich, 1992). On one side, it can be argued that such pay levels are needed to attract and retain top executive talent. The pool of such people is small and competition for them is great. On the other hand, the ratio of executive

[9]TPF&C reports that spending power differences further widen the U.S. lead in pay.

pay to production worker pay has been spoken of as contributing to a "trust gap," described as a situation where employees distrust and resent top management (*Fortune*, 1989). In any case, if a goal in AMT settings is to reduce status differentials, encourage trust, and develop a team orientation among employees, the design of top management pay is likely to receive a good deal of scrutiny from employees who want to see if management is doing its part.

4.8 CONCLUSION

A recurring theme is that no single pay program is capable of achieving the entire set of complex objectives faced in AMT (and other) environments. Thus, in establishing individual differences in pay, some combination of plans (e.g., merit pay, team awards, and profit sharing) may be chosen to balance multiple and sometimes conflicting objectives. Moreover, the success of decisions in any one major area may be significantly influenced by decisions in others. Welbourne and Gomez-Mejia (1991), for example, discuss the potential futility of using team awards to encourage cooperation among individuals, while at the same time having a hierarchical pay structure that encourages individuals to compete for promotion opportunities.

In general, the attributes of AMT settings will often call for a significantly different approach to the management and utilization of human resources. Compensation is typically a very important influence on the attitudes and behaviors of employees, whether they be in the executive suite or on the production floor (Gerhart and Milkovich, 1992). Consequently, it is important to continually evaluate what attitudes and behaviors are being encouraged (or implicitly discouraged) by the compensation system. Teamwork, innovation, proactive involvement, and other human resource objectives that are especially important in an AMT environment, will otherwise be elusive.

REFERENCES

Adler, P. S. (1988). Managing flexible automation. *California Management Review*, **30**(3), 34–56.

Bretz, R. D., Ash, R. A, and Dreher, G. F. (1989). Do people make the place? An examination of the attraction-selection-attrition hypothesis. *Personnel Psychology*, **42**, 561–581.

Bretz, R. D., Milkovich, G. T., and Read, W. (1992). The current state of performance appraisal research and practice: Concerns, directions, and implications. *Journal of Management*, **18**, 321–352.

Business Week (1990, December 10). Farewell, fast track, pp. 192–200.

Business Week (1992, January 20). Motown's fat cats, pp. 34–35.

Campion, M. A., and Berger, C. J. (1990). Conceptual integration and empirical test of job design and compensation relationships. *Personnel Psychology*, **43**, 525–553.

Capdevielle, P. (1989). International comparisons of hourly compensation costs. *Monthly Labor Review*, **112**, June, 10–12.

Cederblom, D., and Lounsbury, J. W. (1980). An investigation of user acceptance of peer evaluations. *Personnel Psychology*, **33**, 567–579.

Chelius, J., and Smith, R. S. (1990). Profit sharing and employment stability. *Industrial and Labor Relations Review*, **43**, 256S–273S.

CompFlash (1990, March). U.S. executives outearn counterparts in other countries by a large margin, p. 7.

Conference Board. (1990). Variable pay: New performance rewards. Research Bulletin No. 246.

Conte, M. A., and Svejnar, J. (1990). The performance effects of employee ownership plans, in *Paying for Productivity* A. S. Blinder, Ed., Brookings Institution, Washington, D.C., pp. 245–294.

Cummings, T., and Blumberg, M. (1987). Advanced manufacturing technology and work design, in *The Human Side of Advanced Manufacturing Technology*, T. D. Wall, C. W. Clegg, and N. J. Kemp, Eds., Wiley, Chichester.

Deming, W. E. (1986). *Out of the Crisis*, Center for Advanced Engineering Study, Massachusetts Institute of Technology, Cambridge, MA.

DeNisi, A. S., and Mitchell, J. L. (1978). An analysis of peer ratings as predictors and criterion measures and a proposed new application. *Academy of Management Review*, **3**, 369–374.

Downey, R. G., Medland, F. F., and Yates, L. G. (1976). Evaluation of a peer rating system for predicting subsequent promotion of senior military officers. *Journal of Applied Psychology*, **61**, 206–209.

Fein, M. (1981). *Improshare: An Alternative to Traditional Managing*. Institute of Industrial Engineers, Norcross, GA.

Folger, R., and Konovsky, M. A. (1989). Effects of procedural and distributive justice on reactions to pay raise decisions. *Academy of Management Journal*, **32**, 115–130.

Fortune. (1989, December 4). The trust gap, 56–78.

Freeman, R. B., and Medoff, J. L. (1984). *What Do Unions Do*? Basic Books, New York.

Gerhart, B. (1990). Voluntary Turnover, Job Performance, Salary Growth, and Labor Market Conditions. Center for Advanced Human Resource Studies, Working Paper, No. 90-12.

Gerhart, B. (1991). Employment Stability under Different Managerial Compensation Systems. Center for Advanced Human Resource Studies, Cornell University, Working Paper 91-02.

Gerhart, B., and Milkovich, G. T. (1990). Organizational differences in managerial compensation and financial performance. *Academy of Management Journal*, **33**, 663–691.

Gerhart, B., and Milkovich, G. T. (1992). Employee compensation: Research and practice, in *Handbook of Industrial and Organizational Psychology*, M. D. Dunnette and L. M. Hough, Eds., Consulting Psychologists Press, Palo Alto, CA.

Gupta, Y. P. (1989). Human aspects of flexible manufacturing systems. *Production and Inventory Management Journal*, **30**(2), 30–36.

Gupta, N., Jenkins, D., and Curington, W. (1986). Paying for knowledge: Myths and realities. *National Productivity Review*, Spring, 107–123.

Hammer, T. H. (1988). New developments in profit sharing, gainsharing, and employee ownership, in *Productivity in Organizations*, J. P. Campbell, R. J. Campbell, and Associates, Eds., Jossey-Bass, San Francisco.

Harrison, M. (1990). *Advanced Manufacturing Technology Management*, Pitman Publishing, London.

Hayes, R. H., and Jaikumar, R. (1988). Manufacturing's crisis: New technologies, obsolete organizations. *Harvard Business Review*, **66**(5), 77–85.

Hayes, R. H., and Wheelwright, S. C. (1984). *Restoring Our Competitive Edge*. Wiley, New York.

Industry Week (1990, June 18). The pay revolt brews, pp. 28–36.

Judge, T. A., and Bretz, R. D. (1992). The effects of values on job choice decisions. *Journal of Applied Psychology*, **77**, 261–271.

Kane, J. S., and Lawler, E. E. (1978). Methods of peer assessment. *Psychological Bulletin*, **85**, 555–586.

Kanter, R. M. (1989). *When Giants Learn to Dance*, Simon and Schuster, New York.

Kaufman, R. T. (1992). The effects of Improshare on productivity. *Industrial and Labor Relations Review*, **45**, 311–322.

Keefe, J. H., and Katz, H. C. (1990). Job classifications and plant performance in the auto industry. *Industrial Relations*, **29**, 111–118.

Kruse, D. L. (1991). Profit-sharing and employment variability: Microeconomic evidence on the Weitzman theory. *Industrial and Labor Relations Review*, **44**, 437–453.

Larcker, D. (1983). The association between performance plan adoption and corporate capital investment. *Journal of Accounting and Economics*, **5**, 3–30.

Lawler, E. E. III. (1990). *Strategic Pay*, Jossey-Bass, San Francisco.

Ledford, G. E., Jr. (1991). The design of skill-based programs, in *The Compensation Handbook*, 3rd ed., M. L. Rock and L. A. Berger, Eds., McGraw-Hill, New York.

Lenz, J. E. (1989). *Flexible manufacturing*, Marcel Dekker, New York.

Love, K. G. (1981). Comparison of peer assessment methods: Reliability, validity, friendship bias, and user reaction. *Journal of Applied Psychology*, **66**, 451–457.

Majchrzak, A. (1988). *The Human Side of Factory Automation*, Jossey-Bass, San Francisco.

McEvoy, G. M., and Buller, P. F. (1987). User acceptance of peer appraisals in an industrial setting. *Psychological Bulletin*, **40**, 785–797.

Milkovich, G. T., and Wigdor, A. K. (1991). *Pay for Performance*, National Academy Press, Washington, D.C.

Mitchell, D. J. B., Lewin, D., and Lawler, E. E. III. (1990). Alternative pay systems, firm performance, and productivity, in *Paying for Productivity*. A. S. Blinder, Ed., Brookings Institution, Washington, D.C.

Organ, D. W. (1988). *Organizational Citizenship Behavior: The Good Soldier Syndrome*, Lexington Books, Lexington, MA.

Pritchard, R. D., Jones, S. D., Roth, P. L., Stuebing, K. K., and Ekeberg, S. E. (1988). Effects of group feedback, goal setting, and incentives on organizational productivity (monograph). *Journal of Applied Psychology*, **73**, 337–358.

Rappaport A. (1978). Executive incentives versus corporate growth. *Harvard Business Review*, **56**(4), 81–88.

Rice, B. (1985). Performance review: The job nobody likes. *Psychology Today*, **19**(9), 30–36.

Ross, T. L., and Ross, R. A. (1991). Gain sharing: Sharing improved performance, in *The Compensation Handbook*, 3rd ed., H. L. Rock and L. A. Berger, Eds., McGraw-Hill, New York.

Santora, J. E. (1991). Du Pont returns to the drawing board. *Personnel Journal*, February, 34–36.

Scholtes, P. R. (1987). *An Elaboration on Deming's Teachings on Performance Appraisal*. Joiner Associates, Madison, WI.

Schuler, R. S., and Harris, D. L. (1991). Deming quality improvement: Implications for human resource management as illustrated in a small company, *Human Resource Planning*, **14**(3), 191–208.

Schuster, M. H. (March 1990). Gainsharing: Current Issues and Research Needs. Workshop, School of Industrial and Labor Relations, Cornell University.

Schwab, D. P. (1980). Job evaluation and pay-setting: Concepts and practices, in *Comparable Worth: Issues and Alternatives*. E. R. Livernash, Ed., Equal Employment Advisory Council, Washington, D.C.

Smith, P. C. (1976). Behavior, results, and organizational effectiveness: The problem of criteria, in *Handbook of Industrial and Organizational Psychology*, M. D. Dunnette, Ed., Rand McNally, Chicago.

Wagner, J. A. III, Rubin, P., and Callahan, T. J. (1988). Incentive payment and nonmanagerial productivity: An interrupted time series analysis of magnitude and trend. *Organizational Behavior and Human Decision Processes*, **42**, 47–74.

Wall, T. D., Corbett, J. M., Martin, R., Clegg, C. W., and Jackson, P. R. (1990). Advanced manufacturing technology, work design, and performance: A change study. *Journal of Applied Psychology*, **75**, 691–697.

Walton, R. E., and Susman, G. I. (1987). People policies for the new machines. *Harvard Business Review*, **87**(2), 98–106.

Weitzman, M. L. (1984). *The Share Economy*. Harvard University Press, Cambridge, MA.

Weitzman, M. L., and Kruse, D. L. (1990). Profit sharing and productivity, in *Paying for Productivity*, in A. S. Blinder, Ed., Brookings Institute, Washington, D.C.

Welbourne, T. M., and Gomez-Mejia, L. R. (1991). Team incentives in the workplace, in *The Compensation Handbook*, 3rd ed., M. L. Rock and L. A. Berger, Eds., McGraw-Hill, New York.

Wexley, K. N., and Klimoski, R. (1984). Performance appraisal: An update. *Research in Personnel and Human Resource Management*, **2**, 35–79.

Womack, J. P., Jones, D. T., and Roos, D. (1990). *The Machine That Changed the World*. Rawson Associates, New York.

CHAPTER 5

HUMAN ASPECTS OF TOTAL QUALITY MANAGEMENT

W. A. J. GOLOMSKI
W. A. Golomski & Associates and
Graduate School of Business
University of Chicago
Chicago, Illinois

5.1 INTRODUCTION

Three questions are implied when considering advanced manufacturing technology (AMT) carefully:

1. What do we mean by "human" when there is such a great range of human talent, personality, and motivation?
2. Which of the gurus has the final definition of "total quality management"?
3. What is AMT?

Fortunately for me, others have defined AMT in earlier chapters. What is considered advanced in the 1990s will not be a decade from now. The degree of implementation will vary greatly.

A favorite approach of many speakers and writers is to find an isolated

Organization and Management of Advanced Manufacturing. Edited by Waldemar Karwowski and Gavriel Salvendy.
ISBN 0-471-55508-8

aberration, assume that it occurs generally, and then suggest that the solution is a major breakthrough. There are some who believe that is what is happening with quality. Of course, this is not the case.

In this chapter the concepts of total quality management (TQM) are discussed first, then the human aspects of TQM are reviewed, and finally how the TQM concepts are modified in an AMT environment is discussed.

In 1924, Walter Shewhart of AT&T introduced two important concepts of quality to the world: (1) Processes should be looked at from the point of view of preventing defects and not just detecting them, and (2) processes have two kinds of causes of variation: random causes due to management actions and assignable causes due to operators or workers in a process.

Most causes are due to management decisions made on supplier selection, maintenance, equipment chosen, process design, product design, employee selection, employee training and leadership. These so-called *random causes* are what W. Edwards Deming (1981) calls *common causes*. They account for at least 80 percent of business process and operations problems. Nonetheless, most managements fantasize that the major problems are with first-line workers and their motivation.

The work of Shewhart and George Edwards of AT&T eventually led to the field of quality control and quality assurance. These were professional engineering and management activities. The official definitions of quality control and quality assurance quoted below are found in ANSI/ASQC Q90-1987, Quality Management and Quality Assurance Standards:

Quality System: The organizational structure, responsibilities, procedures, processes, and resources for implementing quality management.

Quality Control: The operational techniques and activities that are used to fulfill requirements for quality.

Quality Assurance: All those planned and systematic actions necessary to provide adequate confidence that a product or service will satisfy given quality requirements.

5.2 THE TRANSITION TO TOTAL QUALITY MANAGEMENT

During the 1970s quality management became a vice-presidential position separate from manufacturing or operations. However, its main concern was with the quality of the products and services sold or bought. It became apparent that the use of sampling plans to protect companies was no longer effective. With performance rates economically tolerating only a few parts defective per million, or even billion, management had to change. The changes were primarily in the electronics, telecommunications, defense, and aerospace industries. Some pharmaceutical and food operations also fit into that high-performance category and were improved to these high-performance rates in the 1950s. The transition in concept and organizational structure moved from

(1) inspection to (2) quality control to (3) quality assurance to (4) continuous improvement to (5) total quality management. Some see a transition to paradigm innovation.

Several changes in the approach to quality occurred. These changes can be summarized as follows:

1. All work should be viewed as being composed of processes.
2. Some processes fail because of complexity and should be simplified.
3. Processes can be brought into a state of *statistical control* and then be made *capable* of meeting specifications.
4. There can be continuous improvement in all processes.
5. Small groups, properly trained, using principles from sociology and psychology, can be more effective at continuous improvement than experts.
6. Strategic and annual quality improvement plans are helpful in aligning quality improvement activities with each other and the organization purpose. The customer defines quality.
7. There is no best way, but not any way will do. National and international efforts have produced standards for a quality management system that permits a variety of quality systems.

A parallel approach started in the late 1950s with the book *Total Quality Control* by A. V. Feigenbaum. This was extended by the Japanese to develop companywide quality control. This is called "quality control in everything that we do." It is a management systems approach to total quality control but does not have the social and behavioral science components that more recent formulations have.

5.3 WHAT IS TQM?

Total quality management is philosophy, a set of concepts, and a collection of methods for continuously improving organizations. The improvement is focused on customer satisfaction, process improvement, employee participation and involvement, and a strategic quality plan. There is a linkage between the long-term vision of the company and how it can be supported by TQM. In some cases quality, cost, and customer satisfaction are the central issues for the corporation around which everything else is built.

There are a variety of other definitions of TQM. Here's a short one used by W. A. Golomski & Associates in our 1992 training manuals and a longer one from the U.S. Department of Defense.

Total Quality Management (TQM)

> Total Quality Management is a system and a sustained improvement process for improving quality in everything that we do. It is concerned with satisfying both internal and external customers. But does this mean at any cost? No! We will aim for the *rational expectations* of customers over the likely use of the product or service.

Some would disagree with the word "rational."

Whenever a new concept emerges, the tendency is to say that it is just a relabeling of old concepts or it is something that the reader or listener has been using for years. Its somewhat like reading a book on abnormal psychology and wondering which of the many abnormalities you have. There usually are elements of prior approaches in new ones.

> Total Quality Management (TQM) is both a philosophy and a set of guiding principles that represent the foundation of a continuously improving organization. TQM is the application of quantitative methods and human resources to improve: (1) the material and services supplied to an organization, (2) all the processes within an organization, (3) the degree to which the needs of the customer are met, now and in the future. TQM integrates fundamental management techniques, existing improvement efforts, and technical tools under a disciplined approach focused on continuous improvement.

The reasons for all the interest is that the usual planning, budgeting, financial, and marketing approaches failed in many organizations. The broadened perspectives of TQM have provided new insights and great practical benefits.

One of the elements of TQM is the role of standards.

Standards

Some people dislike the word "standards." They might think that it has to do with the current controversy on political correctness. In the business sense a standard is an agreed-upon set of behaviors, or practices, or systems, or methods and devices. The more familiar standards include standards of conduct, standards of measurement such as the official inch and centimeter, and standardized physical and chemical tests. Standards facilitate trade. They are the basis of technical, operational, and legal communications.

Two types of quality standards are emphasized here. They are standards of process conditions and standards of outputs of processes.

1. Who is best in class?
2. What are the two basic standards for quality management?

Measuring Progress and Benchmarking

Some of the most important parts of life are the most difficult to measure. If we are thinking about a vocation as a priest, brother, nun, rabbi, imam, or minister, what will that be like? Or, if we are dating and thinking of getting married, how will the intended spouse work out? In quality we usually take a different route. We make three measures of relative progress. (1) How are we improving relative to a base period? This is called baseline measurement. (2) Are we improving as fast as the best of competitors? This is called making a competitive comparison. (3) Are we doing as well as the best in class, regardless of the industry or field? This is called benchmarking.

One of the problems in most organizations is that there are very few measures of quality. If there are any at all, they are usually measures of complaints, or of scrap levels, or of rework. This is a negative way of looking at quality, and it is based on quality measurements of outputs. We would rather have measures of customer satisfaction and measures early enough in a work process to prevent or minimize problems.

Quality measures are of two types: those while an operation is doing on and those after the end of work. If we were to start a quality improvement effort now, one of the first things we would want to do is to identify various output measures, and more importantly, quality measures within processes. These starting levels are *baseline measures*. They are measures from which to start trend analysis to measure progress of process outputs. We can also have measures of control limit width and patterns within the control limits. But, there is always a nagging doubt. Who in our industry is doing a better job on these same measures? That is why we want *competitive comparisons*. What if we find that someone else is doing better? That should give us a greater sense of urgency and drive us to develop better plans. The counter-argument is that we should always be striving for continuous improvement whether competitors are driving us to do so or not.

There is another possibility: Some organizations in another industry, or field, might be doing better than we are. One of our competitors might find this out before we do. Why not find out who is best in class (BIC) and find that out first? This is called *benchmarking*. Benchmarking studies are difficult for small companies. Trade associations or university business research organizations are the best to conduct such studies. Second, consulting companies, foundations, and large corporations are other possibilities.

A problem with quality measures, which we also have with any process measure, is how to interpret changes. If we have data such as:

J	F	M	A	M	
87	85	88	88(*P*)	. . .	(*P*) partial data

we might be chasing random variation if we are not careful. In January we had 87 on a 100-point scale, and hoped for something better. In February we thought we had slipped and gave a stern talk to the employees and had a worried look. Fortunately, the March data brought smiles on everyone's face and we felt good! April data are only partial as of the end of the first shift today. At least it seems that we did not slip. You felt in an expansive mood and congratulated your employees. Now what was really going on? These data are actually part of a data series starting in 1990. The variation shown is what you could expect from a process in statistical control. Month to month differences are quite likely to occur. Giving stern talks or praising employees under these conditions does no good. The process has not changed. You are fooling yourself and misinterpreting data variation. Actually the long-run variation of this "process" produced data that fluctuated from 84 to 90 while the "process" was in control. These variations were from a process in control.

National Quality Awards

The oldest, and most famous of the national quality awards is the Deming Prize in Japan. Since this award was started in 1950, different award categories have been added, including those based on company size, and a category for individuals. Two years ago even an international category was developed for those outside Japan. So far, a company in the United States and one in Taiwan have won the award in that category.

National quality awards are administered by the government, as in the case of the Malcolm Baldrige National Quality Award in the United States, or by a professional, not-for-profit organization, as in Japan. There the administration is by the Japanese Union of Scientists and Engineers. In the United States over 20 state and local awards have been started. The awards are certificates, or plaques, or medals. There is no cash prize for the winners; however, public recognition helps their business. The State of Florida, the University of Miami, and some other organizations sponsor the Sterling Award, which is open to those in government, industry, and not-for-profit organizations. The Rochester Institute of Technology and the *U.S.A. Today* newspaper sponsor the RIT/U.S.A. Today Quality Cap award for teams of employees who achieve great breakthroughs. Awards may be given to individuals in the future.

All of the awards have key concepts. For example, Malcolm Baldrige National Quality Award Criteria are built upon the following values and concepts:

- Customer-driven quality
- Design quality and prevention
- Leadership
- Long-range outlook
- Continuous improvement
- Management by fact

- Full participation
- Partnership development
- Public responsibility

It is important to note the role of people in "full participation" of the work force.

5.4 IS TQM DIFFERENT?

There are many people in companies who say, "I've been using TQM for years." On careful investigation we find that not to be true. At least one of several elements are missing. Among the more common omissions are:

1. Lack of a strategic quality plan and an annual quality improvement plan that covers *all* products, services, and internal departments.
2. Lack of education and training every year for each and every employee. This includes how to identify opportunities for improvement, how to diagnose them, how to come up with solutions, how to implement the solution, and how to make further improvements.
3. Inadequate use of quality measures and statistical methods to measure progress and to measure process control.
4. Lack of objective customer satisfaction measures by geographical area, industry grouping of customers, and by product (service) type or group.
5. Lack of benchmarking of all business processes and products (services) that are the best, regardless of field, and comparing their level and growth to yours.

Certainly, some organizations had some TQM concepts in place in their quality assurance systems. But, these are static systems based on conformance to specifications. That's not good enough. If there is less variation within specification limits, subassemblies and products usually perform better. Suppliers are compared in terms of how little variation they have relative to specification limits.

5.5 THE HUMAN ASPECTS OF TQM

Eric Baum, the famous social psychologist, discusses the importance of affiliating with an organization. By this he means that if there is a strong nexus between our personal development goals and values and the goals and values of the organization, the employee *affiliates* with the organization. However, if during the rites of initiation to the organization, the veterans stamp out any vestiges of prior success, or status one wishes to gain as a result of prior work, disaffiliation can occur. One then just works in the organization but does not affiliate with it. If the rite of initiation is supportive, nonthreatening, or even the individual develops himself or herself, the organization benefits.

Individuals who affiliate with an organization want to help it; however, desire is not enough. In the first few pages of this chapter the transition to TQM was discussed. One of the concepts that many organizations missed was continuous improvement. As a result, they consider continuous improvement a part of TQM. Continuous improvement, as a concept, has several steps.

- Recognizing opportunities and problems
- Diagnosing them
- Finding solutions
- Implementing solutions
- Making sure they work
- Repeating these steps

All of these require education, training, and coaching.

5.6 THE HUMAN RELATIONS APPROACHES

Employee involvement and participative management in the past was quite different from TQM today. Employees were sometimes brought in for a discussion on a problem with an executive or manager. When that person received the inputs of the group he or she made a decision on what to do. No power was released. No discussion among equals took place. The joy of implementation might have been delegated by the executive or manager to others than those who gave inputs.

Another type of participation was that of Japanese quality control circles. In the United States these were employee groups at the bottom of the organization. They usually identified the problems they wanted to work on. Upper management gave approval. Employees worked on an approved project and presented the proposed solution to upper management for approval to proceed toward implementation. A complaint by some union leaders was that management was involved in too many steps. They, in effect sat on both sides of the table. The weight of the past kept them from empowerment. Another problem that was identified by middle management was that there was no role for middle management. Eventually, middle management killed off quality circles, as they were called in the United States. They are still thriving in Japan. A different form of small-group activity has emerged in the United States.

5.7 TQM APPROACH

In the TQM approach management contributes resources and direction but does not dominate. Small groups are encouraged to identify opportunities and discuss them with a cross-functional steering committee for prioritization. Projects are worked on and implemented unless costs are increased or there is a need for capital expenditures. Employees are empowered and work within acceptable guidelines.

Recent applications have occurred in schools, the government, in business, and in not-for-profit organizations. But there is another reason for survival. Robert Cole, of Berkeley, studied small groups in the United States, Sweden, and Japan. There were several reasons for failure. Among the two most important were the lack of alignment with the strategic plan of the organization and lack of leadership skills within teams.

The first needs some explanation. Support will not continue for a change process unless the activity or approach is part of a generally accepted process or procedure. Teams are able to sustain interest and survive for two or three years, but then die because of lack of support. General systems theory indicates that subsystems that are disjoint from the main system will not survive. The second problem is that of teams wanting to be democratic but still needing someone to encourage creativity to flow within the guidelines decided for the team. Teams can wander and not stay on track. This is especially true of teams of knowledge workers.

Senior management must set the "guiding principles" within which people work as individuals and in groups. But establishing them and posting them will not have any effect if the senior officers do not show a sustained interest in them. That is one way to make sure that they are followed. But even the phrase "sustained interest" might not suggest sufficient direction. It might suggest a passive interest rather than involvement. We do not expect a mandate but, rather, discussions in which people are convinced. This is then followed by what some people call "walk the talk." This means demonstrating what you preach through action.

Here are a few of the guiding principles of the Ford Motor Company circulated to all employees and suppliers. They could be used by a much smaller company.

1. **Quality Comes First**

   To achieve satisfaction, the quality of our products and services must be our number one priority.

2. **Customers Are the Focus of Everything We Do**

   Our work must be done with our customers in mind, providing better products and services than our competition.

3. **Continuous Improvement Is Essential to Our Success**

We must strive for excellence in everything we do: in our products, in their safety and value—and in our services, our human relations, our competitiveness, and our profitability.

4. **Employee Involvement Is Our Way of Life**

   We are a team. We must treat each other with trust and respect.

5. **Dealers and Suppliers Are Our Partners**

   The company must maintain mutually beneficial relationships with dealers, suppliers, and other business associates.

6. **Integrity Is Never Compromised**

   The conduct of our Company worldwide must be pursued in a manner that is socially responsible and commands respect for its integrity and for its positive contributions to society. Our doors are open to men and women alike without discrimination and without regard to ethnic origin or personal beliefs.

Individual behavior differs in various cultures of the world. It differs by gender; it differs by age; it differs by class. All of these differences are due to historical cultural patterns. In most cultures women are not expected to be leaders of companies, although they might be leaders of nations. This has nothing to do with temperament, stamina, genetics, or intellectual capacity. Women are not expected to supervise men in factories no matter how competent they are. Those who do rise through the ranks find it quite difficult because the support of fellow workers is not offered openly. When women of ambition and ability find opportunities for advancement closed, it is hard for them to be enthusiastic and motivated. There are no studies done on factory or office workers that show that women perform better under male supervision. Everything is based on opinions and anecdotes.

One of the interesting observations to make in any cafeteria or lunchroom is to observe who sits with whom. Generally, people sit with the same group each day. If there are a variety of nationalities involved, who are new to the country, people group themselves that way. Males and females sit at the same table. However, if people were born in the country in which they are working, males and females usually sit separately. The usual exception is that of professionals and managers, unless there are religious prohibitions. If there are different races involved, they might sit separately.

It is important for those who manage the quality effort in companies to try to get those with diverse differences to work together as one team. Another goal is to get people to work together as one team rather than to emphasize their supposed differences in status.

There are two basic changes in TQM from the human resources point of view that are new. We are looking for heroic teams and not just emphasizing individual performance. This has many implications to performance appraisal and incentive systems. Second, we view "quality" as the continuously learning organization.

5.8 RELATIONSHIP TO AMT

Advanced manufacturing technology organizations require employees who can think in terms of systems. They must be flexible enough to work with many new product introductions and many process improvements. Change becomes a way of life to them. This is not a natural talent.

Stress can occur if what you know is constantly challenged because of changes. Stress is reduced if employees are part of the process. Recently, professionals at two companies provided me with examples. Some approach the matter of change through using sociotechnical concepts, such as at the General Motors Saturn plant and at the new Miller Brewery in Ohio. Both use advanced manufacturing technology. Both invested thousands of hours in making sure that the human systems are not subverted by the technical system. Among the important considerations are:

- Can another person be seen at all times?
- May I override a process or am I a process watcher?
- Is there an opportunity for me to feel good about my contribution?
- Are suggestions for improvement solicited during the product and process design stages?
- Are suggestions welcomed after processes are in place?
- Do engineers, managers, and operators act as if they are one team?
- Do processes take ergonomic considerations into account?
- Am I ever allowed to talk to a customer or supplier?

High-tech companies usually believe that they can innovate problems away and dislike quality programs. Only after disasters occur do they question whether or not TQM will help. Processes are automated that should not be. These companies often have the unfounded belief in the ability of the group to be invincible.

A recent study by the Indiana Labor Management Council (1991) in the United States is of interest to us here. Employers were asked to report on the effectiveness of three broad social technical systems categories. In Table 5.1 "Technology" means equipment; "Technical Systems" means TQM and various industrial engineering approaches; and "Social Systems" are described in Table 5.2. The importance of social systems, some of which are covered in this chapter, is apparent. The social systems studied are listed in Table 5.2. The footnote in Table 5.1 draws our attention to the second entry in each cell. Correlations with a significance below .05 are found for more entries for social systems than for technical systems or technology. None of the

TABLE 5.1 Relationships Among STS Variables and Outcome Variables

Outcomes Variables	Social Systems	Technical Systems	Technology
a. Domestic market share	.1123* .054	.0321 .582	.0391 .503
b. Foreign market share	.2139 .001	.1880 .002	.1172 .060
c. Overall sales	.0821 .156	.0443 .445	.0331 .567
d. Product quality	.2365 .001	.0450 .436	.1624 .005
e. New and improved market introductions	.1834 .002	.0494 .401	.0526 .371
f. Productivity	.2881 .000	.0600 .300	.1855 .011
g. Organization costs	-.1241 .033	-.0468 .422	-.0879 .131
h. Employee skill and knowledge	.2229 .000	.1529 .008	.2785 .000
i. Profitability	.1672 .004	.0499 .389	.1154 .046

* NOTE: The top entry is the correlation coefficient and the lower entry is the level of significance. Any correlation with a significance below .05 can be interpreted as meaning there is a relationship between the two variables. If the level of significance is above .05 there is probably no relationship between the variables.

correlations are high, but there is a relationship between profitability, products, quality, and productivity with both social systems and technology but not technical systems.

Groups Affected

A way to classify major groups of people affected by AMT is in terms of operators, engineers, and managers.

Operators become not machine tenders but process tenders and maintenance overseers. The scope of the work changes. One has to be able to think in terms of the interrelationship between parts of a process. A big question facing managers is whether operators should only be taught what to do, not why they should do it in that way. Employees with different skills are needed more for computer-integrated processes than for traditional processes. As the tactile senses of jobs are lost, skills are totally mediated. The worst was expected in the chemical industry a couple of decades ago, but the effect on operators was not as severe as expected.

In a study of 36 companies with AMT in three different industry types satisfaction varied based on the complexity of the operator's job. (See Table 5.3)

TABLE 5.2 Performance Improvement Strategies: Social System

Type of Strategy	No. Companies Using	% Companies	Average % Workers Affected	Average Years In Use
Job sharing	38	12.3	25.4	4.9
Job rotation/ cross-training	180	58.4	46.0	9.7
Team building	105	34.1	48.8	3.1
Self-managing teams	65	21.i	35.6	3.3
Redesigning the role of supervisors and middle management	78	25.3	55.2	2.6
Performance appraisal	88	28.6	66.1	3.1
Gainsharing	52	16.9	69.4	5.2
Pay for knowledge	71	24.10	44.7	10.0
White collar teamwork	76	24.7	49.9	5.1
Four-day workweek	41	13.3	39.5	3.7
Flextime	46	14.9	24.3	3.5
Employee suggestion programs	161	52.3	84.9	8.6
Team-based rewards	50	16.2	54.9	5.0

Manufacturing Competitiveness Frontiers
July 2, 1991

One of the chemical companies had a low satisfaction score mainly because of a poor training program and management style. (See Tables 5.4 and 5.5.) In these same companies there were none fewer managers. Their satisfaction level was generally lower than that of their operators. The table codes are the same and the companies are in the same rows as in Table 5.3. Measurements were made 30 days after start-up.

Engineers expressed greater satisfaction than managers and operators. One of the companies in the South had a group with strong religious fervor that settled with their ratings being higher across all occupational levels. Another, also in code B, III, had lower ratings across all occupational levels. Their process was not working well, and there was much blame being passed around by senior management. There was a great deal of stress.

One could also have looked at the satisfaction levels, by occupation type, relative to the percent change in complexity since the prior generation of production facility. Additional measures were made at one year after the facilities were initiated. A much more extensive set of measurements was made that will take a year or more to analyze.

In studies of this type there is always the question of confounding variables. One can attempt to adjust for years of experience, education level, and years

TABLE 5.3 Operator Satisfaction Based on Level of Interface by Industry Type

	Complexity		
Satisfaction	I	II	III
A	72	67	71
	81	63	68
	65	84	68
	74	65	75
B	69	70	66
	67	71	80
	78	45	74
	73	68	67
C	77	72	72
	76	80	72
	80	74	78
	81	73	75

Complexity Code: I Operator Must Review Data Sheets On Monitor And Adjust Process

II Analysts Review Output And Tell Operator Amount Of Adjustment

III Operator Observes Warning Signal; Needs Approval To Adjust

Identity Code: A Electronic And Electro-Mechanical
B Chemical
C Discrete Parts

with work at similar complexity. In these operations overriding the system was discouraged. This was especially true in plastics operations where the resin could set and shut the operation down. Two companies were deleted from the study because of internal bickering as a result of job reclassification.

The Influence of Appraisal

W. Edwards Deming (1981), the famous consultant in statistics, makes a distinction between *enumerative studies* and *analytical studies*. These distinctions grew out of his work on both the census and on process variability. In enumerative studies in the census and in marketing research you start with a frame or set whose size and characteristics you wish to measure. A subset of this set is used to estimate the properties and/or value of the whole. This is done for carloads of coal and for production lots of product. Oftentimes accept/reject criteria are used or the lot is accepted at a price on a predetermined scale. The same approaches are used by managers in evaluating performance of people relative to goals over a discrete period of time, for example, a year. In both cases there are problems with whether appropriate sampling plans and appropriate measures are used. In the case of measuring the performance of people, the so-called approach of pay for performance is inappropriate.

TABLE 5.4 Supervisor Satisfaction Based on Level of Interface by Industry Type

	Complexity		
Satisfaction	I	II	III
A	55	53	54
	59	47	51
	51	82	53
	68	50	56
B	54	51	56
	52	49	70
	58	32	59
	55	53	54
C	61	55	54
	60	67	50
	67	57	55
	58	56	61

Complexity Code: I Operator Must Review Data Sheets On Monitor And Adjust Process

II Analysts Review Output And Tell Operator Amount Of Adjustment

III Operator Observes Warning Signal; Needs Approval To Adjust

Identity Code: A Electronic And Electro-Mechanical
B Chemical
C Discrete Parts

When you have a process, such as a production process, you want to study what is going on during a given period of time or frame. Based on a cause-and-effect analysis, a decision is made on whether to leave the process running or to intervene. This generally causes problems when you assume that the past is a predictor of the future. To make such assertions you must know about the dynamics or circumstances that were present during the frame. If they will be the same in the future that is very convenient. Improvements in the process come about from studying the system that produced the product or service. Accumulating the output, and studying it will not help improve the system. Information is lost if one does an enumerative study. If one preserves the sequence within which a study of a process is made, more information is gathered. When diagnosing a process, one not only collects data periodically, but also records the events taking place that could affect that process. If a low of raw material or components from a new vendor were started in a process, that event would be logged. A comparison of the data patterns for the characteristics of importance would be made with that of other lots. An analytical study would also be made to see if the process was in *control* and *capable* of meeting specifications. Causes of variability in the process are due to management (common causes) and to the operator (special causes).

There are two issues here. Is AMT capable of providing feedback and feedforward signals based on a process in control? What influence does an

TABLE 5.5 Engineer Satisfaction Based on Level of Interface by Industry Type

	Complexity		
Satisfaction	I	II	III
A	88	80	85
	93	83	79
	80	94	79
	79	77	81
B	87	79	72
	87	75	87
	89	61	80
	81	72	76
C	84	76	78
	86	85	85
	90	78	90
	89	83	84

Complexity Code: I Operator Must Review Data Sheets On Monitor And Adjust Process

II Analysts Review Output And Tell Operator Amount Of Adjustment

III Operator Observes Warning Signal; Needs Approval To Adjust

Identity Code: A Electronic And Electro-Mechanical
B Chemical
C Discrete Parts

operator have in AMT when the process was designed to reduce or eliminate human intervention? Performance appraisal systems usually assume that the work of an employee is subject to an enumerative study, when, in fact, an analytical study should be made. As a result employee effectiveness is decreased by an inappropriate appraisal system. Appraisal systems in an AMT environment make the situation worse. Lower work satisfaction measures usually occur. An exception is when the managerial style is one of fixing the blame when process upsets occur. Under AMT they occur less often, but fewer of the causes of upsets are due to the operator or manager. The unfairness of the appraisal system becomes more evident.

Technology Diffusion

Rosenberg believes that in complex technologies with interactions between many subsystems it is difficult to anticipate exactly how they will perform. It is important to learn by doing. Not all of the learning comes from technology transfer because the user will understand the equipment better than the manufacturer. Although TQM is well diffused, the application within a given system has to be recreated because the developer does not know how to do it.

Organizational Learning

Our work has also been on the effect of employee turnover rates, learning, organizational structure, and complexity. In some studies we have separated learning rates and performance level after a given period of time. In the case of all complexity levels, teams do better than hierarchies for fully trained people. They do about 20 percent less well at high complexity than low. Teams, which are common in TQM, perform better by almost 30 percent when compared to hierarchies and when experienced workers are hired after being interviewed by the team. Teams with new members with low experience do poorly. The implications to new workers who do not have the opportunity to gain experience in a technical school is not pleasant. Satisfaction goes down, in AMT, the greater the complexity. The greater the opportunity for the employee to feel useful, the better the quality and employee satisfaction.

BIBLIOGRAPHY

Adas, M. (1989). *Machines as the Measure of Men: Science, Technology, and Ideologies of Western Dominance*, Cornell University Press, Ithaca, NY.

Baum, H. S. (1990). *Organizational Membership, Personal Development in the Workplace*, State University of New York Press, Albany, pp. 43–44, 97.

Cole, R. E. (1991). *Different Quality Paradigms and Their Implications for Organizational Learning*, presentation at the Conference, Japan in a Global Economy—A European Perspective, Stockholm School of Economics, September 5–6.

Deming, W. E. (1981). Transformation of western-style management, in *Productivity through Performance Appraisal*, K. K. Shethy and V. M. Buehler, Eds., Addison-Wesley, Reading, MA.

Indiana Labor Management Council (1991). *Midwest Manufacturing Company Survey of Quality Productivity Improvement Strategies*, Indiana Labor Management Council, Indianapolis.

Peterson, D. E., and Hillkirk, J. (1991). *A Better Idea: Redefining the Way Americans Work*, Houghton Mifflin Company, Boston.

Rosenberg, N. (1982). *Inside the Black Box: Technology and Economics*, Cambridge University Press, Cambridge, MA.

Salvendy, G., Ed., (1992). *Handbook of Industrial Engineering*, Wiley, New York.

Shea, G. P., and Guzzo, R. A. (1987). *Group effectiveness: What really matters? Sloan Management Review*, Spring, 25–31.

CHAPTER 6

DESIGNING THE HUMAN INFRASTRUCTURE FOR TECHNOLOGY*

J. K. LIKER
Industrial and Operations Engineering
University of Michigan
Ann Arbor, Michigan

A. MAJCHRZAK
Institute for Safety and Systems Management and
Industrial & Systems Engineering
University of Southern California
Los Angeles, California

6.1 INTRODUCTION

Digital Equipment Corporation (DEC) installed a flexible manufacturing system (FMS) and reported that the technical issues were the easy ones to work out. The major problems were human resource issues. Employees at every level lacked awareness of new technology and its implications, feared newness,

*The first author would like to thank Ford Motor Company for its support of research on participatory ergonomics. The second author would like to thank her current sponsor, the National Center for Manufacturing Sciences as well as past sponsor Digital Equipment Corporation for their support.

Organization and Management of Advanced Manufacturing. Edited by Waldemar Karwowski and Gavriel Salvendy.
ISBN 0-471-55508-8

and relied on getting the product out and not using the FMS effectively. Little training was provided as "engineering superstars" responsible for installing the equipment believed that people in the plant could not possibly understand the new technology (Majchrzak, 1988).

At Westinghouse a Fanuc robot was designed to user specifications but never used. A case study of the failed installation revealed the following sequence of events (Knight, 1985). A project team began design of the robot at the request of upper-level manufacturing managers. Shop floor supervisors were not consulted during the design work. When the robot arrived, several technical problems needed to be worked out during production runs, but shop floor personnel helped as little as possible and essentially circumvented the efforts of the project team staff.

A young information systems specialist at Ford developed an automatic inspection system linked to a personal computer (PC). The system, installed on the camshaft line of a Ford automotive engine plant, produced statistical process control charts based on measurements automatically taken by in-line gages. The system automatically produced hundreds of charts each day; however, supervisors and workers on the camshaft line ignored the output, relying instead on the original manual gages. These supervisors and workers never asked for the information from the system and had no commitment to using system outputs. The plant manager finally used the system to find out how many defects were on the camshaft line and used the data as a negative performance review for the area superintendent. One year later the young information systems specialist returned to find the entire system disconnected and the PC used as a simple spreadsheet. In other words an automatic inspection system initially intended to help superintendents and workers not only failed to provide information seen as useful but the information it did provide was used against them.

These and other examples illustrate the fallacy of assuming that implementation of technology is simply a matter of good technical design, a little training, and perhaps rewriting a few job descriptions. A 1984 study by the Congressional Office of Technology Assessment (OTA) concluded that "the main stumbling blocks in the near future for implementation of programmable automation technology are not technical, but rather are barriers of cost, organization of the factory, availability of appropriate skills, and social effects of these technologies" (Office of Technology Assessment, 1984, p. 94).

Despite these examples and evidence of several national studies since the 1984 OTA report, managers too often still seem to ignore the human aspects of human–technical systems. For example, Butera (1984) reports that in 19 case studies of the implementation of automated equipment worldwide, most managers failed to use any structured approach to considering human resource issues and simply reacted to problems as they occurred. Moreover, when the human consequences of new flexible technology are considered, often only reduced labor costs are examined (Ayres and Miller, 1983; Majchrzak, 1988).

There are three major reasons for considering these human resource issues early in the process of designing and selecting technical systems:

1. *To Avoid Resistance to Change*: If these human considerations are ignored, people will resist change, perhaps even refusing to use or sabatoging the technology. Even if the technology is implemented, demoralization of the work force may follow leading to increased absenteeism and turnover (Liker et al., 1987).
2. *To Respond to Unpredictable Events*: Advanced manufacturing technologies are by definition new and unproven and can break down in unpredictable ways. Thus, smooth operation of the technology requires operator commitment to learning and responding quickly to novel situations (Hirschhorn, 1984).
3. *To Maximize Total System Effectiveness*: The technology is only part of a broader sociotechnical system. Unless the social and technical systems are compatible and effectively designed, parts of the total system will inadvertently inhibit effective functioning of other parts, such as is often found in "work-around procedures" (Taylor and Asadorian, 1985).

In this chapter we argue that the design of the "human infrastructure" is as important to the overall success of human–technical systems as the design of the technical system. Much attention has been paid recently to the concurrent design of products and manufacturing processes to jointly optimize product functionality and manufacturability (Susman and Dean, 1992). Yet, there has been much less attention to the concurrent design of the human infrastructure and technical systems. As with any design problem, effective design solutions depend on an effective design process. It is the design *process* by which human and technical infrastructures become compatible that is the focus of this chapter.

There has been much written on the characteristics of various best practice human infrastructures for different technologies. For example, a chapter in the *Handbook of Industrial Engineering* (Majchrzak, 1991) discusses the design elements of the human infrastructure and reviews the literature on the best design features to incorporate for programmable automation—the *content* of the human infrastructure design. There is much less written on the *process* by which a human infrastructure compatible to technical needs is designed. Moreover, there is no one agreed-upon approach to the design of the human infrastructure of technology.

Thus, as the focus of our chapter, we have chosen to describe the process of compatible human infrastructure design by first proposing a general philosophical perspective to designing compatible human infrastructures—an open-system view of a sociotechnical system. Second, we briefly review the best practice human infrastructures for advanced manufacturing technology and show how they fit (and fail to fit) within the open-systems framework. Third, we argue that we need a general approach or methodology for designing the human infrastructure and suggest some desirable characteristics of such an approach. Fourth, we discuss four alternative design methodologies and the degree to which they have these desirable characteristics. We conclude with a preliminary synthesis of these alternative design methodologies.

As an overview of the chapter, the principle points we will make can be summarized as follows:

1. Production systems are sociotechnical systems, and effectiveness depends on arriving at an appropriate fit between the social and technical systems.
2. Research is converging on some general human infrastructure characteristics that constitute "best practice" for complex technological systems. These characteristics provide a useful input for the design of the human intrastructure but are too general to provide specific solutions.
3. Every sociotechnical system is unique and the human infrastructure needs to be *designed* to fit the unique characteristics of that system.
4. All sociotechnical systems are open systems and hence the system must be designed to fit the unique requirements of its external environment. This means the design process should consider environmental constraints and opportunities in designing to achieve the system mission.
5. Participative system design will generally lead to a more effective sociotechnical system and greater user acceptance than a design by experts that is imposed on members of the system.
6. There are a number of design methodologies for the human infrastructure that are based on open-systems thinking. There is considerable overlap between these methodologies and they provide complimentary strengths.

6.2 AN OPEN-SYSTEMS VIEW OF THE HUMAN INFRASTRUCTURE

The paradigm we believe has proven most powerful for understanding human organizations is open-systems theory. Originally developed by a biologist, Von Bertalanffy, for understanding ecological systems, the open-systems paradigm has identified principles that describe how systems are born, grow, and die (Von Bertalanffy, 1956; Katz and Kahn, 1966). The open-systems model has been adopted by many contemporary organizational scholars as the central paradigm in the study of formal organizations. Indeed, contemporary quality methodologies such as total quality control are based on a foundation of open-systems theory (Checkland, 1981; Ishikawa, 1985; Schonberger, 1990).

According to open-systems theory, formal organizations import energy (e.g., raw materials, money, customers) from their environment, transform these inputs through a production process, and then export their products (e.g., physical products, services, information) into their environment. Feedback is necessary so that the organization can adjust its course (positive feedback) as well as make internal adjustments to produce products to specification (negative feedback) (Hanna, 1988).

Figure 6.1 is a representation of an organization as an open system. The figure portrays the internal organization as four interrelated systems. The formal organization is most commonly represented by an organizational chart but includes all formalized rules, job descriptions, reward structures, career paths, and standard operating procedures. The informal organization is not written down but consists of shared understandings among organizational members. Cultural systems and power relations comprise key elements of the informal organization. Individual systems include the skills, knowledge, experiences, attitudes, and needs of all individuals in the organization.

One critical principal in open-systems thinking is that all thriving organizations must maintain a "dynamic equilibrium" between their internal subsystems and their larger external environments. To maintain equilibrium the system must be able to dynamically change itself in response to a change in any subsystem or in the broader environment—local changes reverberate throughout the entire system. Open-systems theorists translate this into the concept of "congruence" or fit. System effectiveness depends on how well the internal subsystems fit with the environment, as well as how well each of the subsystems fit with all of the other subsystems—technical systems, formal systems, informal systems, and individual systems. By human infrastructure, we mean all of these subsystems when they involve humans.

When looked at in this way, it becomes clear that a technical system change cannot be designed in isolation from the subsystems involving humans, and these internal systems must fit with the demands and constraints imposed by the organization's environment. Thus, rephrasing our first principle point, which underlies most of the arguments in this chapter: Effective design of production systems means finding an appropriate fit between the social and technical systems and between these systems and the environment.

6.3 ONE BEST HUMAN INTRASTRUCTURE?

Early efforts to develop a systematic theory of management focused on attempting to codify principles not unlike physical laws. It was assumed that organizations followed scientific laws that, if they were discovered, would lead to a prescriptive science of management. Moreover, it was assumed that the prescription would be in the form of "oughts" that applied to all organizations regardless of their technology and environmental context. For example, those early scholars who are now called classical management theorists argued for formal organizational structures that clearly separated the line organization of doers from the technical staff of thinkers, planners, and controllers. They recommended an extensive division of labor, a strictly enforced chain of command, and clear reporting relationships in which everyone has

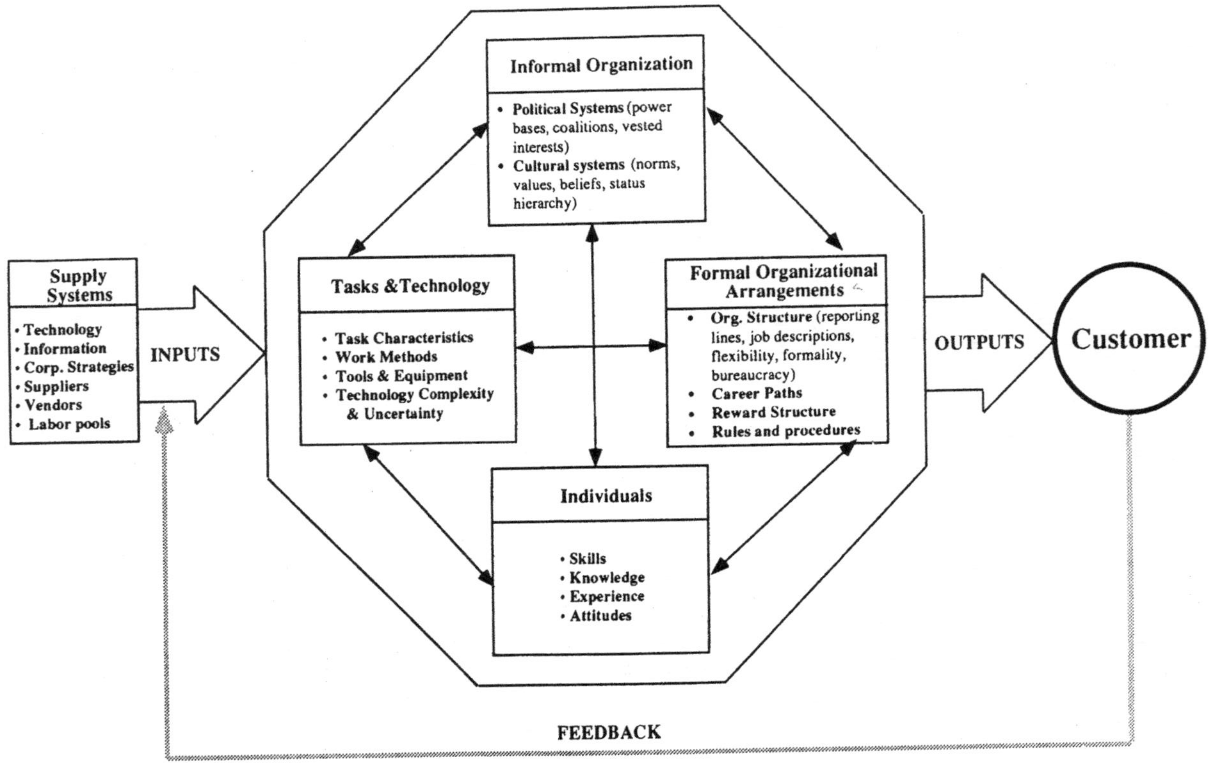

Fig. 6.1. General model of organizational systems. [Adopted from Nadler and Tushman (1980)].

one and only one boss to whom he or she gives complete obedience (cf. Fayol, 1960).

Early studies testing these recommendations of the classicists have demonstrated that under conditions of a large, stable or gradually growing demand for relatively simple products with well-understood, stable product and process technologies, these requirements are right (Burns and Stalker, 1961; Lawrence and Lorsch, 1967). The environment could be ignored and a focus on internal standard operating procedures and efficiencies was very effective.

Unfortunately, the world is not so conveniently predictable and stable. The environment does not always behave as we wish. These same studies (Burns and Stalker, 1961; Lawrence and Lorsch, 1967) discovered that when the environment is changing rapidly and in unpredictable ways and when core tasks and technology are unpredictable and nonroutine, the most effective form of organization could be characterized by what Burns and Stalker called an "organic" form. Organic organizations can be described as loosely structured, flexible organizations in which people are committed to a common purpose, and those doing the work take immediate, nonprescribed, action when the situation demands without always going through the formal chain of command. The main characteristics of organic organizations are:

1. Organizational relationships are characterized by a network structure, not a strict hierarchy.
2. Commitments by workers are to tasks and to progress, not to obedience.
3. Tasks are adjusted and redefined through interaction, not standardization.
4. Managers communicate information and advice, not instructions and decisions.
5. Expertise and contribution of special knowledge are valued, not hierarchical position.
6. Production decisions are made by people closest to the production process.

This was the beginning of contemporary "contingency" theories of organization that argued, based on open-systems theory, that the "best" way to organize depends on the specific situation facing the organization. Two key contingency variables are the organization's technology and the external environment.

Perrow (1967) provides one useful framework for classifying at a very general level the core tasks of an organization. His model consists of two dimensions: the analyzability and variety of the core tasks of the organization. *Analyzability* of a task refers to the degree to which tasks can be reduced to a clear set of mechanical steps that can be documented through standard operating procedures or the knowledge needed can be easily looked up in a manual. In nonanalyzable work novel problems often occur and the correct

	Variety: Low	Variety: High
Low Analyzability	CRAFT Descriptor: Low exceptions with non-standardized tasks Organization: Mostly Organic	NONROUTINE Descriptor: High exceptions with non-standardized tasks Organization: Organic
High Analyzability	ROUTINE Descriptor: Low exceptions with standardized tasks Organization: Mechanistic	ENGINEERING Descriptor: High exceptions with standardized tasks Organization: Mostly Mechanistic

Fig. 6.2. Perrow's (1967) characterization of core tasks and their organizational implications.

response cannot simply be looked up in a book. The cause of the problem or the solution is not obvious and employees must rely on accumulated knowledge and judgment. *Variety* of a task refers to the number of exceptions encountered in one's work. Tasks are high on variety if there are a large number of unexpected and novel events to which employees need to respond. Perrow uses this classification scheme to specify which organizational design is most suitable for the organization.

As shown in Fig. 6.2, tasks that are high on analyzability and low on variety are called *routine*. For example, in a production system that mass produces simple gears with highly reliable automation that rarely breaks down, the tasks of the machine operator are likely to be very routine. In this case a highly mechanistic organization is very effective. By contrast, tasks that are difficult to analyze and high on variety are *nonroutine* and are most effectively managed through highly *organic* organizational structures. Organizations in which the core tasks are high on analyzability and high on variety (called *engineering systems*), as well as those that are low on analyzability and low on variety (called *craft systems*) are best organized as some hybrid of mechanistic and organic organizations. Craft systems should be mostly organic, while engineering systems should be mostly mechanistic.

There is mounting evidence that many advanced manufacturing technologies, such as computer-integrated manufacturing and flexible manufacturing cells, create tasks that can be characterized as nonroutine since they are less analyzable and experience more exceptions than traditional automation (Hull and Collins, 1987). For example, when a flexible manufacturing cell is designed to accommodate a wide range of products, it must be able to flexibly adapt to changing environmental demands (e.g., changing product mix, unique products). To obtain the potential for flexibility from the technology, products may need to be rapidly customized to meet unique customer needs. This rapid customization requires rapid setups, which create the potential for unpredictable equipment breakdowns and setup problems. As a result, a more organic organization is needed in which humans operating the systems can respond quickly to novel problems.

The implication of Perrow's contingency framework is not that all flexible manufacturing system installations should use their potential for flexibility and thus require organic organizations. However, when this complex new technology *is* used to make varied products, the organization needs to be designed to have more organic characteristics rather than mechanistic characteristics. Many observers of contemporary advanced technologies have come to this conclusion. For example, a National Research Council study of 16 plants with model installations of advanced manufacturing technology (AMT) found that human resource policies that can best be described as organic seemed to account for the success of these firms (Walton and Susman, 1987). Other studies have come to a similar conclusion (Dean and Snell, 1991; Hirschhorn, 1984; Majchrzak, 1988; Pasmore, 1988). Thus, there seems to be a growing consensus on a new "one best way" for advanced computer-integrated systems. These more organic human infrastructure best-practice design features for advanced manufacturing technology can be summarized as follows:

1. *Team-based Organization.* The shop floor is divided into production areas in which teams, including workers and core support staff (e.g., maintenance, engineers), manage all day-to-day operations with little management interference and varying degrees of standardized work procedures. Many traditional management responsibilities such as scheduling, vacation planning, and employee selection are handled by the teams. These are also often called autonomous work groups or self-managing teams.
2. *Broad Job Classifications.* The many highly specialized jobs in traditional plants are replaced by a small number of broad classifications that encompass a large part of the production process. Production workers typically learn all of the production jobs in their area through job rotation, perform much of their own maintenance, and do their own quality control checks.
3. *Participative Management.* The goal is to push all daily operating decisions down to the people who operate the machines. Workers participate as much as possible in decisions that directly effect them or the operation of their immediate production area, including decisions about new technology. The role of first-line team leaders is to facilitate team decision making, not to make decisions for the team.
4. *Flat Hierarchies.* Since teams are largely self-managing, few layers of management are needed to operate the organization.
5. *Rewards for Learning and Commitment.* Pay-for-knowledge systems

are an example of one way to reward workers for acquiring new skills. In general, the organization should encourage continual learning, which includes allowing people to take risks and make mistakes without punishment. Commitment to the general mission of the organization is encouraged through group bonus plans like profit sharing or gain sharing.

6. *Continuous Improvement.* Personal learning and continual (often small) improvements to the production system are used to stay ahead of the competition by fully utilizing human talents and interest.
7. *Training and Education.* Large investments in training, as well as broader education, of the work force are required. This goes beyond teaching people narrow technical tasks and includes broader education in how to solve problems, understand the business context, and work effectively in teams.
8. *Open Communication Channels.* Open communication is needed in all directions including lateral communications with suppliers and customers, extensive information sharing by management, and upward communication from work teams to management since individuals at all levels are making decisions and tasks that cut across the organization are continually being redefined as conditions change. Coordination is handled through direct communication rather than through rules and standard operating procedures.
9. *Flexible Operating Procedures.* Rather than the extensive formal standard operating procedures of mechanistic systems, flexible rules and regulations are used in order to ensure that team members can flexibly respond to novel situations.
10. *Employment Stability.* To reap the investments of a highly trained, committed work force and ensure that workers improve production processes that force continual redefinitions of their own jobs, job security is needed. The promise of job security also sends a message to employees that they are valued and cements the mutual interdependence between employer and employee.
11. *Information and Discretion to the Source.* Real-time information and empowerment to act is provided directly to those who can take action rather than being hoarded by technical staff and management.

While these general characteristics seem to fit with many complex technological systems in the rapidly changing environments of our postindustrial society, there are limitations to such lists of prescriptions. The first limitation is that the prescriptions are becoming a new "one best way," which violates the open-systems assumption of seeking a unique fit for each organization internally and with its environment. These prescriptions assume the organic organization is best for all modern-day circumstances. In fact, there is large variation in the stability of the environment and the routineness of technologies. For example, G.M.–Toyota's Nummi plant uses highly specified tasks (with time standards to the second) while DEC's Enfield plant uses broadly defined tasks (Adler, 1991; Perry, 1984).

The second limitation is that the prescriptions are too general. For example, knowing that training and education are important does not tell the designers of a sociotechnical system how much training and education to provide, to whom, or on what topics. For example, in an extensive review of AMT training programs, Majchrzak (1988) found that some organizations could manage with 160 hours of training for AMT operators, while other organizations needed 2000 hours of technical and human resource training. Thus, the best practice organic characteristics are of limited practical value to managers faced with decisions about the human infrastructure they need for their particular installation of new technology.

We doubt that it is useful to continue to delineate characteristics of best practice human infrastructures. Rather, we believe what is needed are practical design methodologies that allow designers to develop detailed technical and organizational specifications taking into account the fit between subsystems and with the larger environment—a system that is right for their specific combination of circumstances.

6.4 CONCURRENT ENGINEERING OF ORGANIZATION AND TECHNOLOGY

The traditional design process for the human infrastructure of technology has a parallel in product/process design. Traditionally, product/process design proceeded sequentially. Products were designed first by specialized product design staff who worried about product functionality, and, after some prototype testing that proved this functionality, the design was shown to manufacturing process designers who critiqued the design from the vantage point of manufacturability. In those critiques, product designers were provided a list of complaints about features that would be hard to manufacture and manufacturing requests for changes in the design. These complaints led to an iterative process of making *minor* refinements to the product design; however, major product design decisions were typically not altered and the design modifications really amount to searching for a local optimum within the original global design decisions (Ward, 1990).

Contemporary design processes attempt to design the product and process simultaneously, or at least with overlap (Clark and Fujimoto, 1991). Techniques referred to as design for manufacturability (Susman and Dean, 1992), concurrent engineering (Winner et al., 1988), or simultaneous development (Industrial Development Division, 1991) emphasize the concurrent design of the product and process. But, according to an Institute for Defense Analysis report, concurrent engineering is more than simultaneous design of products and processes:

> Concurrent Engineering is a systematic approach to the integrated, concurrent design of products and their related processes, including manufacture and support. This approach is intended to cause the developers, from the outset, to consider all elements of the product life cycle from conception through disposal, including quality, cost, schedule, and user requirements. (Winner et al., 1988, p. V)

Concurrent engineering methods recognize that the product, manufacturing processes, end use, and servicing represent a system of interrelated parts. As in open-systems theory, changing one part of the system impacts other parts. Thus, product designers need to simultaneously consider such factors as vendor sources, the implications of alternative manufacturing processes, the impact of product design decisions on manufacturing ease, cost, quality, the ability of the end user to service the part, and so forth. There has been a great deal of work on methodologies to support this systems approach to design of products and processes. And there is evidence that effective utilization of these approaches can lead to impressive improvements in reduced lead time, reduced cost, and improved quality (Shumaker, 1990).

We can draw a parallel to the traditional serial process used to design the human intrastructure of technology. Traditionally, process engineers designed and selected technologies without considering the humans in the system. Then an industrial engineer might be brought in to figure out the staffing requirements and design the jobs—from a purely technical perspective. How should tasks be combined into jobs? How many people were required? What types of tools and aids were needed to perform the work? The industrial engineer might ask for some modifications to the equipment so that jobs could be performed more efficiently (e.g., changing a fixture so a part could be inserted more easily). At that point someone in personnel might get involved in thinking about training. Thus, we see a series of handoffs much like the handoff between product and process design. In both cases—product/process design and human infrastructure design—there are similar problems:

1. *Sequential vs. Simultaneous*: Designing the product first and fitting the process to the product parameters is similar to designing the complete technical system first and then fitting people into that system.
2. *Functional vs. Cross Function*: As a result of the sequential process, different components of the design are developed based only on narrow functional perspectives. Just as the product engineer is not considering manufacturability issues in product design, the manufacturing process engineer is not considering human infrastructure issues in manufacturing system design.
3. *Focus on Components Rather Than an Integrated Whole*: In both cases, parts of the system are independently designed in isolation and then

forced-fit together. There is little attention paid at the outset to dependencies and interactions between components.

Finally, "designing for manufacturability" was historically purely a matter of intuition. If the product engineer happened to have a feel for manufacturing (e.g., through experience in manufacturing), the product might be designed for manufacturability. Similarly, designing for human–technical system integration has been mostly a matter of intuition. More recently, product design tools for "designing for assembly" or "designing for manufacturability" have been developed to more systematically bring downstream considerations upstream to the concept design stage of new products. Majchrzak and Gasser (1992a) argue that similar tools are needed and feasible to design sociotechnical systems to achieve a fit between technology and the human infrastructure.

We can get some ideas about the features of an effective human infrastructure design approach from what has been learned about concurrent engineering, as well as gain some insights into the potential barriers to concurrent technology–human infrastructure design. Susman and Dean (1992) suggest three sets of factors are needed for effective concurrent engineering: effective cross-functional teams, integrative mechanisms, and technical tools. The most common method of concurrent engineering in the United States is the use of cross-functional teams that have representatives from different specialty groups that understand each of the key aspects of the product and process (e.g., product design, analysis, process design, finance, machine operation, marketing, vendors) (Clark and Fujimoto, 1991). According to Susman and Dean, cross-functional teamwork must be supported by integrative mechanisms (e.g., common reports, colocation of the project team, combining separate departments into one, rewarding teamwork). Additionally, technical tools have been developed to assist the designers in considering the downstream implications of product design decisions, such as design for assembly programs, as well as the use of computer-aided design (CAD) tools to enhance communication among different functional specialties.

Some barriers to effectively implementing concurrent engineering for product/process design are discussed by Liker and Fleischer (1992). They identify three sets of factors: mechanistic organizational structures, political barriers, and cultural barriers. Mechanistic structures based on functional departmentation lead to the creation of rigid boundaries separating product engineers and process engineers. These often result in political conflict resulting from conflicting interdepartmental goals. For example, product engineers have the goal of designing new performance features that add functionality to the product, while manufacturing engineers have the goal of optimizing use of currently owned or off-the-shelf manufacturing systems that may not be well suited to these new functional features. Product engineers want tolerances to be very tight to ensure fit between parts, while manufacturing engineers want loose tolerances that are forgiving of tool wear and manufacturing errors. These organizational boundaries often become embodied in distinct beliefs, values, and assumptions that can be thought of as incompatible subcultures. For example, product engineers are typically college graduates who work in clean offices, are relatively high status, and value ideas and sophisticated analytic techniques, while manufacturing engineers often come up through the ranks of production or the skilled trades, enjoy "getting their hands dirty," and value concrete machines that make concrete parts.

Similarly, concurrent engineering the human infrastructure of technology means bringing together different functional perspectives that are traditionally separated into different functionally homogenous organizational units in large bureaucratic organizations (e.g., manufacturing engineering, industrial engineering, personnel). To do this may require the use of cross-functional teams, but according to Susman and Dean (1992) we need more than that. There may be some need to provide supporting organizational structures (e.g., new reward systems) to encourage teamwork across boundaries, or perhaps even the creation of new departments that bring together multiple functions (e.g., a human infrastructure design department). Moreover, design tools are needed to systematize consideration of human infrastructure features and development of the design. Liker and Fleischer (1992) warn that prevailing political and cultural forces that maintain distinctive functional identities and conflicting goals between functions must be overcome for true cross-functional design to occur. This begins to suggest the characteristics necessary for an effective human infrastructure design process.

6.5 FEATURES OF A CONTEMPORARY CONCURRENT DESIGN PROCESS

Based on the open-systems paradigm, the research on the relationship between technology and organization, and evolving concurrent engineering methodologies, we have identified nine features that any design process should have to be effective in designing the human infrastructure of technology.

1. *Integrated Human Infrastructure Design*: The design process should recognize that the human infrastructure is a complex system of interrelated parts and provide a means for considering the fit between the formal organization, informal organization, individual characteristics, and linking mechanisms to the environment.
2. *Concurrent Human Infrastructure and Technical System Design*: The design process should facilitate concurrent design of the human infrastructure and technology, rather than serial design. This means the design process should provide analytic tools for understanding the likely implications of any particular technical solution for people and the organization, as well as the likely implications of different organizational arrangements for technology design.
3. *Fit with Environment*: The design process should consider the larger

environment of the system and the requirements and constraints imposed on the human infrastructure and technology by this environment and the systems mission.

4. *Specificity*: The design process should provide information on implications that are precise enough so that engineers can alter specific technical designs and organizational planners can modify organizational scenarios as the design process proceeds.
5. *Multiple-Scenario Generation*: The design process should facilitate the development of multiple design scenarios that can then be compared.
6. *Design Evaluation and Refinement*: The design process should facilitate systematic comparison among these scenarios, not unlike the prototyping processes in product design. An interative design procedure should then be used to gradually fine tune the design based on "prototyping" and "testing" design concepts.
7. *Life-Cycle User Involvement*: The design process should provide a process for involving potential users of the new technology throughout the design and implementation process.
8. *Facilitating Design Team Learning*: The design process should increase the knowledge of the planners of the sociotechnical system about issues relevant to organizational–technology integration.
9. *Facilitating Design Modification*: The design process should enable redesign as circumstances, such as new technology, change. This implies the ability to capture the original design reasoning so that implications of new sociotechnical implications can be considered in light of that reasoning.

Features 1 to 4 all have to do with the content of the design requirements. Basically, they say that the design process should produce an integrated sociotechnical design in accordance with open-systems principles at a level of specificity required to actually implement the design. Features 5 to 9 all have to do with the process of design—considering multiple scenarios and their associated trade-offs, evaluating and refining the design, involving the right people in the design process, facilitating design team learning, and facilitating later design modification and improvement.

Given these features, are there design process methodologies that have all of these features? Clearly, current methods used by most firms today for human infrastructure design do not have all of these features. While traditional manufacturing system design processes lead to specific sets of requirements (feature 4), they are rarely cross functional and integrative, user involvement is rare, the designers learn little about the broader implications of their functional designs for total system functioning, and there is no systematic approach to human infrastructure design modification. We have identified four existing design process methods that show promise for the future. These four were picked because they are all based on a systems perspective and include many of the characteristics of a concurrent design process. We also tried to identify diverse approaches that seemed to offer compensatory strengths and weaknesses. The four methods are sociotechnical systems design, participatory ergonomics, human-centered human factors design, and computer models of technology, organization, and people integration.

6.6 SOCIOTECHNICAL SYSTEMS DESIGN

The Sociotechnical systems (STS) approach has been used to successfully redesign organizations for more than 35 years (Pasmore, 1988). Pioneering work began at the Tavistock Institute in London. In studies of the introduction of new technology to coal mines beginning in 1949, Emery and Trist (1978) discovered that the most successful installations (in terms of productivity, safety, and low absenteeism) were those in which teams of workers performed the coal mining instead of workers assigned to individual specialized jobs. These teams were not unlike the teams of miners who historically went down into a mine with pick axes in the early years of mining and seemed to form a natural work unit. They argued that the social systems and technological systems were intimately connected and changes in one necessarily had implications for the other. The early STS approach provided a social system solution to managing technological change—the use of autonomous work groups.

Sociotechnical systems eventually evolved into a methodology for design rooted in open-systems theory that included analysis methods for the organization's environment, social system, and technical system, along with a set of principles of design to integrate the social and technical systems. Nonetheless, the resulting solutions almost invariably retained the core ingredient of autonomous work groups. We will begin by discussing the analysis methods and then move to some design principles.

The STS Methodology

The STS methodology has been refined since the early Tavistock days (cf. Cherns, 1976; Cummings, 1976; Davis and Taylor, 1979). The methodology has been used in many industrial plants to design blue-collar work with great success in business outcomes and employee quality of work life. Pava (1986) and Pasmore (1988) have developed modifications of the analysis methods for nonroutine work in service organizations. Contemporary versions of the methodology are detailed in Pasmore (1988) and Taylor and Asadorian (1985).

Most applications of STS are to the design of new plants based on routine manufacturing operations ranging from the production of dog food (Walton, 1982) to weaving (Miller, 1975) to printed circuit boards (Taylor and Asadorian, 1985). The methodology generally proceeds through a common series of steps, though the particular details of the analysis method vary from consultant to consultant. Generally the analysis is based on the current ("as is") system, which might be in an existing plant, and reveals the weaknesses of that traditional system. A common process, similar to that described by Pasmore (1988), is as follows:

1. *Define Scope of System Redesign*: This step involves defining the boundaries around the system to be analyzed and redesigned (or designed). Generally this decision is made by top leadership in the company.
2. *Form Steering Committee*: Generally a steering committee is established that includes top management of the system and union leaders (if relevant).
3. *Environmental Scan*: This focuses on analyzing the current and expected future external environment. What are the relevant aspects of the environment that will influence the functioning of the bounded system? These can include many sectors such as suppliers, customers, competitors, government, the local community, the labor pool, and technological developments. What are expected challenges and constraints in the future?
4. *Mission Statement*: A mission statement is developed that includes the business purpose and operating philosophy of the plant. The philosophy includes statements of values and ways of managing people in the system. The importance of designing to the system purpose is very much stressed by Taylor and Asadorian (1985) who argue this was missing in many early STS formulations.
5. *Educate Organizational Members*: Education is an ongoing process and may have begun with the steering committee and at this point may focus on educating potential members of a design team. Education continues throughout the process eventually educating all system members in the concepts of STS.
6. *Create Change Structure*: Typically a design team is established (through invitations to volunteer) that represents a diagonal slice of the organization (representing all levels and functions). In some cases the design team would be established before step 3 and work on the environmental analysis and mission statement, and in other cases this is presented to the design team by the steering committee. The design team has primary responsibility for analysis and developing design concepts, which they present to the steering committee.
7. *Technical System Analysis*: This typically begins by identifying a small number of key "unit operations" in which there is a significant state change in the product and then focuses on potential "variances" in the product that can lead to defects. Variances are deviations from specifications. A variance matrix lists the potential variances for each unit operation and then identifies relationships between variances, for ex-

ample, which variances will themselves cause later downstream variances? Then a variance control table is constructed that considers where in the current system (as it is traditionally organized) each variance is observed and where and how it is controlled. Information and control requirements are identified to control these variances.

8. *Social System Analysis*: A common approach to analysis of the social system is to construct a "focal role network." The analyst picks a key role in the production process, such as the production worker, and then analyzes their patterns of communication with other key actors in the system. This enables the analyst to identify interdependencies between roles as they are currently defined in the existing system. Another approach is to use the four key functions that the sociologist Talcott Parsons (1951) theorized were necessary for the success of any social system—called the GAIL model. GAIL stands for goal attainment, adaptation (to change), integration (between key roles), and latency or pattern maintenance. The GAIL analysis consists of asking the degree to which the current system performs these functions effectively. Other analysis methods include surveys to determine the degree to which work is designed so it is intrinsically motivating to workers (cf. Hackman and Oldham, 1980).
9. *Formulate Design Proposals*: In this step, often referred to as "joint optimization" as discussed below, the design team develops alternative design scenarios, presents them to the steering committee, perhaps shares them with other key people in the organization, and develops a final "provisional" design. It is regarded as provisional as it is always subject to refinement, even after it is implemented. Thus, it is a living design that is expected to be continually improved.
10. *Implement Recommended Changes*: This involves a great deal of education and is sometimes begun on a trial basis in a pilot area. The design is always subject to modification through input from system members and experience.
11. *Evaluate Changes/Redesign*: This is an onging process of continuous improvement of the social and technical system.

Joint Optimization

Once the mission is defined and the analysis is complete, the actual design objective is to "jointly optimize" the social and technical system considering the fit between the people and technology. While the analysis methods in steps 7 and 8 are relatively rigorous and systematic, this design step is unstructured and creative. There is no systematic method for joint optimization. Generally, the design is informed by best practice case examples (i.e., autonomous work groups) viewed either through videotapes or design team visits to successful sociotechnically designed plants and a set of general principles defining characteristics of jointly optimized systems.

One widely used set of principles for designing jointly optimized sociotechnical systems was developed by Cherns (1976). A compatible, but much more specific, set of principles were described by Davis and Wacker (1982):

1. Self-maintaining work teams are the basic building blocks for achieving joint optimization.
2. Those within a team should be responsible for controlling variances that occur in performing work.
3. Those within a team should have access to all the information they need to solve the unit's problems and assess its performance.
4. The boundaries between teams should be between main transformation processes rather than dividing a core transformation process between several teams.
5. Bounded organizational teams should control a complete process, product, or subdivision of a product.
6. The product should be measured at the boundaries to provide feedback to the teams.
7. Coordination between activities and people should be left to people within the team leaving integration to boundary managers.
8. Members of a team should develop face-to-face relationships in carrying out the work.
9. Requisite skills needed to perform assigned work and to maintain the technical and social systems of the unit should be within the team.
10. The need for external control and coercion should be minimized.
11. Boundary management should be handled by managers and supervisors (e.g., planning for environmental changes, training, extra resources from outside).

Cherns (1976) adds several important principles. One is *compatibility with organizational objectives*, which means the system design should fit the system purpose. A second is *design incompletion*, which means the design is always evolving and subject to improvement.

Case Example: Romeo Engine Plant

The Romeo Engine Plant, manufacturer of V8 engines designed initially for the 1991 Lincoln Town Car, is the first Ford Motor Company plant in the United States to implement the team concept plantwide, and it did so using the STS methodology (Liker and Sanborn, 1991). Formally, an operating agreement was signed with the UAW in July, 1987, designating Romeo, formerly a tractor plant, as the site for building the new engine. By the summer of 1987 a full-time launch team included approximately 30 people, including managers of production, quality, finance, and personnel, UAW leaders, a training coordinator, key process engineers, and a small number of hourly skilled trades and production employees. This team was soon joined by an outside consultant who, along with a full-time internal consultant, led the design effort. This group performed the complete analysis and developed a provisional design consisting of work teams in all major areas of the engine plant.

The launch team decided to involve machine operators, assemblers, and maintenance employees in the selection and build phases of the equipment being purchased for their lines, as well as the processes being considered for assembling engines. This decision allowed the three areas of engineering, production, and maintenance to have input not only into equipment and process selection and design but training provided by vendors as well. The same production employees and process engineers who purchased the equipment were also to be the operators, thereby creating a sense of ownership and responsibility. From the viewpoint of production workers, this meant that many of them were hired long before they would have been in a traditional launch and, as a result, they had input into key design and installation decisions.

A skeleton crew of initial salaried and hourly members were assigned to each production team by early 1988 and had a great deal of authority over the design, purchase, and installation of equipment in their production areas. Team managers had responsibility for budgets, technology, and production planning decisions in their areas. The teams were trained in interpersonal skills and team building and were expected to make all major decisions using consensus decision making. Most decisions about the selection of major equipment vendors and some parts vendors had already been made by the launch team. But the production teams had considerable autonomy and even influenced changes to the basic organizational design of the plant. The overall flow of the design and implementation process is presented in Fig. 6.3.

Some of the design features of the Romeo Plant that resulted from this process were (Liker and Sanborn, 1991):

1. *Self-Managing Work Teams*: A typical machining team is shown in Fig. 6.4. The team manager is a salaried position, but there are also hourly team coordinators who have certain leadership responsibilities. Each team runs like a small business with responsibility for production, scheduling, and quality control and operates as a profit center.
2. *Flat Hierarchies*: While a traditional engine plant at Ford had seven hierarchical levels, Romeo has four.
3. *Pay-for-Knowledge*: There is a small bonus for learning multiple jobs within the team.
4. *Team Facilities*: Each team has its own physically distinct area with a

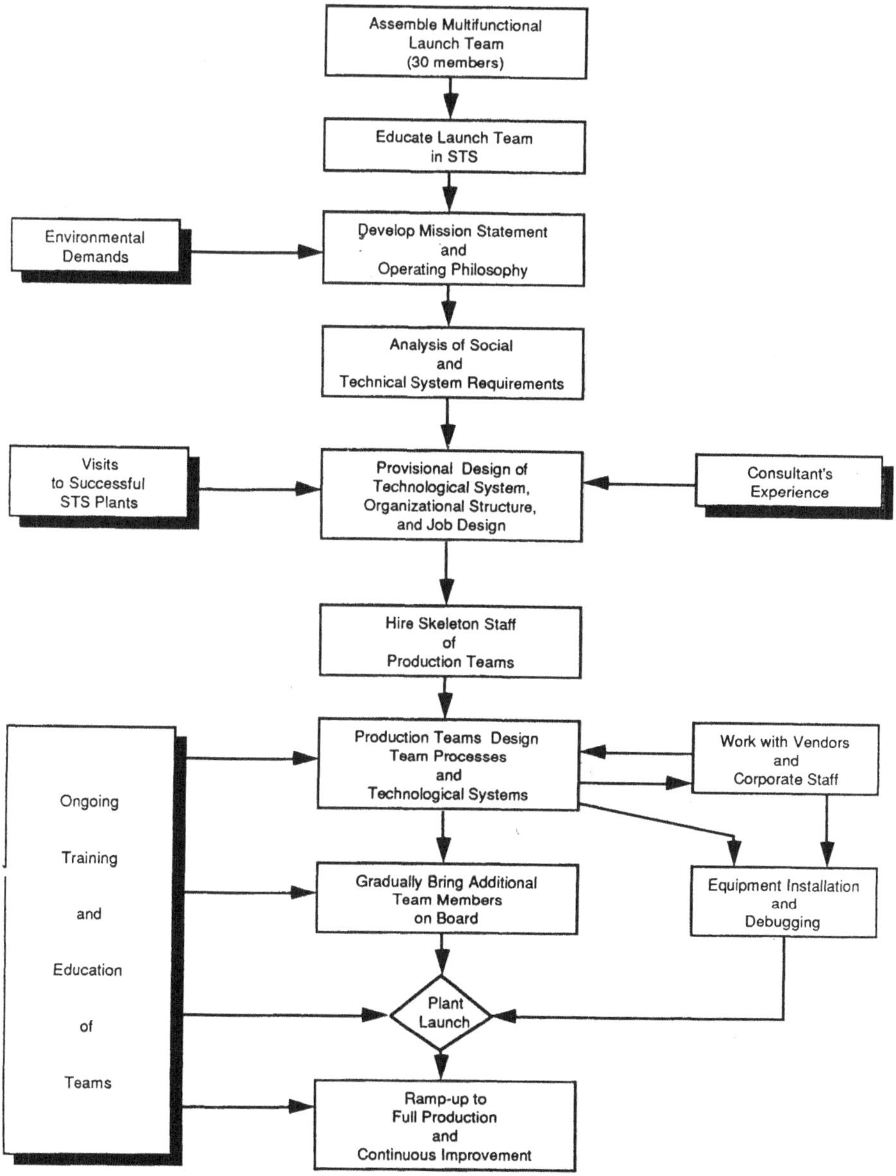

Fig. 6.3. Ford Romeo Engine Plant design and implementation process.

two-story building next to the work area that includes lockers, offices, meeting rooms, and computers.

5. *User-Friendly Information Systems*: The extensive information systems were developed by analysts who worked with the teams. For example, individual pagers are available to all workers for communication and

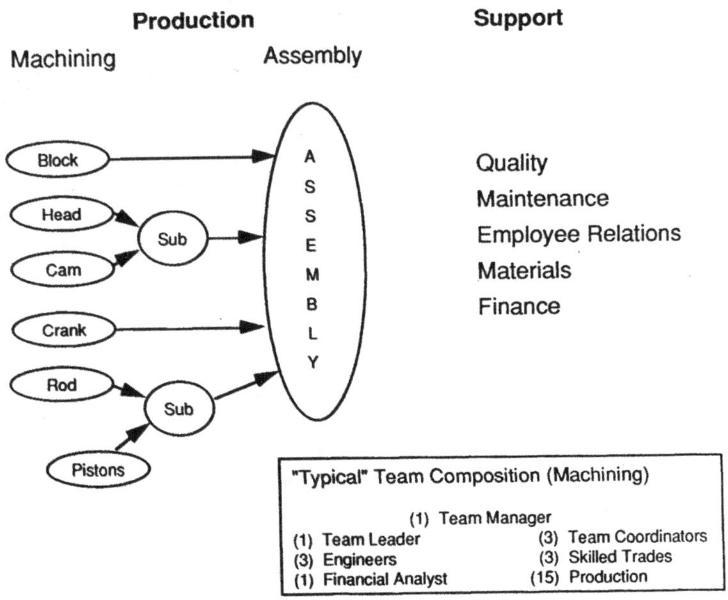

Fig. 6.4. Team structure at Romeo.

in some cases the machines can page the individual when he or she is producing defects. This frees up workers to wander from their machines for personal reasons, meetings, and so forth. The personal computers accessible to all workers include detailed information about productivity of every machine, time worked for each employee, quality levels by team and for the plant as a whole, and so forth. In some cases workers directly input their start and stop times into the computer rather than punching in or reporting to their supervisor.

6. *Broad Education and Training:* To provide ongoing education and training Ford (with funds from the state of Michigan) built a 11,600 square foot learning center costing almost one-half million dollars. The total cost of the center, training, and team building is estimated at $17.8 million. Education, including technical training, interpersonal skills, and business is required of all employees. Most courses are taught by Romeo employees. For example, the plant manager teaches Productivity 101, the UAW leader teaches Union History, and the quality manager teaches Quality 101. Technical training, administered largely by equipment vendors, has to meet strict standards for the development of training materials and approach documented by Romeo. In cases where the training did not meet these standards Romeo refused to pay for equipment until the training packages were adequate. Each employee spends a minimum of 14 work weeks on training and then is required to take two weeks of budgeted training per year.

7. *Extensive Environmental Contact*: When possible, arrangements were made for team members, hourly and salaried, to visit assembly plants where the engine is installed and vendors who supply parts.
8. *Ergonomic Workplace Design*: Though early team members were generally not trained in ergonomic analysis methods, when they visited vendors to look at equipment and when equipment was being installed, they often made suggestions on improving the ergonomics (e.g., moving a dial so it was readable, placing a button where it would not accidentally be hit, placing controls in easy reach of operators, etc.).
9. *Participative Management*: All major decisions are made participatively using the teams, task forces, and standing joint committees.

The result of the first year of operation of Romeo Engine in 1991 was the best launch that Ford ever had for an engine plant in terms of on-time start-up, quality, and productivity. As an example of the quality at launch we can look at defect rates. The average Ford engine plant in the Romeo launch year had 60 defects/1000 engines. Rome launched at 37, which was the best in the company even compared to established plants. This was close to the Japanese average of 25 and considerably better than General Motors' best engines. This also represented 80 percent fewer defects than the prior four launches of Ford engine plants. And repairs/1000 engines continually improved in the first year of operation starting at 3 engine repairs/1000 in the first month of launch (i.e., engines repaired by dealers after sale of the car) and dropping to less than 1 repair/1000 toward the end of the first year of operation.

STS as a Design Methodology

It is clear that STS as a design methodology has an enviable track record. It also has many of the desirable design features that we identified earlier for effective concurrent organizational–technical integration. The sociotechnical systems methodology is explicitly based on the open-systems model and emphasizes the fit between subsystems and the environment. Specificity is provided by visits to successful STS designed plants and the specific knowledge of people who work in the system and influence the design of their work area. Design teams are encouraged to consider alternative scenarios and evaluate the alternatives relative to the analysis they conducted. For example, at Romeo each design team was trained to identify potential failure modes for each step in the machining or assembly process in their area and develop action plans for avoiding these failure modes. This often led to redesign of the equipment before launch. There is a high level of user participation, and in the process of analysis and design the design team learns a great deal about the system. If the design team does a good job of documenting the analysis they perform, that documentation can always be referenced when considering design modifications.

On the other hand the actual design process of joint optimization is very unstructured, and the methodology provides little guidance on the process for ensuring that the social and technical systems will be matched. Thus, feature 2 (concurrently designing technology and human infrastructure) is not facilitated in any systematic way but rather relies on the intuition of the design team. The design principles provided to the design team are very vague and tend to lead to a single solution regardless of the analysis—autonomous work groups with a supportive social system that fits the organic model of organization. Visits to other STS plants by design team members also reinforce this particular design solution, and new designs are rarely more than minor variations on the general team-based theme. Moreover, it is common for the design to focus on social system characteristics (e.g., teams, participative management, open communication, pay for knowledge) without influencing the design of the technical system (though at Romeo there were some technical system modifications). Thus, the STS methodology does not provide specific tools for concurrent human infrastructure and technical system design, multiple-scenario generation, design evaluation and refinement, and facilitating design modification. These features are mainly left to the intuition of design team members whose awareness of the importance of these issues is raised through education and a detailed critique of an existing traditional system.

6.7 PARTICIPATORY ERGONOMICS

Participatory ergonomics is the application of the tools and methods of ergonomics by people who perform the work (Noro and Imada, 1991). They are often assisted by outside ergonomics experts, but the "expert" role is that of a trainer, facilitator, and resource rather than an analyst, solution developer, and implementer. There is evidence that ergonomic analysis and redesign combined with user participation is more likely to lead to user acceptance than when analysis and design are the strict province of experts alone who recommend changes (Orta-Anes, 1991).

Analysis Methods

Ergonomics has traditionally focused on job design, specifically the design of individual jobs. The biomechanical aspects that focus on the design of repetitive manual work have largely been the province of industrial engineering, while the cognitive aspects combining industrial engineering with psychology is often referred to as *human factors* engineering. The focus is on the human–machine interface.

Tools to analyze repetitive manual work are based on human physiology and cognitive science and can help to achieve a fit between human capabilities and job design to minimize physical and psychological stress on the operator and improve job performance (Chaffin and Andersson, 1984; Armstrong,

1986). For example, biomechanical models can be used to identify the stress on the operator of various work postures, and the job can then be designed to reduce this stress to acceptable levels (Chaffin and Andersson, 1984). Job design often includes significant modifications to machinery and equipment (e.g., relocation of buttons, changes in conveyor heights, redesign of material handling methods to eliminate working overhead).

One body of techniques focuses on cumulative trauma disorders (CTDs) defined by Armstrong (1986) as disorders of the soft tissue (e.g., nerves, muscles, and tendons) that are caused or aggravated by repeated exertions and movements of the body. They are considered chronic injuries (as opposed to acute injuries) since they develop over a period of weeks, months, or years as a result of the accumulation of repeated motions. Some risk factors that are known to be associated with the incidence of CTDs are repeated or sustained exertions, awkward postures, exposure to extreme vibration and temperatures, usage of gloves, and mechanical stress.

There are no straightforward techniques for quantifying the extent of the exposure to CTD risk factors and comparing measures to agreed-upon standards for acceptable levels of exposure. Armstrong et al. (1986) have developed a laborious job analysis technique that consists of two steps. The first step is a work content analysis based on the traditional techniques of time and motion studies documenting each task the operator performs. This is often done through frame-by-frame viewing of videotapes. The second step is to go through each work element and rate it for occupational risk factors. Simplified methods based on checklists of risk factors rated by trained observers have also been used, though not validated (Joseph, 1986).

A biomechanical model has been developed and computerized to measure the degree of stress on the spinal column in manual lifting tasks (Chaffin and Andersson, 1984; Chaffin and Baker, 1970). The computerized version is based on two-dimensional movement (lifting within a single plane) based on a static-strength model (it does not incorporate the cumulative effects of repeated lifting). The analyst inputs the worker's posture when lifting (manipulating a graphical stick figure), and the weight of the object and the program presents the back force along with control limits on acceptable force for men and women. By manipulating the design the analyst can ask what-if questions about the effects of different job designs.

Ergonomics analysis tools are much more developed than design tools. The analysis tools can alert the designer to risk factors that need attention. After redesign the designer can test the extent of risk factors associated with the new design. However, there are no tools that explicitly suggest ways to modify the workplace design to reduce risk factors. One attempt at such a tool was a CAD system that designed graphical representations of people (Evans and Chaffin, 1986); however, this tool has never gone beyond the prototype stage.

The analysis approaches described are just a few examples from the rich set of ergonomics tools developed through decades of research. Most have been developed to be used by experts and are not particularly user friendly. In order to enable users to actively participate in the analysis and design of their own workplace, all published cases of participatory ergonomics have used simplified analysis procedures. Such devices as checklists (Kogi, 1991), questionnaires (Nagamachi, 1991), and simple diagrams of workers in workplaces (Liker et al., 1991; Nagamachi, 1991) replace relatively complex computer programs. This leads to a reduction in the accuracy of analysis but facilitates user participation and ultimate acceptance. The result is an informed, intuitive design of the workplace.

Examples of Participatory Ergonomics in Practice

Numerous examples of the use of a participative approach to ergonomics in a number of different companies are reported in a compilation of studies by Noro and Imada (1991). Mechanisms for participation include the use of cross-functional task forces in automotive plants, combined with individual meetings with operators whose jobs are to be redesigned (Liker et al., 1991), the use of quality circles in Japanese plants (Nagamachi, 1991; Liker et al., 1989), legislated health and safety committees in Australian office environments (Rawling, 1991), and design workshops in which individuals are systematically led through a process of redesigning their own workplaces (Wilson, 1991). We will summarize one example here—the use of cross-functional task forces in automotive plants.

The use of cross-functional teams to systematically redesign worker's jobs in automotive plants and parts plants has been described by Liker et al. (1991) and Orta-Anes (1991). The process used to implement these programs has been as follows:

1. *Define Program Scope and Goals*: What parts of the plant will be the focus of the program and what are the goals of the program? In unionized plants this should be decided in partnership with union leaders.
2. *Assemble Cross-Functional Team(s)*: Identify members and leaders of the teams. In a large plant it is often best to establish separate teams for separate areas of the plant. All major functions who are affected by or can influence the implementation of work redesign should be included on the teams.
3. *Train Team Members*: The focus is on the philosophy of fitting jobs to people and specific tools and techniques for analyzing and redesigning jobs. This is most effective in a workshop setting in which hands-on experience is built into the training (Liker et al., 1990).
4. *Evaluate and Modify Workstations*: Liker et al. (1991), based on two cases, recommend starting in an area of the plant where there are clear ergonomics problems that appear to be solvable without great resources. Early success experiences can motivate continuation and expansion of the program.

5. *Expand and Refine Process*: This can mean moving from a pilot area to other areas in the plant, improving on group process, providing more in-depth training in ergonomics, or introducing additional analysis tools.

In the case of a stamping plant for automotive body parts (Liker et al., 1991) two teams were set up—one in the stamping area and one in the welding area. Team members included representatives from industrial engineering, plant health and safety, human resources, the union, process engineering, tooling engineering, plant medical, hourly operators, and plant maintenance. A university graduate student was available as a resource at all meetings to provide expert advice and run computer models. The teams evaluated workplaces based on a combination of computer models, checklists, and a rating form based on a diagram of a worker in a workplace. The rating form, developed by an hourly team member to simplify job analysis, is shown in Fig. 6.5. The teams brainstormed solutions and developed a refined solution. This was often implemented by maintenance representatives on the team after management approval. Over a 2.5-year period 79 jobs were redesigned. Independent ergonomic evaluations after implementation showed that in almost all cases there was a significant reduction in ergonomic risk factors, though there was no attempt to evaluate actual reductions in employee injuries.

Using a similar program in two engine plants Orta-Anes (1991) found that employee attitudes toward the job improved after implementation of ergonomics changes by committees. Also, there was a significant reduction in reported soreness by workers after the ergonomics changes. Orta-Anes (1991) also found that resistance to the changes was more likely by workers who were not involved in the evaluations and redesigns of their jobs compared to workers who were involved in the analysis and design process.

Participatory Ergonomics as a Design Methodology

Ergonomists have focused mainly on the individual–technical system match and have developed strong tools for evaluating the fit between workplaces and human physiology and cognitive capabilities (see Fig. 6.1). Thus, participatory ergonomics, which applies these methods, only addresses part of the issue of fit between subsystems. Job design is highly focused on technical issues, treating the human as a biomechanical and information processing machine. There is no attempt to consider the intrinsic motivation afforded by the job design or the supporting organizational structures or broader environment. An evolving branch of ergonomics, macroergonomics (Hendrick, 1985), argues that ergonomics must take into account the big picture of management and organizational systems. However, macroergonomics at this point has not developed a methodology for design to incorporate these broader systems.

On the other hand, ergonomics adds much to our understanding of the human–machine interface and provides a wealth of knowledge and analytic

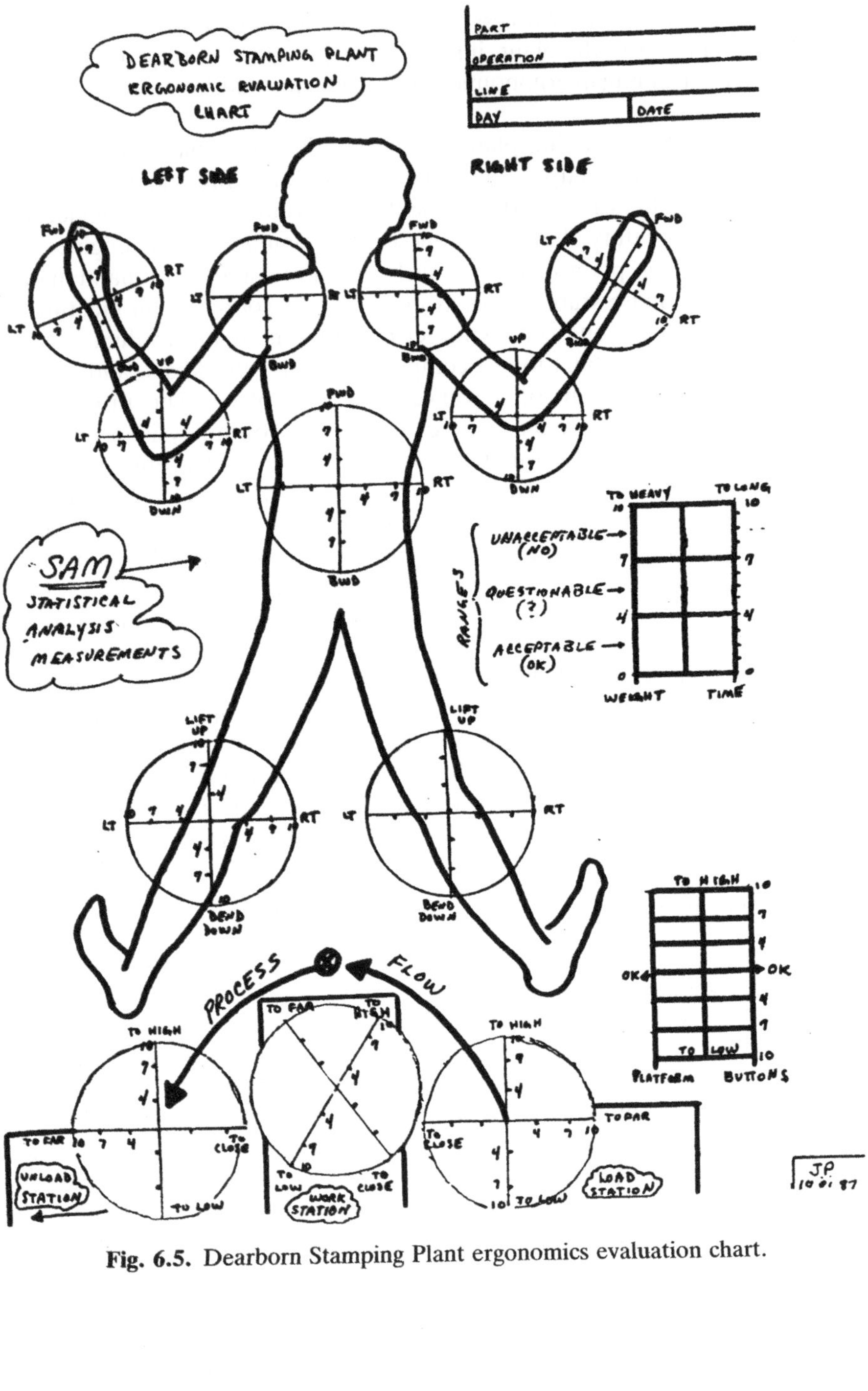

Fig. 6.5. Dearborn Stamping Plant ergonomics evaluation chart.

methods. For the evaluation of some types of risk factors (e.g., static lower-back stress), the analysis results are very specific and suggest specific solutions. Multiple scenarios can be quickly evaluated, particularly with computerized analysis software. Participatory ergonomics typically provides for user participation and cross-functional design teams learn a great deal about ergonomics issues they had not previously considered. Finally, the systematic analysis methods provide the opportunity to evaluate the implications of technology changes and modify job designs.

6.8 HUMAN-CENTERED HUMAN FACTORS DESIGN

The human-centered design approach is both a philosophical orientation and a general design approach. Proponents are primarily in Western Europe. Demonstration projects and evolution of the approach have been funded primarily as part of the European Strategic Programme for Research and Development in Information Technology (ESPRIT). Several companies, including Rolls Royce, as well as several universities, including Bremen University, in Germany, Technical University of Denmark, and UMIST in Britain have been involved.

The approach has been primarily applied to semiautonomous group working practices such as product-focused production and design islands. However, the approach need not be limited to those production applications.

Design Philosophy

In designing a technology, the approach adheres to the following design philosophy:

- System design involves the design of the technology, design of the social system, design of the manner in which the technology and social system will be used, and the design of the design process itself (Eason, 1988).
- The design of technologies is the design of tools that allow users to function as people who build mental models and formulate strategies to deal with changing situations rather than the design of hardware/software independent of how the user is involved (Kidd, 1992).
- Users' skills are a desirable feature for ensuring system performance, not a necessary evil to avoid (Badham and Schallock, 1991).
- System design is an evolutionary process involving user participation and extensive piloting and continuous change even after implementation (Eason, 1988).

System Specifications

This philosophy suggests such a dramatic departure from traditional design processes that human-centered design proponents have felt the need to further clarify their design philosophy by proposing a set of system specifications. These specifications, while general, help to guide system designers in a direction not normally considered in the traditional design process. A sample of these system specifications culled from some of the human-centered design literature include:

- Systems should be designed so that the users maintain control, flexibility, and choice over all operations, as well as choice over which functions are automated (Corbett, 1985).
- Systems should be designed to maximize users' involvement in the planning, execution, evaluation, and improvement of tasks.
- Systems should evolve as users' skills evolve (Murphy, 1989).
- Systems should be designed to support cooperative work, not just individual work (Havn, 1989).
- Systems should be designed for a broader notion of user acceptance and usability than just user-friendly interfaces; rather, a range of design issues should be considered including system functionalities and the user's relationship with the technology (Kidd, 1991).
- Decisions about which functions are allocated to human versus machine should be based not solely on machine capabilities but on other factors such as psychological importance for humans performing the function, predictability of the function, criticality, and scope for experimentation and learning (Kidd, 1992).
- Human–machine interface should allow for (a) interaction between the operator and the computer, (b) operator being able to "see" the internal processes of the computer software in order to facilitate the development of a schema, (c) people learning structures rather than isolated pairings, (d) the system not performing in ways unexpected by the operator in light of the operator's knowledge of the present state of the system, (e) provide information feedforward to inform the operator of the likely consequences of his or her actions, (f) human override, and (g) operators to have the freedom to trade-off requirements and resource limits by shifting operating strategies at their discretion (Corbett, 1985).

Design Process

The human-centered design approach has also suggested a design process for incorporating these specifications into the implemented design. Ken Eason (1988) has perhaps elaborated a design process in the most detail, although others (such as at the Technical University of Denmark) have offered suggestions for aspects of the design process. For example, Havn (1989) suggests that the IDEF modeling procedure be used as an excellent starting point for graphically representing information flow and thus talking about the social/behavioral requirements needed to support such an information flow.

In brief, Eason's design process, called *user-centered design*, suggests an evolutionary system design process. By evolutionary, Eason means that the system incrementally evolves, with the pace of evolution determined by users as they gradually acquire the experience and knowledge needed to understand options and their implications.

In addition, Eason believes that not only must the technical system evolve over time, but the evolution must be influenced so the evolving social system complements the evolving technical system, the patterns in which the system will be used, and the skills and process used by the design team itself. Thus, for example, just as the tool design process proceeds from prototype development to operational systems design and installation, the design of how the system can be used proceeds from prototypical experimentation with the tool, through initial use of the tool, to diffusion of the tool throughout the organization.

On the design of technical systems for human use, Eason suggests that the design of the technical system must follow after the design of jobs. To design the technical system, Eason suggests that features must be established with user participation for four categories of factors: functionality, usability, user acceptability, and organizational acceptability of the system. To ensure functionality, Eason suggests that the detailed task analysis necessary as part of any job design process should be used to specify functionality requirements for the technical system. To ensure usability, the design process should be one that first explicitly specifies various categories of usability including match to the user's perceptual-motor skills, level of user's expertise, match to the user's cognitive skills, and the extent to which the system should create a learning environment. To ensure user acceptability, Eason suggests that the design process should explicate the features for acceptability, including degree of access and pacing of the system, as well as the degree of choice in the system, the assumed knowledge, and the degree to which the system would be used within a cooperative work mode. Finally, criteria should be specified for organizational acceptability such as the degree of flexibility of use and the organizational culture and norms that should be designed into the system. These criteria should be set, according to Eason, with user participation. Once this range of criteria are identified, Eason suggests an intensive use of prototyping, trials, and real-life scenarios flowing in an incremental evolutionary fashion.

Example Demonstrations

Several demonstration projects have been completed thus far. In one, a shop floor monitor and controller were developed as a human-centered computer-assisted planning (CAP) software package. The main thrust of the human-centered CAP work was the development of a workstation to assist with scheduling within a production island. While many features of the system are similar to those of most scheduling systems (e.g., such as priority setting,

TABLE 6.1 Characteristics of a Human-centered Island at Sealectro Ltd.

PRODUCT
Manufactures contact pin/insulating bush subassemblies for a wide range of Sealectro's connectors.
SAMPLE PROCESSES
Metal turning, metal secondary operations, heat treatment, assembly, expoxy resin injection, plating.
BATCHES
Repeated single cycle period batch control, i.e., only those parts which are required for assembly in the next time period will be manufactured in the current time period.
DECISION-MAKING
Quasi-autonomous factory-within-a-factory with the island teams maintaining full authority and responsibility for all aspects of island operations including micro-scheduling, manufacturing engineering, and product quality.
CULTURAL NORMS
• Personnel will be encouraged to develop new skills. • All functions are embraced within multi-disciplinary teams. • Quality is the island's responsibility. • Performance of team is based on output quality and due date performance.

identifying resource needs to perform a machining operation, etc.), a few features of the CAP software clearly stand out as human-centered (Murphy, 1989):

- All scheduling decisions remain with the operators.
- A manual override facility to the automatic resource allocator, which freezes particular operations.
- A means of testing whether an improved machining sequence to manufacture a particular part would lead to an improvement in machine utilization.
- A means of recording particularly successful sequencings that will be used for training less skilled operators.

As another example, the characteristics of a human-centered island at Sealectro Ltd. is presented in Table 6.1. Note the similarities to the team approach in the Romeo engine case.

Human-Centered Human Factors Design as a Design Methodology

This approach incorporates most of our suggested features for a concurrent organization–technology design process. The human-centered approach is based on the open-systems model and considers issues of fit between subsystems and with the broader environment (features 1 to 3). By primarily designing technical systems for small-sized production islands (Badham and

Schallock, 1991), the human-centered approach can foster the organizational changes typically associated with a switch from traditional to cellular manufacturing concurrently with the design of technical systems for those cells. On the other hand, from the examples given here, it can be seen that the emphasis tends to be on technical system modifications to enable decision making and control by the human operator—mainly focusing on the individual–technical system interface. There is less attention given to broader formal and informal organization issues.

By explicitly applying the human-centered approach to the design of specific technical systems, this approach meets our fourth feature of specificity. The features of multiple-scenario generation and design evaluation and refinement is only facilitated in a sequential fashion, as different scenarios are prototyped one after another and surfaced to the user group. While there is no inherent barrier in the human-centered design process to surfacing alternative prototypes concurrently, the average length of the design cycle with human-centered design (12 months or more) precludes the serious simultaneous exploration of alternatives.

The human-centered approach, particularly as elaborated by Eason, provides an excellent means for involving users in the entire life-cycle development of the new technology from design through implementation. Moreover, by viewing the technical system as a tool that allows users to function as people who build mental models (not simply as tool operators), the designers' role shifts from simply providing functionalities to helping users redefine their work with the technology. Clearly a design process embodying this view would also facilitate design team learning. Finally, the human-centered approach does reasonably well for the feature of design modification. Design modification is fostered in the human-centered approach by explicitly building into the design the notion that user skills will evolve; thus the system must be flexible enough to accommodate to this skill evolution.

In summary, the human-centered approach incorporates most of our features of an effective design methodology, though it has a somewhat narrow focus. The greatest problem with the approach (and a problem that by most standards may not be critical) is that the approach is indeed cumbersome and slow; as a result, multiple-scenario comparisons on any significant level of detail cannot be surfaced.

6.9 COMPUTER MODELS OF INTEGRATED TECHNOLOGIES, ORGANIZATIONS, AND PEOPLE

A final design process for concurrent organization–technology design has been under development at the University of Southern California. Initial work resulted in the development of a manual checklist procedure for ensuring that the necessary range of appropriate issues were considered in an appropriate sequence when designing an integrated organizational–technological system

(Majchrzak et al., 1991). Experiences from industrial applications of the checklist (at Hewlett-Packard, Douglas Aircraft, and other U.S. manufacturers) indicated that the checklist procedure provided an insufficient number of our features for an effective concurrent organization–technology design. For example, while such a procedure facilitated integration of technology and organization, design evaluation and refinement and highly specific design features were difficult to elicit from such a manual procedure. Thus, work was begun on a computer model of integrated technology–organization designs for different configurations of technologies, business needs, organizational contexts, and environments.

The result of this work to date is an automated knowledge-based design, decision support, and simulation software package that is to be used by people involved in the design and planning of technology for discrete parts manufacturing systems (Majchrzak and Gasser, 1992a). An early prototype version, entitled HITOP–A (Highly Automated Technology, Organization, and People–Automated), used user inputs and theory-based heuristics to generate a supportive human infrastructure and then to evaluate this infrastructure for its likelihood of successful implementation. The general framework is presented in Fig. 6.6. A sample list of user inputs is presented in Table 6.2. The framework was implemented in the prototype using the artificial intelligence (AI) structure depicted in Figure 6.7.

Since that prototype, considerable enhancements have been made to the knowledge base. The automated knowledge base, now called ACTION, can be used during the preparation of capital investment proposals to identify hidden costs and benefits associated with a planned technological change, as well as during the implementation process to identify critical success factors. ACTION incorporates specific decision rules and heuristics, drawn from subject matter experts, the current theoretical literature, current best-practice approaches, and our own formal analyses. These rules and heuristics predict in detail, for a wide range of contexts, technologies, and management values, the future human infrastructure needed to support a proposed technological change. The current experimental domain for the knowledge base is the human infrastructure supporting flexible manufacturing cells, but some initial investigations support the possibility that the embodied domain knowledge generalizes to other types of technologies and business processes.

Uses of the Automated Knowledge Base in the Design Process

An intended use of ACTION is as a support tool for planning modernization projects. Ideally, the tool would be used by all those responsible for the design and implementation of a technology change, including users, engineers, production designers, plant managers, technology/system designers, and organizational development, human resource and/or human factors professionals. Based on our experiences with an early manual checklist, and an early version of the software tool at Texas Instruments, the most effective users of

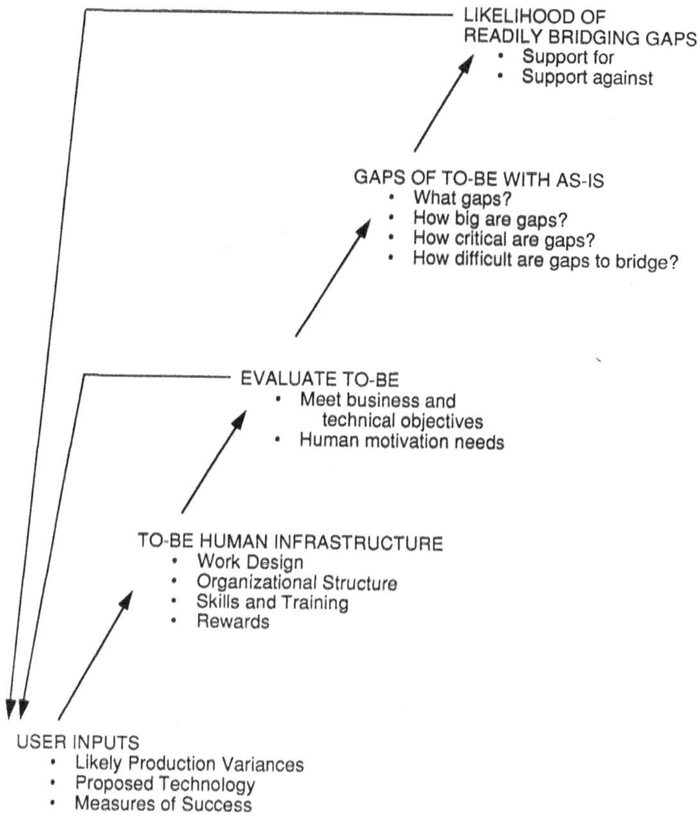

Fig. 6.6. Example in HITOP–A framework.

the tool are those who can span the three domains of engineering system design, manufacturing processes, and human resources either because of their credibility, coordination skills, and/or knowledge breadth (Majchrzak et al., 1991).

For manufacturing systems designers, ACTION is intended to be used in a number of different ways (Majchrzak and Gasser, 1992b). The automated knowledge base can be run when a plant engineer or manager or technology planning group is first considering multiple alternative technology scenarios. ACTION describes the organizational infrastructure (work design, skills, coordination, cultural norms, and performance management structure) that best supports the various alternative scenarios and the extent of fit between each proposed infrastructure and the current organization. With this information, managers, planners, and technology users are then able to decide among alternatives based not just on technology and economic reasons but organizational needs and fit reasons as well. For example, managers might decide that it is more effective to install state-of-the-art flexible equipment before it

TABLE 6.2 Inputs Required of User for HITOP–A.

Inputs	# Options	Examples
Business Philosophies	3	JIT
Management Philosophies	14	Push Discretion Down
Product Specifications	4	# Parts Per Product
Marketplace	7	# Suppliers
Measures of FMC Success	10	Part Mix Flexibility
Production Environment Variances	9	On-Time Arrival of Parts
Cell Load	7	# Batches
Technology Specifications	156	Expected Reliability of MHS
Workforce Characteristics	5	Experience with Teams
User Preferences	2	Preferences Against Teams
Allocation of Tasks to Human, Machine, Outside Cell	28-96	Allocation of Dynamic Rescheduling Tasks

is fully debugged not because the equipment will initially speed up quick changeover times but because the equipment debugging process in which machinists become involved will expand machinists' ability to maintain and repair the equipment once it stabilizes. An alternative way in which the knowledge base could be used is to help designers develop initial implementation plans such as those needed for training programs. A third use of

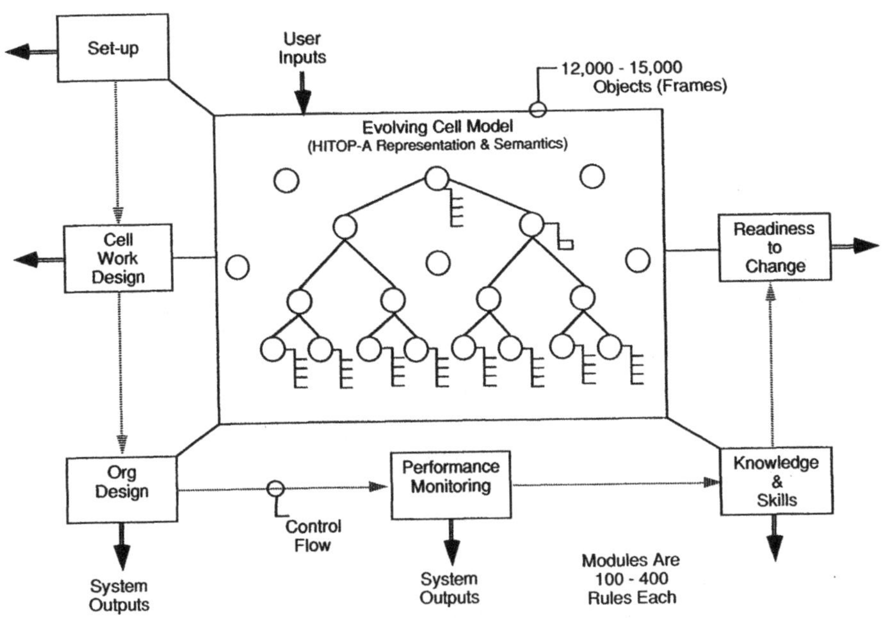

Fig. 6.7. HITOP–A system implementation structure.

ACTION is to open a dialogue among stakeholders about inputs. For example, initial uses of the software indicate that significant time is often spent in determining which values the organization wishes to input into the automated knowledge base. Moreover, when a consensus cannot be generated, the automated knowledge base is often run multiple times to allow for differences in input values to determine if real differences in supportive technology and organizational infrastructures result.

The Automated Knowledge Bases Design Process

ACTION is actually a tool rather than a complete design process. Thus, as with any tool, the precise ways in which the tool is to be used are not defined. Continuation work now being supported by the National Center for Manufacturing Sciences will further delineate best-practice procedures for using the tool. Nevertheless, clearly the tool has some *intended* design methodology features. These design methodology features include:

- The use of the tool as a mechanism for surfacing specifications for a supportive human infrastructure given a set of desired business goals and initial technical design features.
- The user evaluating the outputs of a supportive human infrastructure to determine if the supportive human infrastructure is acceptable to the user community and if not initiating changes in desired business goals or initial technical design features.
- Considerations of the outputs of a supportive human infrastructure as one possible scenario from which further discussion about its plausibility, cause, and desirability can ensue.
- Use of the tool's request for input values as a structured checklist of factors to consider when designing sociotechnically matched systems.
- Use of the tool to generate and compare many alternative scenarios to encourage exploration and experimentation of inputs and outputs not previously identified.

Automated Knowledge Bases as a Design Methodology

Given this list of features of ACTION, when used as a design methodology, the tool compares favorably along many of the features for an effective design methodology. The design of ACTION was guided by the open-systems model and thus has the features of concurrently designing for technology–organization integration and environmental fit. This is done through hundreds of decision rules, as well as by asking the user to answer a number of input questions about the environment. The current version of the tool achieves concurrent organization–technical design with this information; however, concurrency is achieved in a highly cumbersome manner. Outputs of a supportive human infrastructure that are found unacceptable are left to the tool user to determine which inputs (including reconsideration of some technical inputs) need to be altered to generate different outputs. The new version of the tool under development at the time of this writing will overcome this cumbersome nature.

The feature of specificity is accounted for by the tool by providing significant detail for implementation, including, for example, allocation of over 400 tasks to jobs, information about specific coordination activities needed to perform specific tasks, and specific cultural norms likely to yield effective work structures. The tool facilitates the feature of life-cycle user involvement by providing a ready mechanism for structuring a consensus process around input values and even allowing the possibility of examining the implications of selecting one preferred set of inputs over another. The system also facilitates design team learning by identifying a range of issues as inputs that are often ignored by system designers (such as whether or not there is a preference for employment security or discretion and what production variances are likely to remain regardless of efforts to control). Moreover, for the designer to evaluate the outputs about a supportive human infrastructure, the designer must turn to the users to determine if they are willing to live within such a structure. Multiple-scenario generation and design evaluation and modification are obviously fostered due to the automated nature of the output generation. Because of the ease with which detailed scenarios are generated (in a matter of minutes), in theory, an infinite number of design alternatives could be considered. Design modification is also fostered by the implied feedback loop between unacceptable human infrastructures and changes to the initial technical design.

One problem is that the tool does not currently define its use. As we have learned with computer numerical control (CNC) machines, the same machines that deskill workers in one factory can upgrade the skills of the same types of workers in another factory. Thus, it is not the tool alone that determines how it will be used but implementation and management strategies. Thus, any design tool may not be used to its best intentions.

6.10 CONCLUSIONS

The human infrastructure of technology is not something that automatically happens. To achieve an appropriate fit between technology and organization, planners should consciously design the two concurrently. There is no cookbook to achieve this fit and we have argued that merely copying "best-practice" procedures that seem to work for other firms will often lead to suboptimal results. We have argued that open-systems theory provides a powerful framework for the design of the human infrastructure. The theory emphasizes that system elements are interdependent and must be concurrently designed considering fit between subsystems and with the larger environment.

Drawing on knowledge about effective concurrent engineering practice, we identified a set of desirable characteristics of a design process for the human infrastructure of technology. We then identified four existing design approaches that have most of these characteristics. In looking across these design approaches it appears that use of these techniques need not be mutually exclusive. In fact, each has some unique strengths, and a combination of all four approaches may be most effective. We discuss the strengths of each in turn.

The sociotechnical systems perspective is the oldest approach and has evolved into a complete process of analysis, design, and implementation. It is firmly rooted in open-systems theory, and the analysis works through the entire system from external environment, to mission statement, to internal social and technical systems. There is a strong tradition of multifunctional involvement and employee participation in the complete process. Within this tradition has evolved an understanding of the design of autonomous work groups and the use of various linking structures that connect specialized subsystems. Organizational development skills are very important to orchestrate the process of group decision making and consensus building. This can be handled by outside consultants, internal consultants, or some combination of the two. Its weakness is that there is a tendency to apply a single solution—autonomous work groups—to all situations, and there seems to be a tendency of focusing mainly on social system design without influencing the design of the technical system.

Participatory ergonomics builds on the long tradition of ergonomics research and adds the involvement of users in the design of their own workplaces. Often simplified analysis procedures are developed to enable users to evaluate their own jobs and understand where the design of their workplaces is currently deficient. There is evidence that ergonomics changes that are designed with user participation are more likely to be accepted than changes imposed by experts or management. The strength of this approach is in fact in the design of technical systems to fit people's physiological and cognitive capacities—an area where STS is weak. On the other hand, this approach does not lead to the broad social system changes where STS is strongest.

Human-centered human factors design is a relatively new approach and attempts to design technological systems, primarily computer-controlled systems, to fit human capabilities and needs. The emphasis is not only on cognitive capabilities but also on the broader motivational needs of humans. For example, human–computer allocations consider the need of humans to have discretion and continue to grow as they master new skills in their job. The advocates of this approach tend to be well trained in human factors methods, systems thinking, and computer system development. The strength of this approach is in the design of the human–computer interface. The current examples, however, suggest that this approach, like participatory ergonomics, is not likely to lead to the broad social system changes of STS. Again this method adds a dimension lacking in the other approaches.

Computer modeling of integrated technologies, organizations, and people is an ambitious attempt to codify what is known about the fit between technology and social systems in the form of decision support systems (DSS). The DSS is not designed to recommend one best solution but to interact with the design team to ask probing questions and suggest alternative solutions. The early anecdotal evidence suggests that the DSS succeeds in causing design teams to think about issues and alternatives that they might not otherwise consider. It points out implications that are not obvious. One possible vision is that this computer model will grow to the point where it will replace the human facilitator in STS, build in all of the methods and knowledge of ergonomics, and build in all of the analytic reasoning of the human-centered human factors design approach, thus subsuming all of these methods. This seems unlikely, at least for the foreseeable future. Thus, we can view this method as one additional tool to be used as part of a broader process that incorporates the critical features for an effective design process.

The effective operation of a sociotechnical system is fraught with uncertainty and unpredictable problems. It would be foolhardy to assume that these can all be anticipated in advance and designed in any deterministic way by system planners. However, the tools and approaches we have described here can help to effectively design a system capable of identifying and reacting to new problems and opportunities and to learn from experience. The design itself must evolve and grow as humans in the system evolve and grow. If there is anything we know about the human infrastructure of the complex technologies we are creating, it is that the human infrastructure must be a process of discovery and learning.

REFERENCES

Adler, P. S. (1991). The Learning Bureaucracy: New United Motors Manufacturing, Inc. Unpublished paper, L.A., CA., University of Southern California.

Armstrong, T. J. (1986). Ergonomics and cumulative trauma disorders. *Hand Clinics*, **3**, 553–565.

Armstrong, T. J., Radwin, R. G., Hansen, D. J., and Kennedy, K. W. (1986). Repetitive trauma disorders: Job evaluation and design. *Human Factors*, **28**, 325–336.

Ayres, R. U., and Miller, S. M. (1985). *Robotics: Application and Social Implications*. Harper & Row, New York.

Badham, R., and Schallock, B. (1991). Human factors in CIM: A human-centered perspective from Europe. *International Journal of Human Factors in Manufacturing*, **1**(2), 121–142.

Burns, T., and Stalker, G. (1961). *The Management of Innovation*, Tavistock, London.

Butera, F. (1984). Designing work in automated systems: A review of case studies, in *Automation and Work Design*, F. Butera and J. E. Thurman, Eds., Elsevier Science, New York.

Chaffin, D., and Andersson, G. (1984). *Occupational Biomechanics*, 2nd ed. Wiley, New York.

Chaffin, D., and Baker, W. H. (1970). A biomechanical model for analysis of symmetric sagittal plane lifting, *AIIE Transactions*, **2**(1), 16–27.

Checkland, P. (1981). *Systems Thinking, Systems Practice*, Wiley, New York.

Cherns, A. (1976). The principles of sociotechnical design. *Human Relations*, **29**, 783–792.

Clark, K. B., and Fujimoto, T. (1991). *Product Development Performance*, Harvard Business School Press, Boston.

Corbett, J. M. (1985). Prospective work design of a human-centered CNC lathe. *Behavior and Information Technology*, **4**(3), 201–214.

Cummings, T. (1976). Sociotechnical systems: An intervention strategy, in *Current Issues and Strategies in Organization Development*. W. Burke, Ed., Human Science Press, New York.

Davis, L., and Taylor, J., Eds. (1979). *Design of Jobs*, Goodyear, Santa Monica, CA.

Davis, L., and Wacker, G. (1982). *Handbook of Industrial Engineering*, Salvendy, Ed., Wiley, New York.

Dean, J. W., and Snell, S. A. (1991). Integrated manufacturing and job design: Moderating effects of organizational inertia. *Academy of Management Journal*, **34**(4), 776–804.

Eason, K. (1988). *Information Technology and Organizational Change*, Taylor & Francis, London.

Emery, F., and Trist, E. (1978). Analytical model for sociotechnical systems, in *Sociotechnical Systems: A Sourcebook*, W. Pasmore and J. Sherwood, Eds., University Associates, San Diego.

Evans, S. M., and Chaffin, D. B. (1986). Using interactive visual displays to present ergonomic information in workspace design, in *Trends in Ergonomics/Human Factors III*, W. Karwowski, Ed., Elsevier, Amsterdam, pp. 17–24.

Fayol, H. (1960). General principles of management, in *Classics in Management*, H. F. Merrill, Ed., American Management Association, New York, p. 236.

Hackman, J., and Oldham, G. (1980). *Work Redesign*, Addison Wesley, Reading, MA.

Hanna, D. P. (1988). *Designing Organizations for High Performance*, Addison-Wesley, Reading, MA.

Havn, E. (1989). Creating a Human-Centered CIM System. Paper presented at CIM-Europe SIG 3 Workshop for ESPRIT CIM, BIBA Bremen, September.

Hirschhorn, L. (1984). *Beyond Mechanization: Work and Technology in a Post-industrial Age*. MIT Press, Cambridge, MA.

Hull, F. M., and Collins, P. D. (1987). High-technology batch production systems: Woodward's missing type, *Academy of Management Journal*, **30**, 786–797.

Industrial Development Division (1991). Product Development Study: A Key to World Class Manufacture of Automotive Bodies, Vol. I. Prepared for Auto/Steel Partnership Program, University of Michigan, Ann Arbor.

Ishikawa, K. (1985). *What is Total Quality Control? The Japanese Way*, Prentice-Hall, Englewood Cliffs, NJ.

Joseph, B. S. (1986). A Participative Ergonomic Control Program in a U.S. Automotive Plant: Evaluation and Implications, Ph.D. Dissertation, University of Michigan, Ann Arbor.

Katz, D., and Kahn, R. (1966). *The Social Psychology of Organizations*, Wiley, New York.

Kidd, P. T. (1991). Workshop-Oriented IT Tools for Skill-based Automated Manufacturing. Proceedings of the Open University Conference on Information Technology and People, Milton Keynes, United Kingdom, April.

Kidd, P. T. (1992). Design of human-centred robotic systems, in *Human-Robot Interaction*, M. Rahimi and W. Karwowski, Eds., Taylor and Francis, London, pp. 225–241.

Knight, K. A. (1985). Socio-Technical Engineering: An Approach to Technology Integration. Paper presented at the Society of Manufacturing Engineering Autofact Conference, Detroit, Mich., November.

Kogi, K. (1991). Participatory training for low-cost improvements in small enterprises in developing countries, in *Participatory Ergonomics*, K. Noro and A. Imada, Eds., Taylor & Francis, London.

Lawrence, P., and Lorsch, J. (1967). *Organization and Environment*. Harvard University Press, Cambridge, MA.

Liker, J. K., and Sanborn, L. (1990). Launch of the Ford Romeo Engine Plant: Team management in practice, unpublished manuscript, University of Michigan, Ann Arbor.

Liker, J. K., and Fleischer, M. (1992). Organizational context barriers to design for manufacturability, in *Integrating Design and Manufacturing for Competitive Advantage*, G. I. Susman and J. Dean, Eds., Oxford University Press, Oxford, England.

Liker, J. K., Roitman, D., and Roskies, E. (1987). Changing everything all at once: Work life and technological change, *Sloan Management Review*, **28**(4), 29–48.

Liker, J. K., Nagamachi, M., and Lifshitz, Y. R. (1989). A comparative analysis of participatory ergonomics programs in U.S. and Japan manufacturing plants. *International Journal of Industrial Ergonomics*, **3**, 185–199.

Liker, J. K., Evans, S. M., Ulin, S. S., and Joseph, B. S. (1990). The strengths and limitations of lecture-based training in the acquisition of ergonomics knowledge and skill. *International Journal of Industrial Ergonomics*, **5**, 147–159.

Liker, J. K., Ulin, S., and Joseph, B. (1991). Participatory ergonomics in two U.S. automotive plants, in *Participatory Ergonomics*, A. Imada and K. Noro, Eds., Taylor & Francis, London.

Liker, J. K., Fleischer, M., and Arnsdorf, D. (1992). Fulfilling the promises of CAD: Integrating technology and organization. *Sloan Management Review*, **33**(3), 74–86.

Majchrzak, A. (1988). *The Human Side of Factory Automation*, Jossey-Bass, San Francisco.

Majchrzak, A. (1991). Management of technological and organizational change, in *Handbook of Industrial Engineering*, 2nd ed., G. Salvendy, Ed., Wiley.

Majchrzak, A., and Gasser, L. (1992a). HITOP-A: A tool to facilitate interdisciplinary manufacturing systems design. *International Journal of Human Factors in Manufacturing*, **2**(3), 255–276.

Majchrzak, A., and Gasser, L. (1992b). On using artificial intelligence to integrate the design of organizational and process change in U.S. manufacturing. *AI and Society*, **5**, 321–338.

Majchrzak, A., Fleischer, M., Roitman, D., and Mokray, J. (1991). *Reference Manual for Performing the HITOP Analysis*, Industrial Technology Institute, Ann Arbor, MI.

Miller, E. (1975). Sociotechnical systems in weaving, 1953–1970: A follow-up study. *Human Relations*, **28**, 349–386.

Murphy, S. (1989). Human-centered CIM. Paper presented at CIM-Europe SIG 3 Workshop for ESPRIT CIM, BIBA Bremen, September.

Nadler, D., and Tushman, M. (1980). A congruence model for diagnosing organizational behavior, in *Organizational Psychology*, I. Rubin and J. McIntyre, Eds., Prentice-Hall, Englewood Cliffs, NJ.

Nagamachi, M. (1991). Application of participatory ergonomics through quality circle activities, in *Participatory Ergonomics*, K. Noro and A. Imada, Eds., Taylor & Francis, London.

Noro, K., and Imada, A., Eds. (1991). *Participatory Ergonomics*, Taylor & Francis, London.

Office of Technology Assessment. (1984). *Computerized Manufacturing Automation*, Library of Congress no. 84-601053. Washington, D.C.: Government Printing Office.

Orta-Anes, L. (1991). Employee Participation in Reduction of Ergonomic Risk Factors: Attitudes and Perception of Effects, doctoral dissertation, University of Michigan, Ann Arbor.

Parsons, T. (1951). *The Social System*, Free Press, New York.

Pasmore, W. A. (1988). *Designing Effective Organizations: The Sociotechnical Systems Perspective*, Wiley, New York.

Pava, C. (1986). Redesigning sociotechnical system design: Concepts and methods for the 1990s. *Journal of Applied Behavioral Science*, **22**(3), 201–222.

Perrow, C. (1967). A framework for the comparative analysis of organizations, *American Sociological Review*, **32**, 194–208.

Perry, B. (1984). Enfield: A High Performance System, unpublished paper, Digital Equipment Corp., Bedford, MA.

Rawling, R. G. (1991). Participative approaches to the design of physical office work environments, in *Participatory Ergonomics*, K. Noro and A. Imada, Eds., Taylor & Francis, London.

Schonberger, R. (1990). *Building a Chain of Customers: Linking Business Functions to Create the World Class Company*, Free Press, New York.

Shumaker, G. C. (1990). Integrated Product Development Program Strategy. Report by Concurrent Engineering Office, Manufacturing Technology Directorate, Wright-Patterson Air Force Base Research and Development Center, Ohio, July.

Susman, G. I., and Dean, J. W. (1992). Development of a model for predicting design for manufacturability effectiveness, in Susman, G. I. (Ed.), *Integrating Design and*

Manufacturing for Competitive Advantage, Oxford University Press, Oxford, England, pp. 207–227.

Taylor, J. C., and Asadorian, R. A. (1985). The Implementation of Excellence: STS Management. *Industrial Management*, July–August, 5–15.

Von Bertalanffy, L. (1956). General System Theory, in *General Systems Yearbook of the Society for General Systems Theory*, Vol. 1, pp. 1–10.

Walton, R. (1982). The Topeka Work System: Optimistic visions, pessimistic hypotheses, and reality, in *The Innovative Organization*, R. Zaeger and M. Rosow, Eds., Pergamon, New York.

Walton, R. E., and Susman, G. I. (1987). People policies for the new machines, *Harvard Business Review*, **2**, 98–106.

Ward, A. (1990). A Recursive Model for Managing the Design Process, presented at the ASME Design Theory and Methodology Conference, Chicago, September 16–19.

Wilson, J. R. (1991). Design decision groups—A participative process for developing countries, in *Participatory Ergonomics*, K. Noro and A. Imada, Eds., Taylor & Francis, London.

Winner, R. I., Pennell, J. P., Bertrand, H. E., and Slusarczuk, M. G. (1988). The Role of Concurrent Engineering in Weapon Systems Acquisition, Institute for Defense Analysis Report R-338, Washington, D.C., December.

CHAPTER 7

SKILL-BASED AUTOMATED MANUFACTURING

P. T. KIDD
Cheshire Henbury Research & Consultancy
Macclesfield, United Kingdom

7.1 INTRODUCTION

Manufacturing industry is facing a competitive environment that is becoming more international and customer driven. Competitive pressures include decreasing production volumes, increasing product variety, shortening of product life cycles, and a reducing number of repeat orders. This business environment leads to opportunities for improvements in profit growth through economies of scope rather than the traditional economies of scale, and this is creating a demand for highly flexible manufacturing systems.

Traditional manufacturing concepts such as fully automated systems, deskilling, division of labor, hierarchical command and control organizations, and so forth seem to be better suited to a low-variety, large-volume economies of scale manufacturing strategy than to a situation where a high degree of flexibility is needed. A high degree of flexibility demands a balanced manufacturing response based on consideration of organization, people, and technology. Moreover, such a balanced manufacturing response can offer man-

Organization and Management of Advanced Manufacturing. Edited by Waldemar Karwowski and Gavriel Salvendy.
ISBN 0-471-55508-8

ufacturing companies strategic advantages by virtue of a much improved capability to rapidly deal with increased complexity, without error. This balanced approach to manufacturing is known as skill-based automated manufacturing.

This chapter presents some of the main concepts of skill-based automated manufacturing systems. The characteristics of traditional manufacturing practices and concepts are first highlighted. The need for a new approach is then outlined. Skill-based automated manufacturing is then presented with reference to its features and the underlying design paradigm. Issues relevant to the design of these systems are also discussed. The design of computer-aided technologies for skill-based automated manufacturing systems is then addressed, and some important design concepts are presented and illustrated by means of examples. The role of knowledge-based systems is considered.

7.2 SKILL-BASED AUTOMATED MANUFACTURING

Traditional Manufacturing

Traditional manufacturing has taken shape over the period since the industrial revolution in the mid-eighteenth century. The traditional factory is based on an 1890s model of factory organization and control. The main characteristics are (Kidd, 1990a,b) centralization of control and decision making, steep hierarchies, functional factory layout, automation, functional specialization, horizontal and vertical division of labor, deskilling, and standardization.

These organizational and control models are based on ideas developed during the late nineteenth century by people such as Frederick Taylor, an American mechanical engineer, who advocated a system of factory management, aspects of which are still in use today. This system is based on vertical division of labor with the separation of thinking and doing and the centralization of all decision making. Taylor believed that all workers should be told exactly what to do and that their working methods should be standardized. He believed that there was only one best way of doing work, which could be scientifically established. His system also prescribed functional specialization (Taylor, 1902, 1907).

Max Weber, another Victorian, developed the concept of the bureaucratic organization, which, like Tayloristic concepts, is still found in industry today. Weber argued that the most efficient type of organization was a bureaucratic one in which (i) the duties and responsibilities of all members of the organization are clearly defined; (ii) the various positions in the organization are arranged hierarchically with each person responsible to a superior and responsible for subordinates, with the exception, of course, of those at the top and bottom of the pyramid; and (iii) an elaborate system of rules governs the manner in which each person carries out his duties.

In traditional manufacturing technology tends to be used in the form of automation and provides a means of replacing people, or reducing their role, or removing the need for their skills. Technology is also often seen as a solution to most if not all problems and is perceived to be the sole source of improved competitiveness.

Associated with traditional manufacturing is an underlying design paradigm. This paradigm has both given birth to, and fostered the concepts of, traditional manufacturing. At the same time, these traditional concepts have helped to shape the design paradigm.

The design paradigm on which traditional manufacturing is based appears to be dominated by technological considerations. People are almost always treated in a mechanistic way and are in most situations regarded as an issue of secondary importance. The worker is often seen as a nuisance, a source of uncertainty, a necessary evil. The design paradigm emphasizes reducing the role of people through automation and deskilling, and control over the work is transferred to computers and to management.

However, there are a number of important aspects of the traditional design paradigm. One of these important aspects is the reductionist approach used by engineers. Reductionism is based on the scientific method and helps engineers to simplify and to decompose problems so that they become manageable and solvable. Another related issue is the belief that optimization at the task level leads to optimization at the enterprise level. Reductionism by itself cannot be justified simply because it allows problems to be solved. It must also lead to solutions that in turn lead to optimum performance at a higher level.

Another important feature of the design paradigm is its monodisciplinary nature (Kidd 1991a). Society is a very complex system. It is characterized by enormous size and massive interconnections between components. So complex is this system that it is impossible to cope with it in this form, and it has been necessary to reduce it into manageable areas that have tended to be examined separately.

As a result knowledge of the world is divided into boxes that are man-made and purely arbitrary. There is no natural law that states that the world should be divided into software engineering, psychology, organization and management science, and so forth. These subjects are man-made, and the divisions between them purely arbitrary and a matter of convenience. In manufacturing the result of this monodisciplinary perspective has been a tendency toward serial engineering.

Finally, economics plays a central role in the design paradigm through the notion of cost efficiency. In manufacturing this attention to cost efficiency is largely seen in terms of the goals that managers pursue when investing in new manufacturing technologies, that is, cost reductions. It also permeates all aspects of the cost accounting system, that is, the investment justification methods used, the performance measurement and control systems, and product costing methods (Kaplan, 1989; Meredith and Suresh, 1986).

The main characteristics of the design paradigm underlying traditional manufacturing can thus be summarized as:

- A reductionist approach
- Monodisciplinary professions and views
- Serial engineering
- Optimization at the task level implies optimization at the enterprise level
- A concentration on technology
- Emphasis on cost reduction

The Need for a Different Approach

There is a growing understanding in industry of the need for a different approach to manufacturing. There are several reasons for this.

Business Issues Over the past 30 years some notable changes have occurred in the marketplace. Price was for many years the most important competitive factor for many manufacturing companies. In this type of economic regime the importance of reducing costs can be understood. In order to remain competitive and profitable, costs had to be continually reduced, and this objective was achieved through mass production (economies of scale), product rationalization, division of labor, and the application of technology to achieve deskilling, increased machine utilization and capacity, and automated production.

During the 1970s and 1980s the dominance of price competitiveness began to wane, and quality and other issues such as delivery performance, customer choice, and so forth became equally important competitive factors. However, this did not mean that the need to reduce costs lost any of its force. Product price was still important, but the right price alone was not enough because customers were also demanding a wide range of high-quality products.

World markets are clearly becoming more international, more dynamic, and more customer driven, and competition on price alone is no longer a viable business strategy for many manufacturing companies. Other nonprice competitive factors have become equally important. During the 1990s there will be an increasing trend toward tailoring of products to meet customer needs. Although this trend has been described as a move from mass production to "one of a kind production" (Browne, 1990), the actual change in market conditions is more complex than this because there will be a whole range of company-specific conditions governing strategic choice (Atkinson, 1990). Basically, the emphasis will be on moving from price competition to quality competition. This can be based on large batches of specialized brand products or on low batches of diversified quality products. These changes in business strategies do not mean that price competition is dead. In fact there is the opportunity for the most innovative companies to offer a diverse range of low-price high-quality products.

The important point about these changes in markets is that they have led to reductions in production volumes, increased product variety, shorter product life cycles (as companies strive to keep their products up to date), and a reduced number of repeat orders. These trends are set to continue throughout the 1990s (Department of Trade and Industry, 1989a). Opportunities for profit growth will not therefore be derived from economies of scale but from economies of scope (Goldhar and Jelinek, 1983), that is, competitive advantage will be derived from flexibility rather than from low costs (Williamson, 1989). Economies of scope are said to exist where the same equipment can produce multiple products more cheaply in combination than separately. The key business issues are therefore to reduce manufacturing lead time (including product development time), improve customer service, reduce inventory, and increase responsiveness against a background of highly variable customer demand patterns and changing specifications.

Markets have become more volatile and dynamic, and to remain competitive in these fast changing markets manufacturing companies have to become highly responsive organizations. To survive they will therefore, in general, have to achieve, without compromise, three potentially conflicting goals: reduced costs, improved quality, and greater flexibility.

Addressing the Wrong Costs In the early 1980s it was pointed out by Ingersoll Engineers (1982) (also see Small, 1983) that

> most of the skills of our manufacturing and industrial engineers, our production managers and our facilities planners have been inappropriately focused on direct labour costs. Yet this cost element has been given so much attention for so long that today it is a small part of the cost of sales. It is almost not worth worrying about anymore. [p. 2]

Direct labor costs in many manufacturing companies ranged between only 5 and 15 percent of total manufacturing costs. Typically, overhead costs often reached 40 percent, with material making up around 50 percent (Department of Trade and Industry, 1989b).

Ingersoll also noted that

> guided by traditional training, the cream of our manufacturing talent continues to pound away at what has become a mere fraction of the cost difference between Japanese and Western products. [p. 2]

Tradition, of course, is a major factor. What has worked in the past will be used over and over again. But this tradition seems to be reinforced by cost accounting methods that indicate that products can be made more cheaply by reducing (often insignificant) direct labor costs. Moreover, traditional cost accounting methods often allocate overhead to a product on the basis of the number of direct labor hours used in its manufacture. So if the number of direct labor hours is reduced, so is the overhead allocated to the product. The product can therefore be made more cheaply!

A survey of over 700 leading firms in the United Kingdom, undertaken in late 1990 by a large firm of accountants and the Manchester School of Management, revealed that, despite the fact that the number of shop floor workers has shrunk dramatically with the advent of automation, 80 percent of factories still use costing methods devised half a century ago when factories were almost 100 percent labor intensive (Dunn, 1991). The real costs in manufacturing lie in materials and overheads, not in direct labor costs (Kaplan, 1989). Reducing direct labor costs is relatively ineffective and often irrelevant.

Organizational Issues With the implementation of new manufacturing technologies has come a growing realization of the importance of organizational change. Many of the new technologies have highlighted the inappropriateness of traditional organizational structures and work practices. In addition, techniques such as just-in-time manufacturing, which is a management philosophy and an organizational concept, have demonstrated that new technology is not the sole source of improved competitiveness.

The benefits of organizational change include shorter throughput times, reduced inventory, improved product quality, more economic operating conditions, and improved responsiveness. For example, by changing the factory to a product-based organizational structure it is possible to shorten throughput times by as much as 80 percent and to reduce in-process inventory by as much as 60 percent (Brödner, 1984). The key point is that these benefits can be achieved without any investment in new technology. These are the results of organizational changes. These changes can also result in simplified material flows, easier production planning and control, and improved job satisfaction that can lead to more highly motivated people.

In Germany one company (see Klingenberg and Kränzle, 1987) reorganized its factory from a traditional process layout to a product-based organization. As a result of organizational and work practice changes, the company achieved reduced costs, improved due date performance, a high speed of response to quantitative and qualitative changes of demand, improved quality, reduced expenditure on computer equipment, and a reduction in breakdowns.

Costs were reduced because there was a net reduction in labor and because less capital was needed to fund the business. Improvements in quality arose because shop floor employees were made responsible for the quality of their work. Defects could therefore be detected at an earlier stage or avoided altogether. This brings further cost savings through reduced rework and scrap.

Improvements in flexibility were achieved because throughput times were reduced. Organizational factors that lengthen throughput times include too many levels in the hierarchy, too many functional departments, complex procedures, and long signal processing times (Bolwijn et al., 1986). Organizational changes that lead to reduced hierarchies, a smaller number of departments, simpler procedures, and faster signal processing can therefore result in shorter throughput times.

The factory planning and control functions can also become much simpler as a result of organizational changes. Systems are no longer needed that attempt to organize the correct schedule of work for every single operation on every single resource at every moment of time. Instead the system only needs to provide factory planners with information allowing each manufacturing cell to be loaded with a sensible volume of work over a period of time. The detailed planning and scheduling can be done by the shop floor employees working in the cells. This results in effective but simpler and cheaper computer systems.

Organizational changes can also provide the basis for achieving continuing improvement in manufacturing. For example, a rejected part, or a half finished item waiting for new tooling, or a part program that can be improved, or a design feature that leads to machining difficulties, represent time and money that is being wasted and competitive opportunities missed (Hayes et al., 1988). Remedying these problems is what continuing improvement in manufacturing is about. This is an issue that concerns everybody, including shop floor employees, but to tap into the creativity of all employees, new organizational structures and new work practices are needed.

Skill-Based Automated Manufacturing

In order to remain competitive in world markets, manufacturing enterprises must take full advantage of the potential for increased productivity, better quality, and greater flexibility provided by more responsive organizational structures, better utilization of human skills and experience, and the power of modern computer-aided technologies. To achieve these goals a skilled, cooperative, and motivated work force is required. Traditional management hierarchies and forms of control must be modernized. More participation by employees, at all levels, in planning, designing, and implementing new technologies and systems is needed. Technical systems need to be designed not just to meet economic and technical goals, but also to satisfy organizational and human requirements.

Skill-based automated manufacturing is the term that has been devised to describe a modern manufacturing response that tries to achieve these objectives. The word "skill," should not be interpreted in a narrow sense but should be seen as a short-hand way of referring to a wide range of human abilities and characteristics that are required in a modern manufacturing environment. The skills referred to are wide and include motor, perceptual, conceptual, discretionary, communication, management, team, leadership, and creative skills. Other human abilities include intelligence, knowledge, creativity, and experience. Of course, the term *skill-based* has been used before by other authors for different purposes. Rasmussen (1983), for example, discusses human behavior in terms of three classes: skilled-based, rule-based, and knowledge-based behavior. In Rasmussen's classification skill-based behavior represents sensory-motor performance; rule-based behavior represents performance guided by rules that worked well during previous similar actions; and knowledge-based behavior represents performance guided by knowledge when dealing with new situations. The sense in which the term *skill-based* is used here is much broader and is therefore very different from that considered by Rasmussen and really covers all three aspects of Rasmussen's scheme.

The approach aims to overcome the belief that technology provides the answers to all problems. Nontechnical problems should not be treated as technical ones. A more sensible approach is required if manufacturing companies are to compete in world markets. Reduced costs, improved quality and greater flexibility are needed, but these must be achieved without trade-off. This cannot be satisfied by technology alone. A balanced manufacturing response is therefore required, not a technological one. This demands that managers develop a new and broader vision of the *factory of the future*: a vision with a future.

Skill-based automated manufacturing is a broad, strategic, business-driven approach that involves a balanced consideration of organization, people, and technology in an integrative way. It is neither technology driven or human centered. It aims to create an environment for the exercise of human skills, judgment, creativity, knowledge, and ingenuity and to make full use of modern computer-based technologies. These two goals are both essential requirements for future success in manufacturing.

Main Characteristics The change from economies of scale to economies of scope requires more than fine tuning of existing manufacturing concepts in order to adjust to the new market conditions. A whole new set of manufacturing concepts are needed in order to develop an appropriate manufacturing response to the emerging market conditions. New manufacturing strategies are therefore needed that should be based on the appropriate use of organization, people, and technology. These new strategies will involve people, working in advanced organizations, supported by technology.

Skill-based automated manufacturing has several important characteristics. First, is the philosophy that competitive advantage is not only derived from technology but also from appropriate and responsive organizational structures, appropriate work practices, and from the skills, knowledge, experience, and so forth of all the people in the enterprise. It is however, the skills, knowledge, experience, and so forth of the people that are most important (people make it happen!), hence the term *skill-based model*. Second, competitive advantage can only be derived from organization, people, and technology if an appropriate manufacturing strategy has been developed to meet business needs and market opportunities. Skill-based automated manufacturing is therefore a business-driven strategic manufacturing response. Third, a balanced approach that considers organization, people, and technology needs be adopted. This is not a simple question of technology centered versus human centered but of serious consideration being given to all three dimensions.

Fourth, factories need to be designed as integrated systems. Integration, however, is not just a technical concept relating to the interfacing of computers. Integration needs to be based on the concept of human and computer-integrated manufacturing (Kidd, 1991b). This involves developing integrated systems based on three types of integration: people integration (people communicating with people), human–computer integration (people interacting with computers), and technological integration (machine interfaced with machine). Fifth, people, unlike machines, have a psychological environment that must be respected and taken into account when designing manufacturing systems and technologies. Issues such as motivation, opportunities for development, job satisfaction, and so forth need to be seriously addressed. A business-driven approach, appropriate technology, responsive organizational structures, and a highly skilled work force all amount to very little if people, at all levels, are not motivated and treated with respect. Sixth, manufacturing systems design is an interdisciplinary activity, and organization and people issues should be addressed in parallel with technical issues (i.e., using a concurrent engineering approach rather than a serial engineering one), with trade-offs being made between cost and performance and other factors such as job satisfaction, user motivation, and so forth.

The requirement for flexibility, adaptation, improved responsiveness, and the need to motivate people and to make use of their skills, judgment, and experience all suggest that organizations, work practices, and technology need to be developed in a way that will allow highly trained people, at all levels in the company, to adapt their work strategies to the variety of situations that they will have to face. This implies that organizations, work practices, and technologies will have to be developed as open systems that will allow people a large degree of freedom to define the mode of operation of the system.

There is a need to implement a new organizational model that will not only be effective and responsive but will also make better use of people and take into account people's need for motivation. A reference model that achieves these goals is based on:

- Integrated factories
- Flat organizational hierarchies
- Cell-based factories
- Decentralization of decision making and control
- Increased competence of shop floor people
- A manufacturing environment conducive to learning
- Multiskilled employees
- Team working
- Empowering of people on the shop floor

- Skill supporting technologies
- Continuing improvement involving all employees

The Design Paradigm This new set of manufacturing concepts involves the consideration of organization, people, and technology within an integrated manufacturing environment and an interdisciplinary manufacturing systems perspective. The design paradigm underlying skill-based automated manufacturing has several major features. These include:

- A balanced manufacturing response based on using organization, people, and technology
- A holistic systems approach
- Concurrent engineering applied to the manufacturing system
- Interdisciplinary design involving the application of insights from the social sciences to the design of the technology and the overall manufacturing system

The design paradigm involves addressing organization and people issues at all stages, from formulation of business strategy right through to manufacturing systems design and implementation (Kidd, 1991b), as illustrated in Fig. 7.1.

The design paradigm underlying skill-based automated manufacturing is closely related to the sociotechnical approach and involves a quite different way of dealing with organization and people issues to that currently used in manufacturing industry. The vast majority of manufacturing companies adopt a serial engineering approach to dealing with these aspects of design. Manufacturing engineers design and implement manufacturing systems, making trade-offs between cost and performance and only then deal with organization and people issues. At this late stage, however, these issues may have turned into problems that have to be dealt with retroactively. It is common to find that these problems cannot be resolved satisfactorily because by this stage many degrees of design freedom have been closed off.

Skill-based automated manufacturing involves a concurrent engineering approach to the design of manufacturing systems and computer-based technologies, which involves addressing organization and people issues right from the very start of the design process and in parallel with technical issues. The approach is an interdisciplinary one where insights from psychology and organizational science are used to shape the system.

This involves, in addition to the usual cost–performance trade-offs, making further trade-offs between technical options, costs, and performance and user acceptance, motivation, user skill requirements, implementation issues, job design, organizational structure, and so forth. The benefits of this approach lie in the area of improved design and implementation processes, better system designs, more appropriate organizational structures, and motivated and engaged employees at all levels in the company.

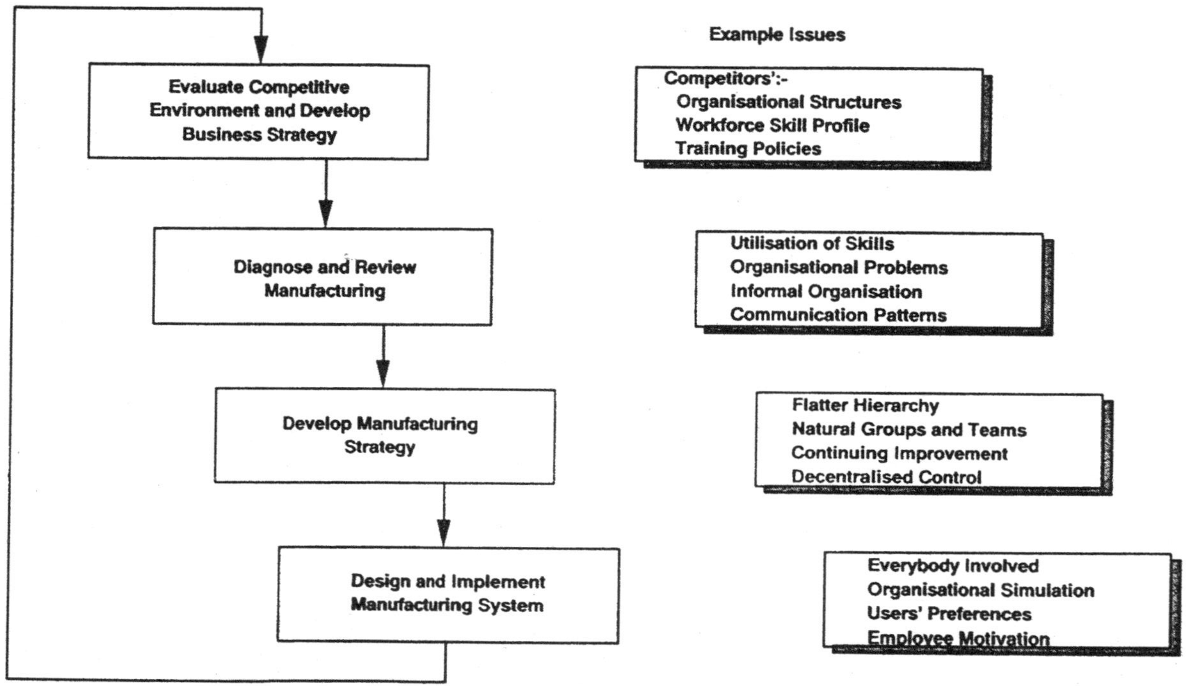

Fig. 7.1 The scope of organization and people issues in manufacturing systems design.

Using a concurrent engineering approach also implies that organization and people issues are considered as being relevant to the design process itself as well as issues that have to be addressed as part of the system design. Reported experiences of the design process (e.g., see Kidd, 1991c; Klein, 1976; Meister, 1987; Perrow, 1983; Rosenbrock, 1989) clearly show that problems are experienced when professional groups and different departments have to collaborate. Concurrent engineering, by definition, should seek to overcome some of these problems, which also means by implication, using insights from psychology and organizational science to shape the design process.

The design paradigm is also based on the spiral model of design (Boehm, 1988). Design experience suggests that it is almost impossible to write a complete and detailed specification using formal and structured analysis methods alone. Some constraints, requirements, and goals are difficult to formulate and often cannot be clearly expressed until a model or a mock-up system has been built. If a specification is written and presented to a client, it is likely that he will accept it. When the system is built, it is likely that he will say that some aspect of the system is not what he wanted or was expecting. There are two basic reasons for this. The first is that some goals and constraints remain tacit and only become explicit when the goals have not been met or the constraints have been violated. The second is that it is often not possible to clearly convey to potential users what the final system will do or look like, using written documents alone. These are two of the main reasons why software often turns out to be inadequate or inappropriate.

During the design of manufacturing systems and computer-aided technologies, there are many uncertainties and risks. Risk areas include, for example, organization and people issues, identification of system requirements, unforeseen problems that arise from new and relatively untried technologies and so forth.

Classical linear models of design (e.g., the waterfall life cycle model) are not suited to design projects that involve substantial risks (and many manufacturing systems projects fall into this category) because they are not sufficiently iterative and because the start of one phase is dependent on the completion of the preceding phase. In projects where there is a significant level of risk, it is important to reduce the risk. This involves identifying the risks and then focusing attention on the high-risk areas, leaving other, less risk-laden aspects, until later.

Thus systems elements with high risk need to be rapidly progressed, at relatively low cost, through various stages of design so that problems can be quickly identified and remedial action taken while the cost of change is still relatively low. This rapid progress can be achieved, not by writing detailed specifications but by a whole range of risk-reducing exercises such as the development of nonfunctioning models, user involvement in the design process, system mock-ups, development of short and precise system definitions and statement of essential requirements, and so forth.

This process is based on several iterations following in rapid succession, involving elements of partial analysis, partial design and partial implementation, at the end of which, the design may well be discarded, or radically changed. As the process of iteration continues, the risk should reduce and an acceptable design should begin to emerge, along with requirements, and so forth, and eventually the design will stabilize, and attention can then be turned to the less risk-laden aspects.

Many of the risk-reducing tools mentioned earlier also act as transitional systems and can help to introduce users to the system while it is still in the very early phases of design, thus giving them an opportunity to both make an input to the design and to learn something about the new system long before it is ready for implementation and use.

Finally, the skill-based design paradigm is not based on a causal mechanistic model of human behavior. Rasmussen (1983) has pointed out that people are not simple deterministic input–output devices, but are goal-oriented creatures who actively select their goals, seeking the relevant information that will enable them to achieve these goals. Peoples' behavior is therefore teleological by nature, that is, it is directed by purpose and intention.

Rasmussen also points to the fact that teleological behavior is not necessarily dependent on feedback during its course but on the experience from previous attempts, that is, the reason for choosing the particular approach. Reasons are the final causes of actions. Thus in the skill-based design paradigm, it is important to design computer-based technologies and manufacturing systems with this teleological model in mind. For this reason, human skills and experience are not just seen as things that are necessary in order to deal with unforeseen events, or with system failures, or to do things that cannot be done by machine, but also as desirable attributes that should not be designed out of systems (Kidd, 1988). Computer-based technologies should therefore be designed as skill-enhancing systems.

7.3 DESIGNING SKILL-BASED SYSTEMS

The HCIM Reference Model

Just as no man is an island, no computer-aided technology is an island. Computer-aided technologies should be designed as part of an integrated manufacturing system. To do this it is first necessary to define what integration means. The *Oxford English Dictionary*'s definition of integrate is "to combine parts into a whole." Computer-integrated manufacturing seems to be concerned with combining technical artifacts (computers and software) into a whole according to criteria that are largely technical in nature.

For the purpose of interdisciplinary design, a much broader vision of integration is needed. Such a vision has been created by Kidd (1991b) with the idea of human and computer-integrated manufacturing (HCIM). The HCIM

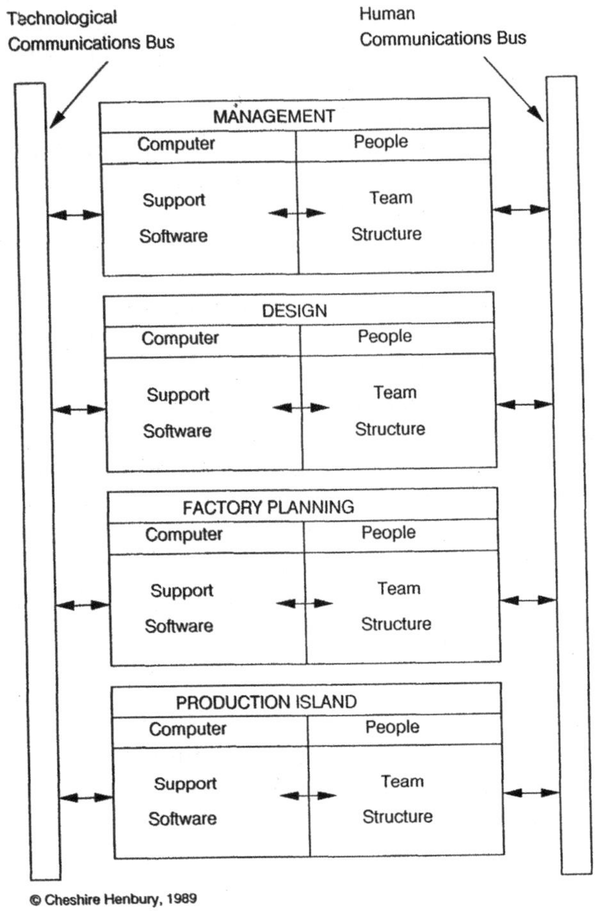

Fig. 7.2 Human–computer integrated manufacturing reference model.

concept of integration is based on the three dimensions mentioned earlier. When designing any computer-aided technology it is necessary to address these three types of integration. More specifically, whether the designer is working in a research laboratory of an Information Technology (IT) vendor company or in a manufacturing company, there is a need to develop technologies so that the three types of integration are enhanced.

Referring to the HCIM architecture shown in Fig. 7.2, the design of the technology should address the overall system architecture. This should be based on considerations of the network of groups, the structure of each group, the integration between groups, the nature of the supporting software, and the technical communication and integration needs between supporting software modules.

These architectures should enable the user company to implement appropriate natural groups, both in offices and in the factory. These natural groups should ideally be designed around complete processes or process flows, or products, and not around individual functions that can lead to inefficiencies and time delays. The designers should also seek to design a technology that will enhance the human communications bus shown in Fig. 7.2. This means avoiding technical barriers to communication (cf. the Traub example considered later in this chapter), seeking opportunities of using the technology to bring about cooperation (cf. the BICC example considered later in this chapter), and using the technology to support cooperative work.

There is also a need to undertake input–output analysis to establish how a given computer-aided technology interacts with the overall system. This is not just a question of technical analysis, for example, data links, but also a question of organizational analysis, for example, who does the numerical control (NC) programming and how can the task be shared between the teams or natural groups.

The Open Systems Concept

The term *open system* comes from general systems theory and is used to describe a system that receives inputs from and sends outputs to the systems environment. The term is very much associated in manufacturing with system architectures based on the International Standards Organisation Open Systems Interconnection model (ESPRIT Consortium AMICE, 1989). The concept of an open system is also used in management science, however, to describe an organizational structure that takes into account and responds to the environment. The idea can, however, be given a much broader interpretation. It can be applied not only to system architectures and organizational structures but also to work practices, human–computer interfaces, and the relationship between people and technologies such as scheduling and control systems, decision support systems, and so forth (Kidd, 1990c,d).

It could be argued that the requirement in manufacturing for flexibility, adaptation, improved responsiveness, and the need to motivate people and to make better use of their skills, judgment, and experience suggests that organizations, work practices, and technologies need to be developed in a way that will allow highly trained people, at all levels in the company, to adapt their work strategies to the variety of situations that they will have to face. This implies that organizations, work practices, and technologies will have to be designed and developed as open systems (Kidd, 1989a).

In this context an open system is one that allows people a large degree of freedom to define the mode of operation of the system and the nature and form of the interaction with the system and its environment. The term basically implies a system that is flexible.

Skill-Enhancing Technologies

Skill-enhancing technologies are an important aspect of skill-based automated manufacturing. Many technologies are skill substituting rather than skill enhancing. A skill-enhancing technology is one that is designed to make human skills and judgment more effective and productive by using insights from psychology and organizational science in an interdisciplinary way.

It is important in this interdisciplinary design process that the technologists who design computer-based systems internalize the values and insights of the social sciences, and use these to influence their creative thinking (Kidd, 1991c). One of the major problems with an interdisciplinary approach to the design of computer-aided technologies is the way in which social science criteria and design methods are used. Social science considerations will not have a major impact if they are not used in a prospective way to guide the subconscious mental processes from which design solutions emerge.

It has rightly been said that the application of social science in a sequential way is unsatisfactory. When design options are shaped by technical and financial considerations and then retrospectively evaluated for the social science implications, the end result is invariably unsatisfactory. Social science issues tend to become secondary and major design changes are not made because it may be too difficult to do so or because the social science critique cannot be operationalized into an alternative design. These criticisms of the conventional social science design approach are, however, only secondary ones. The main point is this. If social science criteria are not used in a prospective manner, in the same way that technical and financial judgment criteria are used, then what is not forthcoming are new ideas, unforeseen possibilities, and new and original ways forward. It is in the areas between disciplines that significant new discoveries are likely to be made.

Examples of Skill-Enhancing Technologies

Example 1: Workshop-Oriented Programming There is a long tradition, dating back in fact to the first NC machine tools developed in the late 1940s, of supporting office-based programming of these machines. There has also been a vigorous debate over the effectiveness of this approach, and there are people who believe that in many cases a more effective way to program these machines is to undertake this work on the shop floor using a skilled machinist.

A survey of 584 U.S. manufacturing plants does provide evidence to support this belief, showing that programming costs of computer numerically controlled (CNC) machines are lower the more management relies on blue-collar workers to do the programming! It seems that the reasons why managers prefer to adopt office-based programming has more to do with the desire to exert control than with efficiency and cost (Kelly and Lan, 1990).

Within Germany there is still a relatively strong craft tradition, and more emphasis seems to be placed on using skilled craft workers to undertake programming than is the case in the United Kingdom and the United States. This is reflected in the emergence of workshop-oriented programming (WOP), which is an example of the development of a skill-based (as opposed to a skill-substituting) technology.

Traub, the German machine tool manufacturer, has developed a WOP system (Hekeler, 1989a, 1990). It has a number of interesting characteristics that are derived from specific design considerations that focused on the users' skills and knowledge and the application of insights from the social sciences.

Many machinists are capable of programming in abstract programming languages such as EXAPT. This is not, however, the area in which the primary skills and knowledge of a machinist lie. These primary skills and knowledge are concerned with metals, the machining process, and how to make things. Workshop-oriented programming has been developed to make the machinist's primary skills and knowledge more productive and more effective, and to avoid the need to develop programming skills, which are seen as secondary and less important.

The way this aim has been achieved is through the development of programming techniques that closely resemble the way that a machinist would produce a component on a conventional manual machine tool using his customary thinking and working practices, while at the same time retaining the same degree of flexibility and freedom of action that a programming language would provide.

The claimed benefits of this approach are quite impressive (Hekeler, 1989b). Evaluation trials undertaken in a factory manufacturing jet turbine and diesel engines suggest that a skilled machinist with no NC experience whatsoever can use the system without aid after 15 minutes of instruction. Within one week it is claimed that these machinist were performing as well as their more experienced colleagues did at the beginning of the trials. The trials also compared the cost of programming and testing for a range of components varying from simple to complex. It is claimed that the results demonstrate that WOP is far more economic than any other type of office or shop-floor-based programming method.

The Traub WOP system has also been designed to provide a standardized programming environment for skilled people, regardless of whether programming directly on the machine, in the foreman's office, or in the planning and programming office. This is an important organizational aspect of the design because it has implications for team working. By using technology that is familiar to both the shop floor and the planning office, it is much easier to bring about cooperation between these parties when a team effort is needed. There are no artificial barriers, created by application of different programming technologies, to hinder the creation of a team to tackle a specific job. This example clearly illustrates how it is possible to assist team working by shaping the technology in an appropriate way.

The Traub system also includes a computer-aided design/computer-aided manufacturing (CAD/CAM) link so that CAD geometry can be transferred, through an interface that converts the drawing into a suitable NC-compatible geometric contour, to the CNC programming system. The architecture of this CAD/CAM system provide flexibility for organizational choice and job design. Unlike many CAD/CAM systems that assume that the NC programming will be done in an office, the Traub system also provides facilities to transfer the CAD geometry directly to the shop floor where it is converted into a NC-compatible geometric contour for direct programming on the machine tool.

The success of the Traub WOP system has encouraged other manufacturers to produce their own WOP systems. However, there are reports from Germany that some WOP systems are not as flexible as Traub's. The reason seems to be that they are not true WOP systems. What appears to have happened is that some manufacturers have added user-friendly graphical interfaces to their abstract EXAPT programming languages. The result of this is that they have restricted the freedom of action of the machinist, which has caused frustration and given WOP a bad image on the shop floor (Hekeler, 1991). This amply illustrates the point that there is more to developing skill-based technologies than simply adding user friendly interfaces to technically oriented systems.

Example 2: Computer-Based Support for Cellular Manufacturing Teams In 1986 BICC Technologies started to address the development of an appropriate manufacturing response to the business conditions of the 1990s. The aims were to reorganize factories to make them more customer responsive, to tap into the creative abilities of all employees, and to develop computer-based technologies to support the organization and the people.

The result was the implementation of small manageable "factories within factories" or cells. Each cell operates in a semiautonomous way in terms of production control and scheduling, machine setting, quality control, and first-line maintenance. These cells can be product based, process based, skill based, or even material based. Responsibility for performance is vested at the cell level, with all members of the cell team contributing their skills and experience to the task at hand.

To support this approach, BICC Technologies developed computer-based support to complement the teams. This computer-based support is called appropriate computer-integrated manufacturing technology. Its purpose, unlike many new computer-based systems, is not to achieve real-time automated scheduling or to deskill the users but to support the scheduling skills and knowledge of the team members. It is a skill-enhancing technology not a skill-substituting technology. It has been designed to fit a particular organizational concept. The organization was not developed to accommodate the technology, which is what some proponents of computer-integrated manufacturing are suggesting needs to happen. This is an example of the way system architectures must be designed to fit the desired organizational structure.

The computer-based tools consists of a number of software modules. Two important ones are the Factory Planner module and the Cell Scheduler module.

The Factory Planner module allows capacity planning for the factory as a whole without the need for detailed and time-consuming scheduling of each operation and resource. This helps to ensure that work is allocated to cells to achieve the customer's specified delivery date and that work in progress is minimized by arranging for production to commence no sooner than necessary. The Factory Planner module uses numerical and graphical displays to highlight bottlenecks and to show overall factory and individual cell loadings. Simulation facilities are also provided to help eliminate bottlenecks and to forecast delivery dates in response to customer inquiries.

The Cell Scheduler module receives orders for production from the Factory Planner module. Scheduling is performed by sequencing orders on each work center, either interactively or by using a number of available algorithms that the team members select based on their scheduling skill and knowledge. The resulting schedules are displayed graphically as a Gantt chart showing operations to be performed by each machine. Alternative order sequences can be simulated and compared using the total time to complete the workload.

An important aspect of the computer system is the requirement for cooperation between the factory planner and the cell teams. This is achieved by the way the technology has been designed to operate and is an example of how the detailed functioning of technology can be shaped by organizational considerations. In this case it was the desire to generate team working between different groups.

In the case of the BICC system this is achieved because the Factory Planner module works on the basis of infinite-capacity backward loading from the due date using period batch control principles and information on the bottleneck resources in each manufacturing cell, while the Cell Scheduler works on the basis of finite-capacity forward scheduling of the work for the current period, plus any backlog, plus the work for the next period. The aim is to ensure that cells are neither overloaded nor underloaded, and this task requires face-to-face cooperation between factory planner and the cell team in order to ensure that this aim is achieved.

Analysis of Examples In the examples considered very little has been said about the human–machine interface (the surface characteristics of the technology). This is because the most interesting aspects of the design lie in the technology behind the user interface and the way in which psychological and organizational issues have been used to influence what have been called the deep system characteristics of the technology (Kidd, 1992). In the case of the WOP system it is evident that people who have tried to copy the WOP concept have not fully appreciated the difference between surface characteristics and deep system characteristics.

These examples are primarily concerned with questions of architecture, what the computer technology does, and how it functions. In both case studies it is possible to detect insights and knowledge from the social sciences applied to the design of detailed aspects of the computer-based technologies.

In the case of the Traub WOP system the programming environment has been standardized to remove technological barriers to cooperation. The programming techniques are based on the methods used by skilled machinist operating conventional manual machine tools. In addition, the architecture of the computer systems includes CAD/CAM integration, which allows programming to be done on the shop floor or in an office.

In the BICC case the computer architecture is based on a decentralized organizational structure of semiautonomous manufacturing cells, and the functionality of the software reflects this organizational design. In addition, however, the details of the scheduling technology are such that team working between the factory planner and the manufacturing cells teams is needed in order to ensure satisfactory loading of the cells.

These two examples illustrate that technical choice can be influenced by insights drawn from the social sciences. There are other examples. Kidd (1991d) has presented a design of a workshop-oriented CNC programming system that is based on consideration of the existing skills and knowledge of the machinist, as well as technical considerations, such as the need to achieve CAD/CAM integration. Kidd (1988, 1989b) has also presented the design of a decision support system where the technical details of the design were influenced by considerations of the user's need for autonomy and control, as well as technical issues.

Open Systems Applied to Human–Computer Integration

One of the things required for interdisciplinary design is a new paradigm that accepts that design is an art but that also concentrates on developing computer-based technologies in which the users' skills, knowledge, judgment, and creativity are viewed as an important aspect of the design.

The makings of such a paradigm can be found in the case studies already described. These case studies illustrate examples where the aim was not to use the computer to eliminate skill and knowledge, but rather to enhance the users' skills and knowledge and to make them more productive and effective. Thus the skill-based design paradigm seeks to combine the power, speed, and accuracy of the computer, with the skills, knowledge, judgment, and creativity of the user. Unfortunately, this goal is not easily achieved because the concept of skill-enhancing computer technology is subject to wide interpretation. The following example, taken from computer-aided control systems design (CACSD) will serve to illustrate the problem.

A Common Problem MacFarlane (1982) has observed that syntheses approaches to design reduce the designers' scope to exercise skill and judgment. This is because a synthesis method requires the designer to specify, usually in detail and in mathematical form, the system constraints and the required performance. The computer then uses this data within an algorithm to produce a design solution.

MacFarlane then goes on to say that a CACSD system should provide a skilled designer with a set of manipulative and interpretative tools that will enable him to build up, modify, and assess a design put together on the basis of the designer's physical reasoning within guidelines laid down by his engineering experience. The designer of a CACSD system therefore needs to consider the way in which the burden of the design work is shared between the designer and computer. The aim, he argues, should be to achieve a situation such that each makes an appropriate contribution to the overall solution, and this involves developing techniques that will lead to a fruitful and effective symbiotic relationship between the computer and designer.

He then observes that the computer is conventionally used for calculation, manipulation, and optimization. He then proposes, by way of an example of his thinking, that in any fully developed CACSD system, the tuning of controller parameters during simulations is best done by a systematic use of appropriate optimization techniques. Although this sounds like an appropriate sharing of design tasks between the designer and computer, the idea does have a number of problems that are not immediately evident.

Basically there are two approaches that can be used to undertake the task of controller tuning during simulation studies. The first is to do what MacFarlane has described and use optimization methods. The second is for the designer to make the adjustments himself.

This manual method can result in a very good learning situation. The designer can experiment and observe results, and the process of making these adjustments can give the designer more insight into the system's performance characteristics, and can give him a feel for the system. This approach also requires the use of knowledge and the exercise of skill and judgment. The prime disadvantage of this approach is the time taken to achieve a satisfactory performance (for a complex system this can be several weeks of effort).

The optimization method is, by comparison with the manual approach, very fast. However, since the computer is doing all the work, there is no learning situation. Furthermore, the designer does not develop any feel for the system, or extend his understanding about the system, and little or no knowledge, skill, or judgment are needed. Moreover, if the optimization algorithm fails to produce a satisfactory result, then there is no indication from the computer why this has happened, and the designer will have gained no insight that he could make use of to resolve the problem.

Although a superficial examination of the optimization approach suggests that the power, speed, and accuracy of the computer are being used to support the designer, what in actual fact exists is the classic human–machine relationship. The computer is doing all the work, and when it fails, the user is left to cope as best he or she can.

The problems that arise in this example stem from two sources. First, the use of the optimization method for tuning controllers is based solely on the criterion of the amount of time taken to produce a result. The manual method is slow, and since the computer is better at examining a large number of options and can do so very quickly, it is concluded that controller tuning is best done by the computer. Jordan (1963) has referred to this kind of thinking as "man–machine comparability." Unfortunately, human–machine comparability ignores the needs of the designer, which in this case are related to learning and the opportunity to exercise skills and judgment.

Second, the design of the system ignores the fact that the algorithm may fail to achieve a satisfactory result. If a design tool sometimes fails to produce a satisfactory result, then the user needs to understand under what conditions it may fail and should have access to alternative tools. If there are no alternative tools or there is no understanding of the range of applicability of the tool, then such a design tool has limited value.

The example considered refers to CACSD. The example itself is not unusual, as the same situation is found elsewhere (e.g., see Slatter et al., 1989). Although MacFarlane is sympathetic to the idea that the user should be in control of the design process, and the computer should support design, rather than undertake automated design, he finds it extremely difficult to break away from the conventional approach of designing computer systems on the basis that either something is done by the computer or it is done by the user.

It appears to be almost a universal fact that there is no new vision of any other way of tackling the human–computer relationship. It is now often stated that people and computers are complementary rather than comparable. But this statement is usually followed by an illustrative comparability statement of the form "computers are fast and consistent but lack flexible reasoning, while human information processing is slower but can handle events which are ill-defined" (e.g., see Clegg et al., 1989; Hwang et al., 1984; Slatter and Craven, 1987). In other words comparability prevails, even when complementarity is preached.

There is one way out of this dilemma—leave the user to determine his or her own relationship with the computer.

User-Defined Human–Computer Relationships The situation discussed illustrates a class of system that can be described as a closed system. Essentially, a closed system is one in which the actions of the user are determined by the designer of the technical system to a degree that is more than required, say, by hardware or software or performance constraints. In other words a closed system is one in which the user–computer relationship is *overdetermined*.

A simple example of a closed system is the human–computer interfaces found on most CNC machine tools. These are determined by the system designers. Kidd (1990d) has demonstrated the concept of an open systems human–computer interface for workshop-oriented CNC systems. This allows the machinist to customize the interface to personal preferences. This involves changing the dialog, the screen layout, and so forth.

Thus a closed system typically restricts the users freedom of action or forces him or her to use the technical system in a particular way that is of the systems designer's choosing and not necessarily of the user's choosing. A closed system also attempts to automate a particular task, but when the system fails, it leaves the user without any computer-based decision support.

An open system is the opposite of the closed system. An open system does not unnecessarily restrict what the user can do. For example, if the user wants to violate some constraint, then the closed system would not allow this, while an open system would (of course, appropriate warnings need to be given), subject to considerations such as safety and potential system failure. More fundamentally, however, an open system is not characterized by the situation described earlier in relation to controller tuning. In that situation controller tuning was either done by the computer acting alone or by the designer acting alone. However, between these two extremes there are an infinite number of possibilities. An open system would allow the designer to make use of any of these possibilities.

In essence, therefore, an open system is one where the relationship between the user and the computer is determined by the user and not the designer of the technical system. The role of the designer of the technical system is to create a system that will satisfy the user's personal preferences, and allow users to work in a way that they find most appropriate given the wide range of problems that the system might be applied to.

Perhaps of greatest importance, however, the open systems philosophy prevents the designers of a technical system from imposing on the system user their own views on what the computer should do and what the user should do. Design is influenced by subjective value judgments (Lawson, 1980), and a good way to minimize the effect of the value judgments made by designers of technical systems is to eliminate as far as possible the opportunity to make these values judgments. The open systems concept is one way of achieving this.

It therefore follows that computer-based systems should not be designed as closed systems, but as open systems, and this means avoiding restrictions, using decision support systems in a way that is highly interactive and that allows sharing of the decision-making process in a way that is unique to each user and to each situation.

A Design Example To understand how an open systems approach can be used to design a human–computer relationship, it is first necessary to introduce the concepts of evaluative and generative decision support systems (Gershwin et al., 1986). An evaluative decision support system is one that takes a set of proposed decisions and evaluates and predicts the results of these decisions. A generative decision support system is one that takes a set of criteria and constraints and generates a set of decisions.

Thus in the example from CACSD, the use of a computer algorithm to undertake controller tuning is a generative approach. The manual tuning method is an evaluative approach. Simulation is an evaluative approach. Automated decision making using artificial intelligence, or computer algorithms, or both, is a generative approach.

The problem that designers seem to face is that they cannot break out of the evaluative versus generative dichotomy. There is no logical reason for this because there is no reason why a system cannot be designed to be both evaluative and generative. Between these two opposite modes of operation there is a continuum of operating possibilities. The only problem is that systems are not designed to operate within this continuum because system designers define the mode of operation of the system in great detail and exclude the possibility of the user determining his or her own operating mode. Often there is no need for them to do this. It is just a matter of tradition. The existence of this continuum has important implications for interdisciplinary design.

During the UMIST interdisciplinary project described in a recent book edited by Rosenbrock (1989), a debate arose concerning the operating mode of a decision support system. The system was primarily operating in a generative mode. The computer came up with a possible set of cutting conditions and the user evaluated these and made any changes. The system then revised its results, keeping the changes specified by the user. In this interactive manner it was possible for the user to utilize the system as a decision support, but control over whether to accept the computers recommendation was left in the hands of the user.

The system operated like a spreadsheet program, and it was also possible to drive the system by defining constraints. The social scientist on the project argued that from the psychologist's perspective the design was unsatisfactory because it meant that the user had to wait for the generative results before he could start to exert control by editing the calculated results. The social scientist thus proposed that a blank table option should be provided so that the user could first enter his own choice of cutting parameters. The computer could then generate any remaining cutting parameters, which might then be edited by the user if so desired. It is, however, possible to design the system to operate in a different way as described below.

Across the bottom of the screen are displayed a set of cutting parameters (depth of cut, feed, and speed) that the machinist has to enter via the keyboard. The computer then estimates the consequences of using these parameters in terms of such constraining variables as power consumption, tool and workpiece deflection, and so forth. These variables are then displayed above the cutting parameter in the form of color-coded bar charts (as described in Kidd, 1989a).

On the keyboard, the machinist is provided with six keys (two keys per cutting parameter). These keys allow the user to increase or decrease any of the three cutting parameters in user-defined steps (say 5 percent as the default value). As soon as any cutting parameter is changed, the bar chart display of constraints (power, etc.) is updated, so the user can see the consequences of his or her action. When the user is satisfied with these cutting parameters, he or she stores them, and then proceeds to specify the next cut (if there is one). The computer gives support by reminding the user of the amount of metal remaining to be removed. The machinist can recall on the screen at any time all the details of the cuts he or she has specified.

This evaluative decision support system allows the machinist to learn how changes to the cutting parameters affect the constraining variables. However, a valid technical and economic objection to this evaluative approach is that there is no guarantee that the cutting parameters will be optimal (in the sense of minimizing cutting time). One way around this problem is to use the computer to advise the user about this situation. This can be achieved in a number of different ways.

One possibility is to offer advice after each cut has been specified or to wait until all the cuts for a given operation have been determined and then to offer advice. This is something that the user can control. The basic principle of operation is as follows. The user specifies the cutting parameters. The computer estimates the cutting time for these parameters. The computer also calculates the optimum cutting parameters. The difference between the machining time is indicated through a graphical display. If the user wants to explore the reasons for the difference, he or she can then call up further data (the optimum cutting conditions and associated bar chart displays of constraining variables, and charts showing the differences between the users choices and the optimum values). The user then has to decide, based on experience, skill, and knowledge whether to accept these optimum values.

The user can also do some experimentation if he or she wants. The user could decide to keep the value of depth of cut originally selected but to use the computer's recommendation of the feed and speed. He or she can immediately see the consequences of these changes. This interaction continues for as long as necessary. The option to generate the computer's recommendations first can also be integrated into the operation of the decision support system. In this case the computer works in a generative mode and comes up with a recommendation the user can change.

Now consider what has been achieved. The computer is now operating in both purely evaluative and generative modes, but there is also a mixture of the two. How the system is used is controlled by the user. The systems designer has provided a decision support system that can be used in numerous ways. From a strictly technical and economic perspective this type of system is not needed. However, if one starts to consider issues such as control and learning, then this sort of system is the result. Thus by using an open systems approach to the design of the human–computer relationship, designers can begin to take account of these wider design issues.

The Role of Knowledge-Based Systems

The question of using knowledge-based systems (e.g., expert systems and neural networks) in decision support tools would seem to be more problematic than using conventional programming techniques because these newer technologies are commonly used in a skill-substituting role. Conventional wisdom would suggest that the rules that an expert uses, say, for example, when scheduling machines in a manufacturing cell, should be elicited from a number of experts and then incorporated into an expert system rule base. The system could then be used to generate a schedule.

Östberg (1988) has described five possible ways that expert systems can be used. The first is the *transactional mode* where the expert system takes over all responsibilities and performs all the tasks through a human facilitator. The second is an *interactive mode* where the human performs the task with the help of the expert system and is in charge during the dialog.

The third is the *commentary mode* where the human performs the task independently and only uses the expert system when in need of a second opinion. The fourth mode is the *teaching mode* where the expert system teaches the human until he or she has reached task proficiency, or serves as a manual for the infrequent task performer. The fifth and final mode is the *supervisory mode* where the expert system is a passive supervisor and only calls attention to illegal human actions or actions likely to interfere with developments in other areas of the system of which the task domain is a part.

The conventional approach to expert systems development is almost exclusively concerned with developing systems that operate in a transactional mode. This is the approach of the closed system school. Essentially the system attempts to mimic the expert knowledge of the user. If the user is an expert already, it tells him what he already knows and does things that he can already do. Thus the user has to deal with a system that contains expert knowledge from the domain of his own expertise, all he has to do is to specify the problem and then the computer will attempt to solve that problem.

The open systems philosophy, however, involves a different approach, which is based on the idea of using knowledge-based techniques, not necessarily just to solve problems, but also to give advice and to support decision making and learning. A knowledge-based system should not be telling people what they already know or doing things they themselves can do but should, instead, be making user's work more productive by bringing to the user advice about things he or she is unsure of or does not know about, and reminding him or her of the possible consequences of decisions. This, of course, is more difficult than just attempting to mimic the actions of the user.

Coombs and Alty (1984) have argued that human experts are rarely required to solve well-defined problems, which is what expert systems operating in the transactional mode are designed to do. Instead they work on problems that involve two or more domains of knowledge. What is required, therefore, is for the expert in one knowledge domain to receive advice from experts in the other knowledge domains, and this involves using expert systems that contain expertise from these other domains.

The open systems approach uses expert systems to support the user rather than to replace him or her and this requires the designer of the expert system to establish what extra skills and knowledge the expert user requires in order to do the job more effectively and to chose an operational mode that leaves the user with freedom to act in the way he or she wants (which may include the transactional mode if that is what a particular user wants).

Analysis usually reveals a number of possibilities. For example, many computer-based system make use of algorithms that are often numerically unreliable, or, there may be a number of procedures that could be used for a given task, each having its own characteristics and range of applicability. The expertise concerning these issues lies in the domain of numerical analysis or operations research and not in the users area of knowledge.

A user may be concerned with scheduling a batch of components in a way that minimizes setup times. As the user works his or her way toward a satisfactory schedule, he or she can use different algorithms. Each algorithm is likely to give different answers depending on its characteristics. What are these characteristics? Which algorithm works best for the particular mix of products to be scheduled? Some of the expertise that could provide answers to these questions and thus assist the user to make more effective use of his or her time and the scheduling system, lies elsewhere, outside the user's area of knowledge.

Knowledge-based systems containing knowledge from a number of knowledge domains could thus assist the user with his or her work. These systems, however, would not only provide a transactional operating mode, but also supervisory, commentary, and learning modes. Thus, such a system would fulfill the requirements mentioned earlier because they would be advising the user about the consequences of his or her decisions and telling him or her about things that the user is unsure of or does not know about, that is, enhancing skills and knowledge.

7.4 CONCLUSIONS AND DISCUSSION

This chapter has presented an overview of a modern approach to manufacturing known as skill-based automated manufacturing. By way of conclusion, a number of important points will be considered.

The first point concerns the issue of the similarity of the skill-based approach to other modern approaches to manufacturing. There are several well-known books written by American authors (e.g., see Dertouzos et al., 1989; Hayes et al., 1988; Peters, 1987; Peters and Waterman, 1982; Savage, 1990) that present both critiques of traditional manufacturing and describe models for future generation manufacturing that are very similar to the skill-based model outlined in this chapter. The main difference between these approaches and skill-based automated manufacturing lies in the emphasis that the skill-based approach places on an interdisciplinary design paradigm and the philosophy of designing computer-based technologies and systems in a way that enhances human skills, creativity, and so forth.

It is fairly clear what industry needs to do in order to survive in the competitive markets of the 1990s. Attention should not be focused on which approach is best but on models for future generation manufacturing that seek to go beyond the technology-driven strategies of the past and address the integration of organization and people issues into manufacturing systems design and that attempt to gain competitive advantage from organization, people, and technology.

The description of skill-based automated manufacturing given in this chapter should be treated as a reference model for future generation manufacturing. The model is one in which the integration of organization and people issues into technical design, using a concurrent engineering design method, is an essential element. While each manufacturing business needs to tailor its own manufacturing system to meet its business needs, using the reference model as a framework and as a guide, the basic principles of dealing with organization and people issues in an interdisciplinary manner from the start of the design process and in parallel with technical issues should be adhered to.

This leads to the second point: The reference model is only one aspect of skill-based automated manufacturing. The process of design is also important. How the system is designed is just as important as the results of the design process. In undertaking design, therefore, ideas such as concurrent engineering, the spiral life cycle, the HCIM reference model, and so forth should be used to help create more effective design processes, which in turn should lead to better systems, both for performance and for people.

REFERENCES

Atkinson, J. (1990). Flexibility and Skill in Manufacturing Establishments, IMS Report No. 180, Institute of Manpower Studies, Brighton, UK.

Boehm, B. W. (1988). A spiral model of software development and enhancement. *Computer*, May, 61–72.

Bolwijn, P. T., Boorsma, J., van Breukelen, Q. H., Brinkman, S., and Kumpe, T. (1986). *Flexible Manufacturing. Integrating Technological and Social Innovation*, Elsevier, Amsterdam.

Brödner, P. (1984). Group technology—A strategy towards higher quality of working life, in *Design of Work in Automated Manufacturing Systems*, T. Martin, Ed., Pergamon Press, Oxford, UK.

Browne, J. (1990). Future integrated manufacturing systems: A business driven approach, in *Organisation, People and Technology in European Manufacturing*, P. T. Kidd, Ed., Interdisciplinary Research for the 1990s. A report prepared by Cheshire Henbury on behalf of the FAST programme, CEC, Brussels.

Clegg, C. W., Ravden, S., Corbett, J. M., and Johnson, G. (1989). Allocation of functions in computer integrated manufacturing: A review and a new method. *Behaviour and Information Technology*, **8**(3), 175–190.

Coombs, M., and Alty, J. (1984). Expert systems: An alternative paradigm, in *Developments in Expert Systems*, M. Coombs, Ed., Academic Press, London, pp. 135–157.

Department of Trade and Industry (1989a). *Manufacturing into the Late 1990's*, HMSO, London, UK.

Department of Trade and Industry (1989b). *Competitive Manufacturing. A Practical*

Approach to the Development of a Manufacturing Strategy, IFS Publications, Kempston, UK.

Dertouzos, M. L., Lester, R. K., and Solow, R. M. (1989). *Made in America. Regaining the Product Edge*, MIT Press, Cambridge, MA.

Dunn, J. (1991). Revenge of the bean counters. *The Engineer*, February 7, p. 26.

ESPRIT Consortium AMICE, Eds. (1989). *Open Systems Architecture for CIM*, Springer-Verlag, Berlin.

Gershwin, S. B., Hildebrant, R. R., Suri, R., and Mitter, S. K. (1986). A control perspective on recent trends in manufacturing systems. *IEEE Control Systems Society Magazine*, **6**(2), 3–15.

Goldhar, J. D., and Jelinek, M. (1983). Plan for economies of scope. *Harvard Business Review*, November–December, 141–148.

Hayes, R. H., Wheelwright, S. C., and Clark, K. B. (1988). *Dynamic Manufacturing. Creating the Learning Organisation*, Free Press, New York.

Hekeler, M. (1989a). *Traub-IPS: The Workshop Oriented Programming System*, Traub AG, Reichenbach/Fils, Germany.

Hekeler, M. (1989b). *Ergebnisse des Pilotanwenders (MTU Motoren-und Turinen-Union Friedrichshafen GmbH)*, Traub AG, Reichenbach/Fils, Germany.

Hekeler, M. (1990). *Workshop Oriented Programming (WOP)—A Standardised Programming Technology for Programmers and Operators*, Traub AG, Reichenbach/Fils, Germany.

Hekeler, M. (1991). Personal correspondence.

Hwang, S-L., Barfield, W., Chang, T-C., and Salvendy, G. (1984). Integration of humans and computers in the operation of flexible manufacturing systems. *International Journal of Production Research*, **22**, 841–856.

Ingersoll Engineers (1982). *It's Not So Much What the Japanese Do . . . It's What We Don't Do*, Ingersoll Engineers, Rugby, UK.

Jordan, N. (1963). Allocation of functions between man and machine in automated systems. *Journal of Applied Psychology*, **47**(3), 161–165.

Kaplan, K. S. (1989). Management accounting for advanced technological environments. *Science*, **245**, 819–823.

Kelly, M. R., and Lan, X. (1990). Does decentralisation of programming responsibilities increase efficiency? An empirical test, in *Human Aspects of Hybrid Automated Systems II*, W. Karwowski and M. Rahimi, Eds., Elsevier, Amsterdam.

Kidd, P. T. (1988). The social shaping of technology: The case of a CNC lathe. *Behaviour and Information Technology*, **7**, 192–204.

Kidd, P. T. (1989a). *Systems Based Approaches to CIM: Questions of Method, Competitiveness and Profitability*, Proc. IEE Seminar on the Systems Engineering Contribution to Increased Profitability, IEE, London.

Kidd, P. T. (1989b). The social and engineering design of computer numerically controlled technology, in *Designing Human-Centred Technology. A Cross-Disciplinary Project in Computer-Aided Manufacturing*, H. H. Rosenbrock, Ed., Springer-Verlag, London.

Kidd, P. T. (1990a). Organisation, people and technology: Towards continuing improvement in manufacturing, in *Computer Integrated Manufacturing*, L. Faria and

W. van Puymbroeck, Eds., Proc. 6th CIM-Europe Annual Conf., Springer-Verlag, London.

Kidd, P. T. (1990b). Organisation, people and technology: Advanced manufacturing in the 1990s. *Computer-Aided Engineering Journal*, **7**(5), 149–153.

Kidd, P. T. (1990c). Designing CAD systems: An open systems approach, in *Computer-Aided Ergonomics*, W. Karwowski, A. M. Genaidy, and S. S. Asfour, Eds., Taylor & Francis, London.

Kidd, P. T. (1990d). An open systems human-computer interface for a workshop oriented CNC lathe, in *Human Aspects of Hybrid Automated Systems II*, W. Karwowski and M. Rahimi, Eds., Elsevier, Amsterdam.

Kidd, P. T. (1991a). *Interdisciplinary Engineering Design: Life Without Boxes*, Proc. IEE Colloquium on Professionalism in Engineering Design: European Perspectives, IEE, London.

Kidd, P. T. (1991b). Design of human-centred robotic systems, in *Human Robot Interaction*, W. Karwowski and M. Rahimi, Eds., Taylor and Francis, London.

Kidd, P. T. (1991c). Human and computer integrated manufacturing: A manufacturing strategy based on organisation people and technology, *International Journal of Human Factors in Manufacturing*, **1**(1), 17–32.

Kidd, P. T. (1991d). Workshop-Oriented IT Tools for Skill-Based Automated Manufacturing, Proc. Open University Conf. on Information Technology and People, Open University, Milton Keynes, UK.

Kidd, P. T. (1992). Interdisciplinary design of skill based computer aided technologies: Interfacing in depth, *International Journal of Human Factors in Manufacturing*, **2**(3), 209–228.

Klein, L. (1976). *A Social Scientist in Industry*, Gower Press, London.

Klingenberg, H., and Kränzle, H. (1987). *Humanisation Pays: Practical Models*, Vol. 2, *Production and Production Control*, RKW, Eschborn.

Lawson, B. (1980). *How Designers Think*, Architectural Press, London.

MacFarlane, A. G. J. (1982). Conceptual Frameworks for Control Systems CAD, Proceedings of 21st IEEE Conference on Decision and Control, Orlando, FL, pp. 538–539.

Meister, D. (1987). Systems design, development and testing, in *Handbook of Human Factors*, G. Salvendy, Ed., Wiley, New York, pp. 17–42.

Meredith, J. R., and Suresh, N. C. (1986). Justification techniques for advanced manufacturing technologies, *International Journal of Production Research*, **24**(5), 1043–1057.

Östberg, O. (1988). Applying expert systems technology: Division of labour and division of knowledge, in *Knowledge, Skill and Artificial Intelligence*, B. Göranzon and I. Josefson, Eds., Springer-Verlag, Berlin.

Perrow, C. (1983). The organisational context of human factors engineering, *Administrative Science Quarterly*, **28**, 521–541.

Peters, T. (1987). *Thriving on Chaos, Handbook for a Management Revolution*, Pan Books, London.

Peters, T. J., and Waterman, R. H. (1982). *In Search of Excellence. Lessons form America's Best-Run Companies*, Harper & Row, New York.

Rasmussen, J. (1983). Skills, rules, and knowledge; signals, signs, and symbols, and other distinctions in human performance models, *IEEE Transactions on Systems, Man and Cybernetics*, SMC-13(3), 257–266.

Rosenbrock, H. H., Ed. (1989). *Designing Human-Centred Technology. A Cross-Disciplinary Project in Computer-Aided Manufacturing*, Springer-Verlag, London.

Savage, C. M. (1990). *Fifth Generation Manufacturing. Integrating Enterprises through Human Networking*, Digital Press, Bedford, MA.

Slatter, R., and Craven, F. (1987). A human-centred approach to FMS design, *FMS Magazine*, **5**(4), 191–194.

Slatter, R., Husband, T. M., Besant, C. B., and Ristic, M. R. (1989). A human-centred approach to the design of advanced manufacturing systems, *Annals of CIRP*, **38**, 461–464.

Small, B. W. (1983). Paying for the technology—making the intangibles, tangible, in *Proceedings of 2nd European Conference on Automated Manufacturing*, B. W. Rooks, Ed., IFS Publications, Kempston, pp. 183–187.

Taylor, F. W. (1902). Shop management, *Transactions American Society of Mechanical Engineers*, **24**, 1337–1480.

Taylor, F. W. (1907). On the art of cutting metals, *Transactions American Society of Mechanical Engineers*, **28**, 31–350.

Williamson, I. P. (1989). Integrated manufacturing: Developing your strategy, in *Computer Integrated Manufacturing*, C. Halatsis and J. Torres, Eds., Proc. 5th CIM-Europe Annual Conf., IFS Publications, Kempston, UK.

CHAPTER 8

SOCIOTECHNICAL/ORGANIZATIONAL DESIGN*

L. KLEIN
The Bayswater Institute
London, England

First, I would like to explore why the idea of sociotechnical design is still news when it has been around for such a long time. After all, we do not keep rediscovering that water boils at 100 degrees Centigrade. So why do we have to keep thinking of new ways of saying that human, technical, and organizational aspects of production systems are inextricably interdependent ("intertwingled," as a colleague put it)?

8.1 THE PROBLEM OF UTOPIAS

An important culprit is the Utopian dream. Once upon a time, there was a splendid conference on operational research and the social sciences. In the middle of a discussion on—I think—operational research and submarine warfare, Fred Emery stood up and announced: "Love and hate aren't that

*Some of the material in this chapter was collected in the course of Exploratory Action No. 5603 of the European Community's ESPRIT Programme, and for the Fraunhofer-Institut für Produktionsanlagen und Konstruktionstechnik, Berlin.

Organization and Management of Advanced Manufacturing. Edited by Waldemar Karwowski and Gavriel Salvendy.
ISBN 0-471-55508-8

guff that Freud talked about. They're a basic defense against informational complexity."

Well, love and hate are outside the scope of this volume, but defenses against informational complexity are very relevant to it. On the one hand, there has been a succession of Utopian visions of tidy, predictable, aesthetically pure, and totally efficient manufacturing. In F. W. Taylor's Utopia, the complexity of individual variation among operators was to be tamed. Ever since the 1950s, the Utopian vision of automation has been that messy, variable, error-prone, union-joining operators would be eliminated altogether. It is a powerful dream, which just will not go away. Quite recently, in the UK's Alvey Programme, a project called "Design to Product" was based on the premise that, with enough support from expert systems, the product designer would be able to arrive at a position where he or she would press a button and everything else downstream would happen automatically, including after-sales service.

On the other hand, there has also been a succession of humanistic, or human-centred Utopias, originating from various parts of the social sciences. One such was the vision generated by Karl Marx, which was based on the community of interests of those whose only marketable product was their labor. Another has been based on the possibilities opened up by the concepts and methods of ergonomics: work systems designed to suit the characteristics of the individual human operator, at any rate those characteristics that are susceptible to quantitative methods of research. Another has been based on the discovery that groups of people can develop solutions to problems and effective ways of working. Another has been based on the values of humanistic psychology, investing everything in trust and openness in personal relationships. I have even come across a Utopian vision of how artificial intelligence (AI) methods can be used to help and comfort the dying: Since doctors do not have time to inform patients fully about their condition, the way to help seriously ill and dying patients is to provide terminals by their beds that they can interrogate.

All such ideas have some basis in reality. But whenever such an ideal or body of ideas can only be sustained at the expense of some relevant data, in other words whenever it becomes a defense against complexity, two things happen: First, it creates the next generation of problems. Then the next generation of observers concentrates on those problems and begins to develop solutions that, in turn, denigrate and omit those parts of the data that had been previously included. Second, the idea gets to be defended in ever more fundamentalist and intolerant ways. If things get bad enough, and enough opposition is mobilized—often to the fundamentalist methods rather than to the idea itself—the idea may get killed off, with the risk of throwing a valuable baby out with the bath water.

For example, one consequence of the pursuit of the automation dream has been that there are indeed fewer operators, but those that remain often do not have necessary hands-on experience of actual operations. One consequence of the pursuit of the autonomous work group dream is that words have changed their meaning (a frequent symptom of fundamentalism): "Innovative" organizations are in some circles currently taken to be those that make use of autonomous work groups, and if you question this use of words, you may be labeled as being "against" autonomy and/or "against" innovation.

So it is vitally important to distinguish between concepts, which have some quality of permanence, and solutions, which need to be continually reviewed and redefined in the light of changing circumstances. The early sociotechnical work contains a mixture of both, and I would like next to summarize and discuss, very briefly, some outcomes of that work.

Concepts

In sociotechnical work, the original coal-mining studies (Trist et al., 1963) led to three major concepts:

1. The technical and social systems are interdependent. They influence each other, in both directions.
2. There is choice in the way one organizes round any given technology. This choice will be influenced, explicitly or implicitly, by values.
3. The work system is an open system.

At the time of these early studies, technology seemed to be taken as given. It is now necessary, therefore, to add another basic concept, namely:

4. There is choice in the way in which technology itself is designed.

I take these to be principles that have some permanence; the way they work out in different situations will vary.

To say that the social and technical aspects of a work system are interdependent is to put the concept at a very high level of generality. The originators of the sociotechnical idea went on to work with it in a number of settings that have acquired status as the classics in this field, and in each of them conceptualization continued. However, little was done to make the concepts operational for use by others. And, at the same time as development and exploration continued, Utopian defenses began to creep in.

Link with Industrial Democracy

The coal-mining studies had, contrary to popular belief, been research studies and not action research or experimentation. The first opportunity to work with sociotechnical concepts in an action framework arose in the context of industrial democracy in Norway.

In 1962 the Norwegian Confederation of Employers and the Trade Union Congress of Norway set up a joint committee to consider questions of industrial democracy. They formulated a program, the general objective of which was to explore the question "under what conditions can more rights and responsibilities be achieved for the individual in the work place?" The work was done by the Work Research Institutes in Oslo, jointly with the Tavistock Institute in London.

The first phase of the program consisted of a study of existing Norwegian and other European experiences with mechanisms that allow formally for employee representation at top management level. The researchers concluded (perhaps predictably) that, whatever such representation achieved for the structural rights and institutionalized conditions under which people worked, they did little to influence their everyday experience and quality of working life. A decade earlier, a similar debate had arisen in Britain in connection with the Labour government's policy to nationalize key industries.

It was therefore decided to move into a second phase, which was a program of planned experimental changes (Emery and Thorsrud, 1969). These set the scene and remain classic demonstrations of the effectiveness with which work groups can develop and use autonomy for the benefit of the production process. However, the work group in such demonstration projects could acquire the position of an élite, with all the problems that this entails, and could also become encapsulated if its relationships with the context were not managed in a systemic way. This has certainly happened where the model was adopted elsewhere. In a follow-up study of the famous Ahmedabad experiment (Rice, 1958), it was found that the experimental shed was still "the experimental shed" 17 years later (Miller, 1983).

Moreover, the enhancement of autonomy at shop floor level does not do away with the need for other forms of representation. Where such forms are already strongly institutionalized, ignoring or denigrating them can have disastrous consequences. In a major project of the German government's Humanization of Work Program, a foreign consultant who did not take seriously enough the legal rights of elected works council representatives not only got the autonomous work group part of the project thrown out but engendered a reaction against autonomous work groups on the part of the trade unions that took several years to wear off.

Development of Criteria for Job Design

In preparation for the second phase of the Norwegian experiments, a good deal of theoretical work was done. General psychological requirements that pertain to the content of a job—its design and meaning in the larger setting—were formulated. These relate to a person's day-to-day work as well as to his or her needs for learning and personal growth. They started from the opportunities a job provides for continuing learning, for experiencing some variety, for decision-making, for social support, for moving toward some sort of desirable future. They then went on to concretize these general needs into more specific design criteria.

However, these criteria do not include anything about working conditions, and they do not refer to any need to explore specific local conditions and cultures. When operators are given a chance to say what characteristics of jobs are important to them, they frequently give priority to characteristics that create problems in their current work situation. Factors such as temperature, lighting, the weights that have to be lifted should not be considered too mundane to be worthy of attention in "innovative" projects. It is also important to leave scope for factors to emerge of which the designer might not have thought. There will often be something that is important in the particular situation that general statements will not have captured; "height above floor level" is an example that comes to mind, from a job design activity in a plant in which galleries featured.

Development of Design Principles for Sociotechnical Systems

In a paper published in 1976, Albert Cherns formulated some principles of sociotechnical design, reconsidering and revising them later (Cherns, 1976, 1987). To reproduce them in abbreviated form inevitably creates the trap of oversimplification, leaving out the qualifications and discussion. But here is an amalgam of the principles and their revision; I am taking them in the order of the second paper (Cherns, 1987), which is slightly different from that of the first and where an additional principle is inserted.

Principle 1: Compatibility The way in which design is done must be compatible with its objectives; a participative social system cannot be created by fiat. If the objective of design is a system capable of self-modification, of adapting to change, and of making the most use of the creative capacities of the individual, then people must be given the opportunity to participate in the design of the jobs they are to perform. Recognizing that this principle cannot be literally adhered to in a situation where the people being designed for are not yet in place, Cherns (1987) pays more attention to the dynamics of the design team itself. For compatibility with objective, the design team must work on its own process and principles of operation, members revealing their assumptions so that they are available for challenge, consulting adequately with the various functions they represent, and reaching decisions by consensus.

Principle 2: Minimal Critical Specification While it may be necessary to be quite precise about what has to be done, it is rarely necessary to be precise about how it is to be done. No more should be specified than is absolutely essential, which requires that what is essential needs to be identified. Here, too, it is vital that experts and design team members reveal their assumptions for challenge.

Principle 3: Variance Control Variance is defined as any unprogrammed event, such as a deviation in quality of raw material, a machine failure, and so forth. A *key variance* is one that critically affects outcome. The principle, which was described in the Cherns (1976) version as "the sociotechnical criterion," is that variances, if they cannot be eliminated, must be controlled as near as possible to the place where they arise. This has far-reaching consequences for functions such as inspection and supervision.

Principle 4: Boundary Location Boundaries between departments are frequently drawn on the grounds of organizational tidyness, such as all machines of a particular type together or all the people working in a particular geographical space, or present at a particular time, responsible to the same supervisor. The principle involved here is that boundaries should not be drawn that impede the sharing of relevant information, knowledge, and learning. It has corollaries for the management of the boundaries between departments: The more the control of the activities within departments becomes the responsibility of the members, the more the role of the supervisor/foreperson/manager is concentrated on the boundary activities—coordinating activities with those of other departments, foreseeing the changes likely to impinge on the team, and so forth. Under favorable circumstances, working groups can acquire and handle a greater degree of autonomy and learn to manage their own boundaries. The role of the supervisor then becomes a "resource" to the working group.

Principle 5: Information Flow This principle states that information systems should be designed to provide information *in the first place* to the point where action on the basis of it will be needed. Cherns sees information in organizations as having three functions: for control, for record, and for action. He sees its use for control as pernicious, being generally abused as an incitement to top management to intervene in the conduct of operations for which their subordinates are and should be responsible. Its use for record he sees as essential, but generally abused. Information for action should be directed first to those whose task it is to act. Properly directed, sophisticated information systems can supply a work team with the right type and amount of feedback to enable them to learn to control the variances that occur within their spheres of responsibility and competence and to anticipate events that are likely to have a bearing on their performance.

Principle 6: Power and Authority Louis E. Davis has coined the term *work authority* to describe two linked concepts. Those who need equipment, materials, or other resources to carry out their responsibilities should have access to them and authority to command them. In return, they accept responsibility for them and for their prudent and economical use. They exercise the power and authority needed to accept responsibility for their performance. This principle is closely associated with principle 5, information flow.

Principle 7: The Multifunctional Principle The traditional form of organization relies very heavily on the redundancy of parts. It requires people to perform highly specialized, fractionated tasks. There is often a rapid turnover of such people, but they are comparatively easily replaced. Disadvantages arise when a range of responses is required; this usually occurs if the environmental demands vary. It then becomes more adaptive and less wasteful for each element to possess more than one function, with some element of redundancy of function.

Principle 8: Support Congruence This principle states that the systems of social support should be designed so as to reinforce the behaviors that the organization structure is designed to elicit. If, for example, the organization is designed on the basis of group or team operation with team responsibility, a payment system incorporating individual members would be incongruent with these objectives. Systems of selection, training, conflict resolution, work measurement, performance assessment, time-keeping, leave allocation, promotion, and separation can all reinforce or contradict the behaviors that are desired. Such contradictions are frequently found, for example, between expressed philosophies of participation and actual systems of work measurement. One may either start with existing policies and see to what extent they may be modified, or adopt a more radical approach, designing an "ideal" system that would permit production units to operate as "mini factories" and profit centers, and then modify that.

Principle 9: Transitional Organization Greenfield production sites pose the issue of managing start-up. The design team and its process need to be seen as a vehicle for transition. The treatment given to those who will not have a part in the new organization, the model of selection of those who will, and the training they receive all demonstrate the reality of the espoused philosophy. The principle of compatibility is nowhere more in evidence than here.

Principle 10: Incompletion or the Forth Bridge Principle The Forth Bridge, in its day an outstanding example of iron technology, required painting to fend off rust. Starting at the Midlothian end, a posse of painters no sooner reached the Fife end than the Midlothian end required painting again. As soon as design is implemented, its consequences indicate the need for redesign. More than that, implementation must begin with the start of design; the principle of compatibility foreshadows this. And with implementation comes evaluation and the need for redesign. The myth of stability is reassuring but dangerous if it leaves us unprepared to review and revise.

It is to be noted that Cherns is throughout limiting the discussion to organization design and is not referring to technology design. However, these principles are, on the whole, formulated in a way that takes account of complexity and the varied and changing environment. They are, indeed, principles rather than stepwise design rules. It is interesting to note that they are now much less well known than simpler formulations.

An Attempt to Define "Sociotechnical Analysis"

During the 1960s, members of the Tavistock Institute had the opportunity to undertake work in Shell UK. They worked with the company first in formulating a philosophy intended to encompass sociotechnical principles and then in applying these principles to refinery organization (Hill, 1971). In the course of this, a method of *sociotechnical analysis* was formulated, in nine steps. These are:

1. Initial scanning
2. Identification of unit operations
3. Identification of key process variances and their interrelationship
4. Analysis of the social system
5. Perception of roles
6. Maintenance system
7. Supply and user systems
8. Refinery environment and development plans
9. Proposals for change

Discussion

Looking at the early sociotechnical work from today's perspective, what I see those researchers doing is going into the field, using their eyes, ears, frames of reference, intuitions, and intelligence, making some kind of sense and order out of the detail and complexity, and coming out with a concept here, a method there, a list of criteria somewhere else. That is the process of developing grounded theory (Glaser and Strauss, 1968), or open-system learning. Nevertheless, even in the course of that process, defenses against complexity began to creep in. And what has happened in many different ways since then is that the learning stopped and one or other item from that process of development has been turned—sometimes by the original researchers themselves—into a solution or formula to be applied in situations that may not match the original ones and where, in particular, the context is different. The phrases "sociotechnical," "sociotechnical theory," or "sociotechnical systems" have become quite widely used and popular in recent years. In some circles, in fact, they are reduced to a cosy and familiar-sounding acronym, "STS," which implies that its meaning is obvious and well known. This is not the case.

The following are examples of the shift that takes place if one does not remain grounded:

1. Autonomous work groups have become in some parts of the field a universal panacea, almost an article of religion. It was indeed a stunning discovery that, given the opportunity, people work things out for themselves. However,
   a. Autonomous work groups are not the only possible solution; it depends on the situation. Sometimes a single workplace needs to be designed. Groups can be very hard on their members. Above all, groups need the opportunity to learn and develop.
   b. The miners in the original coal-mining studies that led to the discovery of autonomous work groups were very familiar with their work. Where new factories or new technologies are being created, the situation may be different.
2. Autonomy is not the only job design criterion and in some situations may not be the first priority. Democratic or participative design methods need to include finding out what is important to the people in the situation, even if this does not match the priorities of the researcher. This must not be taken to mean that I am "against" autonomous work groups as a solution or, heaven forbid, against autonomy! There are audiences to whom the fundamental point—that people need to be in control of their work situation and that they can work things out for themselves—still has to be made. There are, however, other audiences for whom the autonomous work group has become a ritual solution that is sometimes, paradoxically, implemented in a very autocratic way. Moreover, there is now an overwhelming need not to limit sociotechnical design to the shop floor part of the system. To start with the autonomous work group as a preconceived solution may well inhibit a creative approach to the whole system.
3. The method of analysis mentioned above was developed in the context of an oil refinery. But because it has been published, and because it is formulated as a series of consecutive steps, and is therefore apparently easy to follow, it has in some quarters become *the* method of technical analysis (or sociotechnical analysis). There are many interdependencies between the technical and social systems that do not come to light using this method, and experience shows that when people try to force some situations to fit it they get into considerable difficulty.
4. Most importantly, in many cases the hold on the fundamental nature of interdependence has been, rather subtly, lost. There are training courses in "sociotechnical analysis" in which students are taught to analyze the technical system and then the social system, *separately*. It is a basic denial of the main concept, but it sails quite smoothly under the flag of "STS." A consultant introduced a sociotechnical course by saying, "We are not here to discuss technology." In this way the heritage and tradition of the human relations movement, which the sociotech-

nical idea challenges but which is deeply embedded in the professional training of many practitioners, continues.

8.2 THE PROBLEM OF PROFESSIONAL CULTURES

A second culprit in the puzzle of why we have to repeat ourselves so much is the division of reality into academic disciplines on which our education systems are based (really another example of reducing complexity). These perpetuate the cycle of reaction and counterreaction and help to explain why the wheel has to be continually reinvented.

In the social sciences there are no artefacts. It is not possible to explain one's technical position by pointing to an object that incorporates it, and it therefore becomes necessary to give some history of the development that has led up to it, and even of a particular scientist's relation to that development. This chapter of necessity therefore must include something of the author's own developing learning.

For many decades now engineering designers and production engineers have worried about problems of technology implementation. These worries have sometimes led to the commissioning of research about the "human factor," a term that has had a range of different meanings. The technologies involved have changed over time, but the pattern has been remarkably consistent:

In the late 1950s, the Department of Scientific and Industrial Research in the United Kingdom was administering a program of research as part of the Marshall Plan. One of the projects was on "The Human Implications of Work Study." The assumptions behind the project were that productivity was low; work study enhances productivity; people on the shop floor tend to resist work study; if you understand more about that resistance, you may be able to overcome it.

It was very much to the credit of sponsors and steering committees that, when the questions began to be reformulated in the course of the research, they were willing to let this happen. In the course of the research the main function of work study came to be seen as one of social control. It was an attempt to control people's behavior, sometimes down to the most minute second-by-second detail, and people responded to this by reasserting controls of their own: In the short run, in a firm where work study was being newly introduced, there were many arguments and much resistance. In the long run, in a firm where work study had been going on for a long time, people had found ways of coping, of rediscovering areas and ways in which they could themselves exercise control over the situation. From a management perspective this was called "fiddling" or "restrictive practice," although the management of the firm in question was prepared to accept a certain amount as the oil that kept the machinery of the firm running. This statement needs to be qualified: Senior management were able to acknowledge that there was fiddling and to accept it. For middle management, such acknowledgment would have meant having to confront and tackle fiddling, and they preferred to deny it.

There was a time when, as a researcher in that project, I thought I was probably an international expert on fiddling. That this response to control was not a merely British phenomenon was demonstrated during a lecture on the human implications of work study before a German trade union audience, whose reaction was to roll in the aisles with laughter and say, "How did you know?"

A secondary and long-term effect of the work study system was therefore also the creation of two separate and very different cultures in the firm. Piecework presupposes an uninterrupted flow of work. If there are interruptions for reasons outside the control of the operator, piecework becomes clearly unfair and people protest. Therefore, a large part of the function of supervision became to ensure the supply of work, and thus to act as a kind of buffer between the problems of the firm, which in batch production engineering were considerable and were preoccupying everyone except the shop floor operators, and the lives of the operators themselves, which were relatively untroubled (Klein, 1964).

At that time there were many appreciation courses, training courses, and conferences on work study, and these often included a session on the "human implications," usually on the last afternoon. It would be too late to influence the main content, but at the same time it satisfied the requirement that something about people should be included.

A few years later it was the turn of operational research (OR). In the mid-1960s there was a good deal of discussion about operational research and the social sciences; conferences were held and people talked about "human aspects." However, what many OR practitioners understood by cooperation between OR and the social sciences was that social scientists should help to deal with "resistance" and make their projects more acceptable. Requests took the form of "We have devised this entirely rational system, and the irrational people down there won't wear it." With patience one could then persuade them that, if people consistently would not "wear" an idea, there might be some relevant data that had not been taken into account. On that kind of basis one could then work collaboratively.

To give an example: Esso Petroleum Co. was experimenting with having an in-house social scientist, and there was some joint work between the social science function and the OR department on distribution systems (Klein, 1976). The company had distribution terminals around the country, where product was stored and from where it was trucked outside to customers. These might be, for instance, garages or central heating installations. That kind of operation lends itself readily to OR applications. The collaboration began with an OR man proposing a system that included the idea that a truck driver should arrive at work in the morning to find his instructions and route in the cab of the truck and drive off. The truck driver would thus have no kind of overlap with the activities and work systems of the terminal or between his role and other roles there. There would not even be any contact with the order clerks who compiled routes, normally a very frequent occurrence during which, although there could be arguments and problems, information was exchanged and some mutual understanding developed. Most of the rest of the company—or so it sometimes seemed—was worried about why drivers did not seem to be identified with the company, its problems, and its interests.

Clearly, there is a connection between the overlap and contact an operator has with other activities in the organization and the extent to which he or she will be identified with it; once such connections are understood, more appropriate systems can be designed. Clearly, also, the social science contribution to systems analysis and design is not merely about a "humane" or "humanistic" or "anthropocentric" approach to individual work roles. It is systemic, that is, about the interplay between technical and social/organizational factors in the way that systems as a whole function. And computer-integrated manufacturing (CIM), of course, is systemwide.

Every time there is a new technical development, one gets wheeled out at the end of conferences and training courses to talk about "the human implications" of whatever it is (I have even been asked to speak on the human aspects of the unmanned factory!); every time the question being put is, explicitly or implicitly, how to persuade people to accept whatever it is. In the course of more than 30 years of working in the area of human aspects of technology, the technologies have been getting more sophisticated, but the message remains essentially the same: Whenever a piece of equipment or a system is being designed, the consequences are also, explicitly or implicitly, being designed. If the consequences are to be taken seriously, then the time to take account of them is at the design stage. However, it seem that, with every new technical development, with every new project, indeed with every change of personnel within a project, that message has to be restated as if it, too, was new.

The following is an example to illustrate how deeply embedded this problem is. It concerns a company building a new plant that wanted, for once, to get the human and social aspects of the work system taken care of at the design stage. As part of the preparatory work, a study was made in one of the company's older factories, where the product to be made in the new one was already being produced. It was a food product. At one stage the finished items came sliding down a chute at the left of an operator and were carried from left to right along a belt on which they were shuffled into rows, before moving into a wrapping machine, which wrapped them as packets. The operator monitored this process, picking out any broken or jammed items, seeing that the wrapping machine was supplied with paper, and so forth. There were eight of these arrangements spaced along the room, each one involving a chute, a moving belt, a wrapping machine, and an operator.

During the preliminary study, an operator pointed out that it would be much better if the equipment was grouped together in pairs, four pairs of belts running parallel and attended by four pairs of operators, instead of eight single ones. "I'd have a mate, opposite me" she said. "We could talk. And we could help each other out. When I go to the toilet I wouldn't have to switch off, because she could watch mine for a bit." (What actually happened was that, when the supervisor was not looking, operators switched off in order to visit each other for a chat.)

The suggestion constituted a small, not very radical, change to take account of sociotechnical interdependence. Management liked it but came back to report that it had not been possible to persuade the machine tool manufacturer to make the equipment so that some ran from left to right and some from right to left, with the controls in the mirror position, which would have to be the case if operators were to face each other in pairs instead of looking at each other's backs. It was, of course, no kind of technical challenge. But it was a challenge to seeing and acknowledging the operator and her role as a real part of the system, and this hurdle was insuperable. Production engineers see the operator's part in production systems as sets of disembodied tasks, with associated costs. They are not trained to see how such tasks and the experience of doing them combine into roles. So deeply ingrained is the traditional engineer's way of perceiving that even in this company the engineer who went to see the manufacturers to try to persuade them to make the equipment run in both directions thought he was doing so because it might then be possible to have one operator running two belts.

Discussion

By the mid 1980s the pattern described had become so clear that there was a need to try to understand it in its own right, not merely to continue working within it. This led to a recognition of two things: first, one was observing a process of *splitting*, in the psychodynamic sense. This is a means of psychic economy, whereby a complex situation is simplified by all its x characteristics being attributed to one of a pair, and all its y characteristics to the other. While all work systems clearly had technical and human aspects that were interdependent, it was simply easier to see them separately. This was true not only for technologists but also for social scientists, very few social scientists being willing actually to engage with technology instead of generalizing or philosophizing about it.

Second and perhaps most importantly, the splitting had become deeply institutionalized. It pervaded the education system, research, the funding of research, career structures, the literature. Engineers read what engineers have written; social scientists read what social scientists have written. [This analysis is elaborated in a paper on the collaboration between social scientists and engineers (Klein, 1989).]

My aim, indeed the aim of all sociotechnical approaches, had all along been to get the consideration of human and organizational factors integrated with the work of technology development and implementation; I now had a better understanding of why this was so consistently difficult: "Projects" in which social scientists made a contribution to engineering developments, of which by then there had been quite a number, were simply too weak a mechanism to counteract institutionalized splitting. It began to seem more important to tackle the roots of the splitting and search for strategies of integration than to place all the emphasis on collaborative design and development projects, even though these had always been a source of interest and learning. There was a need to seek ways of institutionalizing integrated working, that is, to provide for integration through budgets, through education and training, and, where appropriate, through standards.

In 1986 an opportunity to further such integration presented itself. I was invited to give evidence to a Committee of the European Parliament and included in this evidence a suggestion that should go some way toward dealing with the splitting: It was that technologists applying for development grants should be required, as part of their grant applications, to say what effects they expected their project to have on a number of human and organizational categories. Not that one could expect accurate predictions, but applicants would be forced to give thought to these matters and that, in turn, might feed back into the education system where the splitting had its roots. Some of the members of the European Parliament liked this idea. They modified the categories to include some others they considered important, and in November 1986 the Parliament passed a resolution stating:

> [The European Parliament] believes that applicants for financial support under community technology and research schemes should be required to state what effect they expect the new technology to have on employment, the nature of the work, skills, health and safety at work and older workers for whom retraining is no longer possible . . . (Doc A2-142/86)

In such a framework an explicit contribution from the social sciences would become redundant. So far, however, the resolution has not been implemented.

8.3 THE LINK BETWEEN SOCIOTECHNICAL AND ORGANIZATIONAL—THE ORGANIZATION AS OPEN SOCIOSTRUCTURAL SYSTEM

In the past the work of social scientists in relation to industrial organization has tended to come under two distinct headings: The first was concerned with the workplace, work organization at the level of the production process, and the primary work group. The second was concerned with the wider organization and its environment, and thus involved such things as the relationship between marketing and production, centralization and decentralization, different types of control system, and so forth.

These matters have been studied in many different frames of reference, but I shall mainly be referring to the one I find most convincing in its explanation of empirical reality, in other words valuable as theory, and therefore also most useful in practice, in other words for design and development. This is in effect the same thing—there is nothing as useful as a good theory. Like sociotechnical theory it is also about the interplay between structural and behavioral factors. Structural factors, in the case of the wider organization, means things that, while they may change in the long term, in the short term have to be taken as given: such things as types of ownership, the legal framework, size, characteristics of the market, the type of technology being operated, the control systems. On most of these, there are substantial research findings exploring their links with behavioral or "soft" systems aspects. So, for example, what makes an effective organizational culture has been found to be not a matter of ideology and wishful thinking about how things should be but of appropriate response to the characteristics of the market (Burns and Stalker, 1961). Successful organizational forms, too, have been found to be more a matter of organizing to meet the requirements of a particular type of production process, for example batch, mass, or process production, than of obeying general rules about how to organize (Woodward, 1965). The production process has consequences for behavior so that, for instance, the demands on supervision and middle management are much greater in batch production situations, where many decisions are required in the course of the day, than in volume or process production, where many decisions are already built into the design of the plant; and this in turn has consequences. Batch production also offers more choice in how to organize around it, and in this situation the nature of the control systems (i.e., the way in which targets are set, actuals are measured, and the two compared) may be seen as the intervening variable, themselves having profound effects on attitudes and behavior (Woodward, 1970). In general, organizational structures, cultures, and systems of control are interacting and highly interdependent. Together they can be understood as responses to various types of uncertainty—in markets and in production processes—that organizations have to find ways of negotiating (Galbraith, 1979; Mintzberg, 1980).

With the coming of advanced manufacturing technology (AMT) and particularly computer-integrated manufacturing (CIM), the distinction and boundary between sociotechnical theory and organization theory become a disadvantage. Some professionals have responded to this by widening old definitions, for example, expanding the word "sociotechnical" to encompass other aspects of structure. This carries the danger of fudging the fact that "technical" actually has a meaning, and that choices in technology design have to be made. I prefer to regard the organization as an open sociostructural system in which technology is one of the structural factors.

Operationally, working in this framework is not so difficult. It demands a willingness to start from the situation and go where that takes one, using whatever concepts, insights, methods, or empirical research findings one has available. If one is prepared to do this, then the frame of reference of working with the interdependence between structural or "hard," and behavioral or "soft" factors is very robust and has proved itself in many situations.

The institutionalized splitting described earlier has meant that, in the systems world, there are conventions about how organizations are described. There are aspects of organizational functioning that it is permissible to discuss officially and depict on flowcharts and aspects that, although they are equally real and are recognized by everyone in the situation, nevertheless do not feature in such discussions and records. It is because of this abstraction of some aspects of the reality and suppression of others that attempts to support organizations with, for example, "total integrated management information systems" have turned out not to meet the real needs.

The contribution of industrial social research has been that it has looked at the reality of organizational functioning without such taboos. Much of the research has brought out connections between the structural and behavioral realities.

"Natural Groups"

The impact of information technology in lowering the boundary between shop floor operations and the wider organization has led some organizations to develop creative solutions. One way of structuring an organization (this is an example, not a prescription!) is to examine the product and information flows looking for natural groups of activities, each of which can constitute a single "office" (Oldham, 1989). These will be characterized by a high degree of cohesion and interdependence within the office and a low degree of coupling with the other offices, taking into account both the volume and frequency of the flows. One example of this concept is the three-office model in which there are three natural groups, namely commercial operations, business development, and manufacturing operations. Whereas commercial and manufacturing operations are concerned with the day-to-day functioning of the business, the business development office is charged with ensuring the long-term future viability of the business in all its aspects, whether it be people, products, plant, processes, and so on. Thus the time span of control, that is, the time for the full effect of a decision to become apparent, may well be longer than in the other offices.

The same principles can be applied to manufacturing operations on the basis of material flows. This results in sub-offices, or manufacturing cells, each of which is responsible for the complete manufacture of a family of components or the assembly of a range of products. From the point of view of overall factory control, the cell can be treated as a single activity.

Discussion

In summary, the links between technology and "human factors," in the widest sense, are shown schematically in Fig. 8.1. It is important to note that the

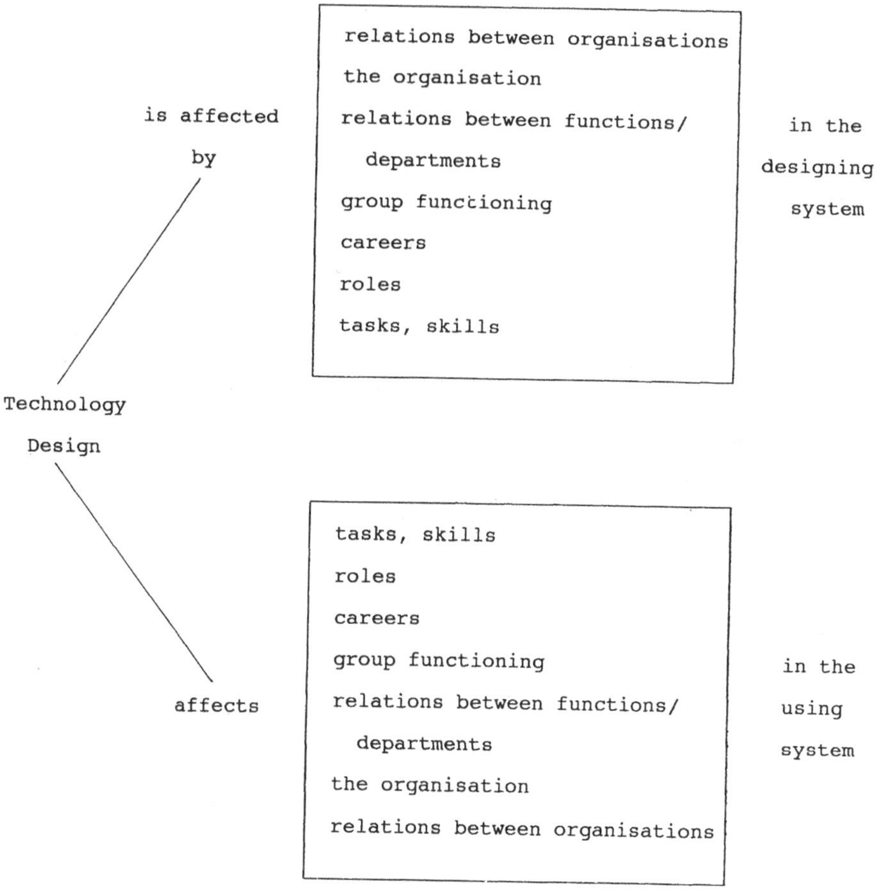

Fig. 8.1. Factors affecting and affected by technology design.

factors that affect individuals and organizations at the output end, that is, as a consequence of technology, apply with equal relevance at the input end, that is, among design teams and the organizations that contain them. In other words, technology is not only an independent variable, having consequences for skills, tasks, roles, values, relationships, careers, group functioning, departments, and the organization in which it is implemented. It is also a dependent variable. It does not come out of thin air but is itself shaped by the skills and understandings, values, tasks, roles, relationships, career paths, and growth and shrinking of departments in the organizations where it is designed.

To give an example: In the course of the design for a new high-speed canning plant, a long and fierce controversy developed about the relative merits of two proposed layouts. One had been put forward by the company's in-house engineering group and the other by the parent company's engineering services group, which had recently been made semi-independent with a remit to sell its services to constituent parts of the group. The debate about the layouts was conducted entirely on grounds of operating costs and the possible work satisfaction of future operators; the factors driving the debate, however, were clearly to do with prestige, career development, and even professional survival of individuals and organizational units at the design end.

Two relatively recent developments have brought into prominence the relevance of relationships *between* organizations as well as the dynamics *within* organizations. At the output end, the most obvious one is the coming of information technology (IT), with its consequential demands on standards, compatibility, and so forth. At the input end an example is the funding of research and development on the basis of groups of organizations or consortia.

8.4 MOVING ON FROM REDESIGN

The early sociotechnical work was concerned with reorganizing, or redesigning, work systems. The crucial component in redesign activities has been harnessing the experience of the operators. All participative design methods involve this, whatever the degree and kind of participation. The crucial next questions, therefore, were first how to devise design methodologies in situations of new design, where operators do not exist yet, and second, how to relate workplace and organization design.

The following are some contributions to this area. They do not form a coherent, comprehensive design strategy. But they may be added to the pool of ideas being considered when a design strategy is being formulated.

Assessing the Constraints

The boundary between new design and redesign is not as clear-cut as the terminology suggests. One may get very excited about the possibilities opened up by the idea of new design and then find that assumptions about the nature of the jobs are already built into the capital appropriation or that key equipment, with its associated effect on work roles, will be bought off the shelf or that the time allocated to development does not allow for enough programming. There is no such thing as a truly green field.

There is therefore a need for an explicit phase of arriving at consensus about what the constraints really are, whereabouts in the design/development process one is located, what options are available, and where one wants to extend them. Do you need to influence equipment manufacturers, and if so, how soon? It may be important to push hard to get apparently firm boundaries extended. In a recent project the timetable set between supplier, user, and funding body did not allow for piloting activities. As a result, the whole development turned into a painful and difficult pilot exercise. But in the end one must also be grateful for constraints. In the present state of our methodological understanding, total freedom is too difficult to handle.

Putting the Infrastructure in Place

As we have seen, if a design or development philosophy is sociotechnical or sociostructural, that is, if it aims at joint or integrated functioning of social and technical aspects, it will be battling against wide-ranging and deeply institutionalized splitting. It is therefore necessary to ensure that the philosophy is supported by a strong infrastructure. It needs to be reflected in training courses, in payment systems, in assessment and promotion criteria, in recruitment, in handover procedures as people move from job to job, and so on, and so on. Consistency and continuity of development are much harder to achieve than an initial enthusiasm for change.

Starting with Values

This idea arose in the course of one of the German government's Humanization of Work projects. Metzger and his colleagues were researching strategies for selecting from alternatives in production design (Metzger, 1975). They evolved a design strategy, as follows:

They got together a design team, including two members of the elected works council. To this team they presented about 10 design criteria for the human aspects of the system. They did not take these criteria from the social science literature but from their own thinking and introspection. They included such things as the opportunity to include more than one task in one's work, the opportunity to move about in the course of the job, the opportunity to communicate with others, and the opportunity to do one's own inspection. They got the design team to give these criteria an importance weighting, by doing paired comparisons: Taking the first two criteria, is 1 more important than 2, 2 more important than 1, or are they equal? Then 2 and 3 as a pair, and so on.

The design team was then presented with a number of different ways of doing a job. In the example I witnessed, which concerned an assembly task, seven different layouts had been prepared. The members of the design team assessed each of the layouts, giving it a score out of 10, on each of the design criteria (opportunity to learn, to communicate, to move about, etc.) By taking these scores, and weighting them according to the results of the previous exercise, they arrived at a quantified "best buy," in human terms, *according to the values of that plant*. This was then put up against the economic best buy, which had been arrived at by normal costing methods, and the outcome negotiated. In the example I saw, the solution picked was the layout that had come second on the human side, because the highest scoring one had been judged too costly.

This design strategy was criticized from a number of points of view: Some people said it was very mechanistic; some asked why two workers' representatives, why not half the team? Some asked why have representatives in the team, why not the operators themselves? Why these particular criteria, why not any number of others? Why these particular layouts, why not any number of possible others?

It is not necessary to make up one's mind on all these questions. The important thing is to grasp the essential core of the strategy: It is possible to make values explicit, to put them on the table for discussion and if necessary negotiation, and to incorporate them into production design.

Although this example concerns redesign, a method that starts with values rather than experience can be adapted for use in a proactive way, provided that some form of representation of the future population is possible. In the design of a new canning plant, where the technology was going to be so new and different that experience in an older plant was not relevant, it was possible to have some trade union representatives on a job design committee, because there were other production activities on the site. The job design committee consisted of management, the engineering group, and representatives from three trade unions. A list of 24 job design criteria was used; they were jumbled up, but essentially fell into four categories: working conditions, social relations on the job, aspects of autonomy, and job content (such things as variety, opportunity to learn, etc.) Members of the job design committee were asked to put these in order of priority, and considerable differences emerged. In order to demonstrate that values can influence design in an explicit way, a criterion that showed both relatively high priority and relatively high consensus—being able to influence one's pace of work—was chosen first to explore with the engineers. Going systematically through the proposed cannery layout from the point of view of pacing revealed that, at the stage where lids were to be put on the cans, an operator would be locked into a 20-second job cycle. Alternative methods were explored and were compared systematically on capital costs, labor costs, and job design aspects. At some capital cost an arrangement leading to a 12-minute cycle was selected and installed (Klein and Eason, 1991).

Starting with Simulation and the Testing of Alternatives

This method was developed by a group of ergonomists who worked at Esso on redesigning the aircraft refuelling facility at London Airport (Shackel and Klein, 1976). Again, although that was a redesign project, the method can be adapted for working proactively.

The situation was that, after landing, aircraft parked at any one of about 100 stands on the long-haul and short-haul "aprons" of the airport. The turnaround time for most aircraft was one hour, but one airline was already trying to reduce this to half an hour. In this time not only did passengers and fuel have to be loaded, but caterers, maintenance engineers, cleaners, and so forth, all needed to park near and work on the plane.

The job of the controller in the fuelling station was to make sure that fuelling trucks reached the aircraft on time. This job had been becoming increasingly difficult: The rate of traffic through the airport was increasing by about 15 percent per year; traffic problems on the ground were increasing in proportion; and the company was very cost conscious and kept to a minimum the controllers' resources of trucks and drivers.

After detailed analysis of the system, redesign suggestions were proposed and were considered in several cycles of discussion with potential users of the system. Then a simulated control room, incorporating the new features, was built in a laboratory, and simulation experiments were designed. The potential users of the system then came and worked in the simulated control room, testing and comparing four different methods of working under three different load conditions (the current load, the load predicted for the following year's peak period, and the anticipated load for 5 years ahead). The intention was that the people who would be operating the system could contribute their experience to its design. They would also be able to try some ideas of the designers that were unfamiliar to them (and which they at first did not like) in safe conditions, to try a number of alternatives before deciding on a solution, and to have some idea of how long that solution would remain viable. The solution arrived at in this way was implemented and, when last reviewed 4 years later, was still in use and well liked.

This work was one of the lines of thought contributing to the development of sociotechnical prototyping in the Cranfield Institute of Technology, United Kingdom. In order to learn about sociotechnical design, it was essential for students to be able to *experience and experiment* with real work systems capable of having alternative forms. Because there were no systems extant and available in this way, Kember and Murray developed an "action simulation" in the form of a work system in which a real engineering product was manufactured. The simulation was "good enough," first, to provide technical options, second, to give the opportunity for different organisational models to be experienced by those who worked it. Third, and most importantly, it was set up in such a way that, once having had some experience, people could then take part in the design process (Murray and Kember, 1991a, 1991b).

Because the students involved were engineers, it was possible to proceed to the next stage, which went part way toward simulating a *flexible manufacturing system* (FMS) before money ran out. An outline history of the whole development is given in Murray (1991).

The Management of Transition

The Cranfield work was, of course, geared toward engineering education, but its relevance for design situations is clear. Both approaches described have in common the fact that, apart from their manifest functions—making values explicit, testing alternatives—they have the latent function of acting as *transitional systems*.

The notion of transition in human life was first put forward by the psychoanalyst D. W. Winnicott (1971). He pointed to the function of teddy bears and all forms of play in facilitating experimentation with new and unknown aspects of the world and the working through of anxieties and conflicts. Throughout the last decades there have been massive and repeated discussions about the problems of technology or system implementation, how to overcome "resistance to change," and so on. It is very often not possible to go directly from A to B; at least as much thought and creativity needs to go into designing the vehicle for getting to B as goes into designing B, and ensuring that the vehicle has the requisite characteristics.

The best illustration I know of a transitional system comes from Harold Bridger: Consider the situation of a young man walking in the park, who sees a young lady he would like to approach. It is not easy, even nowadays, and it is not easy for her to respond without generating possibly wrong assumptions about herself. However, if it happens that she is taking a dog for a walk, the situation is quite different: It is not difficult to say something about the dog, and it is not difficult for her to respond in terms of the dog. In the course of talking about or to the dog, it is then possible to explore whether there is any mileage in going further and, if there is not, to withdraw without loss of face. The dog serves as a transitional system.

I have worked with many forms of transitional system, on one occasion finding myself role playing a product. While I took on the part of the product as it progressed through the production system, exploring the various things that happened to it, members of a design team were able to "play" with alternative production ideas, and thus to arrive at a design solution that had eluded them for a long time.

Institutions such as training courses, negotiations, working parties will have the function of facilitating transition if they create an environment where options can be explored in safety without later repercussions, where people can experiment with roles and behaviors beyond their habitual ones, and where issues can be worked through that the normal working culture may not encourage to surface.

In design situations there are many forms of simulation, prototyping, or piloting that may be used. Such structures may be as simple as cardboard-box prototypes of workstations or as sophisticated as computer-based simulations. But for these to have the function of facilitating transition, certain conditions have to be met. They must be sufficiently flexible to permit alternatives to be demonstrated and sufficiently realistic for the operational realities involved to be experienced or at least imagined. They must also, therefore, incorporate the human and social, or soft systems, aspects in a realistic way, that is, in a way that is not merely the result of wishful thinking. While the idea of simulation is very familiar to system designers, they generally do not include in it simulating the social systems aspects nor the elements of experimenting and working through.

The Issue of Design Rules

In many instances there is a discernible shift from open-ended, grounded diagnosis and action formulation to a more prescriptive and formalized approach to the same problem. Of course, it seems natural that, having invested heavily in a learning and development activity in the first place, one should want to disseminate and institutionalize the outcomes and not repeat what may have been an expensive and laborious process. Project experience or research findings may be encoded in a set of standards for others to follow. Such a standard may refer either to the outcomes to be achieved or the process by which they are to be achieved, or both.

But there are dangers in this approach. If a standard or even a process is applied automatically, without understanding the circumstances in which it was developed and without a diagnostic approach to the circumstances in which it is to be applied, this is likely to result in inappropriate application. It is also important to understand the dynamics involved, which concern the autonomy of those who develop and those who follow such "rules." First, it is one thing to try to make sense of the overwhelming complexity of the environment for oneself by encapsulating it in categories and rules; it is quite another to then reify those categories and rules as if they had some kind of absolute quality. Second, there is a power relationship involved: On the whole people want to devise their own solutions, not follow the solutions of others.

In this context, as has been said, it is necessary to distinguish between design rules and general principles. General principles need a great deal of local, grounded work before one can interpret what a particular principle means in a particular situation. As I understand the hope expressed in the search for design rules, it is that such local work may be minimized. Some aspects of the general principles put forward by Cherns (see earlier discussion), such as those concerning boundary location or information flow, are coming to be generally recognized and incorporated into systems such as group technology or just-in-time methods. In this, incidentally, they demonstrate the possibility of removing the sociotechnical idea from accusations of "do-gooding" or political ideology and formulating it at a general level.

It is important to recognize that the search for all-embracing design rules is at least as much a response to the need on the part of the searcher to gain some control over chaos as a realistic way to progress the subject. A study of what has become of previously formulated design rules should test this hypothesis, if test is needed. Staying with the complexity may in the long run turn out to be the more valuable strategy.

Discussion: The Double Task

The design and development process itself presents a design task. Moreover, if it is accepted that the organization and dynamics of the design process affect outcomes (Fig. 8.1), then these processes also need to be kept under review. Such review needs to go on in parallel with the actual design and development activities themselves and thus constitutes an example of what has been called the "double task."

One possible way of keeping the process actively under review might be an element of longitudinal "accompanying research." This would take the form of tracer studies in an action research framework.

The use of tracer studies in industrial sociology originally derived from medical research, where a "tracer" is sometimes introduced into an organism and studied, at intervals or continuously, on its way through. In this case it consists of taking a key item, usually a product but in this case a process, and tracking it through everything that happens to it in the organization, as well as examining the roles and institutions involved. It has in the past proved very fruitful, not least in that it makes sense to the people in the system. The kind of interview involved in attitude studies would be tedious to experience and inappropriate to carry out repeatedly. This is not true of interviews focused on an object or process, on the lines of "what happened since I saw you last?"

However, the framework would not be that of research but of action research, for three reasons: First, and most obviously, feedback from the work should provide a contribution and not mere post-hoc analysis and critique. Second, as has been the burden of much of this chapter, much relevant knowledge, experience, and research do not find their way into practice. And third, the range of issues is great and will change over time.

Action research can and should be totally rigorous. It is not a matter of intervening in an ad hoc and uncontrolled way but of selecting and agreeing strategies with the participants in knowledge of their consequences, documenting them, and staying in role.

An action researcher attached to a design or development activity might:

Make predictions: If you set it up in this way, there are likely to be consequences of that kind. Is that what you intend?

Help work through dynamics: The dynamics at the input end of design have direct consequences for the output end; in fact, that is often where the origins of poor design decisions are to be found. If design decisions are made for reasons other than design needs, the outcome is bound to be problematic.

Give feedback about patterns that are observed, such as how outcomes differ when some parties are not able to come to meetings.

Offer research findings or experience from other situations.

Offer methodological help, for example, with prototyping or other forms of transitional system, or with methods of evaluation.

Ask questions that stimulate people to consider a wider range of alternatives than might have been considered.

None of the interventions mentioned are "consultancy" in the sense of telling people what to do. It is axiomatic in action research that ownership remains with the client system. The point, in any case, is to find a methodology that counteracts splitting and that does justice to the complexity and dynamic nature of design.

REFERENCES

Burns, T., and Stalker, G. M. (1961). *The Management of Innovation*, Tavistock, London.

Cherns, A. (1976). The principles of sociotechnical design. *Human Relations*, **29**(8), 783–792.

Cherns, A. (1987). Principles of sociotechnical design revisited. *Human Relations*, **40**(3), 153–162.

Emery, F. E., and Thorsrud, E. (1969). *Form and Content in Industrial Democracy*, Tavistock, London.

Galbraith, J. (1979). *Organisation Design*, Addison Wesley, Redding, Mass.

Glaser, B. G., and Strauss, A. L. (1968). *The Discovery of Grounded Theory: Strategies for Qualitative Research*, Weidenfeld and Nicolson, London.

Hill, P. (1971). *Towards a New Philosophy of Management*, Gower Press, London.

Klein, L. (1964). *Multiproducts Ltd. A Case Study in the Social Effects of Rationalised Production*, HMSO, London.

Klein, L. (1976). *A Social Scientist in Industry*, Gower Press, London.

Klein, L. (1989). On the collaboration between social scientists and engineers, in *Designing Human-Centered Technology; A Cross-Disciplinary Project in Computer-Aided Manufacturing*, H. H. Rosenbrock, Ed., Springer-Verlag, London.

Klein, L., and Eason, K. (1991). *Putting Social Science to Work; the Ground between Theory and Use Explored through Case Studies in Organisations*, Cambridge University Press, Cambridge.

Metzger, H., Warnecke, H. J., and Zippe, H. (1975). Neue Formen der Arbeitsstrukturierung im Produktionsbereich. *Z. ind. Fertig*, **65**, 665–670.

Miller, E. (1983). The Ahmedabad Experience revisited; Work Organization in Weaving 1953–70, in *International Yearbook of Organizational Democracy*, Vol. I. C. Crouch and F. Heller, Eds., Wiley, Chichester.

Mintzberg, H. (1980). *The Structuring of Organisations*, Prentice-Hall, Englewood Cliffs, NJ.

Murray, H. (1991). Socio-technical action simulations for engaging with engineering designers, in *The Social Engagement of Social Science, Vol. 2, The Socio-technical Perspective*, E. L. Trist and H. Murray, Eds., University of Pennsylvania Press, Philadelphia.

Murray, H., and Kember, P. (1991a). *Work Organisation for Production Engineers. Handbook*, Cranfield Press, Bedford, UK.

Murray, H. and Kember, P. (1991b). *Work Organisation for Production Engineers. Simulation Manual*, Cranfield Press, Bedford, UK.

Oldham, K. (1989). Unpublished Communication, Birmingham, Lucas Engineering and Systems.

Rice, A. K. (1958). *Productivity and Social Organization: The Ahmedabad Experiment*, Tavistock, London.

Shackel, B., and Klein L. (1976). Esso London Airport refuelling control centre redesign—an ergonomics case study, in *Applied Ergonomics*, **7**(1), 37–45.

Trist, E. L., Higgin, G. W., Murray, H., and Pollock, A. B. (1963). *Organisational Choice*, Tavistock, London.

Winnicott, D. W. (1971). *Playing and Reality*, Tavistock, London.

Woodward, J. (1965). *Industrial Organization: Theory and Practice*, Oxford University Press, London.

Woodward, J., Ed. (1970). *Industrial Organization: Behaviour and Control*, Oxford University Press, London.

CHAPTER 9

MANAGING FOR CONTINUOUS IMPROVEMENT IN QUALITY AND PRODUCTIVITY

J. A. EDOSOMWAN
Johnson & Johnson Associates, Inc.
Fairfax, Virginia

9.1 INTRODUCTION

Advanced manufacturing technologies, such as expert systems and numerical controls tools, are highly sophisticated and complex. As shown in Fig. 9.1, advanced manufacturing technologies can be viewed as assets and means of accomplishing a predetermined result. The technologies play a significant role in improving quality, effectiveness, productivity, efficiency, and profitability. In today's world of increased global competition manufacturing companies must focus on managing and effectively utilize the six manufacturing technology domains (MTD) shown schematically in Fig. 9.2. The elements of MTD include product-driven technology (which is all technologies contained in a product), and problem-solving technology (defined as including all aspects of logical problem-solving concepts and logic designed to help the user analyze and solve a give problem) (Edosomwan, 1989; Porter, 1980). Process-driven technology can be viewed as an asset for improving production processes,

Organization and Management of Advanced Manufacturing. Edited by Waldemar Karwowski and Gavriel Salvendy.
ISBN 0-471-55508-8

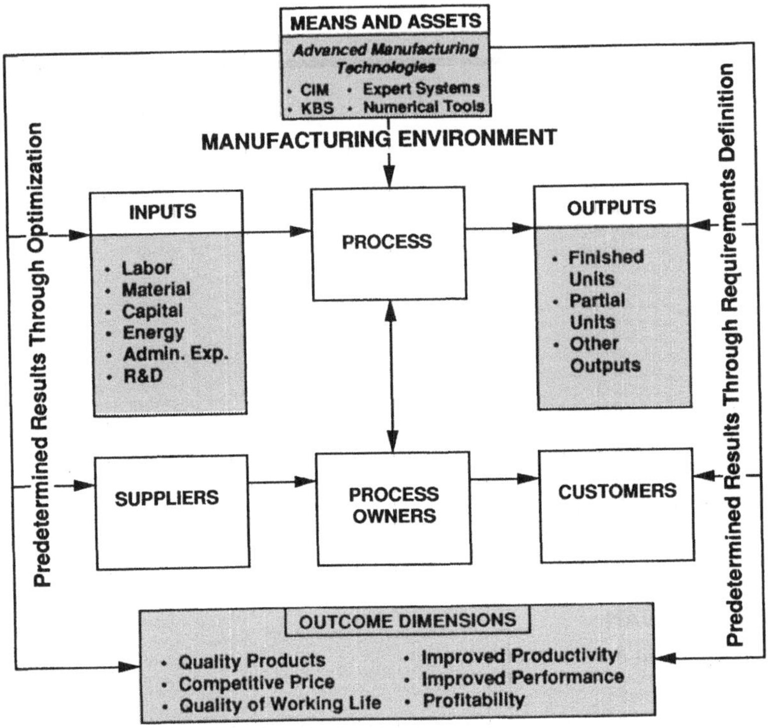

Fig. 9.1 Role of advanced manufacturing technology.

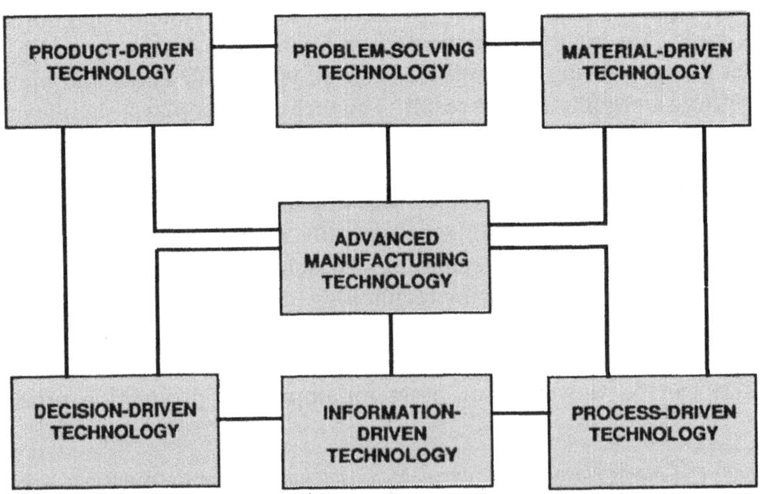

Fig. 9.2 Manufacturing technology domains (MTD).

and the effective optimization of the relationships between productions, inputs, process, and outputs. Materials-driven technology usually evolves out of new product designs dependency and modification of existing materials to suit specific product requirements and needs. Decision-driven technology focuses on optimization of operational, tactical, and strategic aspects of business management. It can be used to improve functional performance, streamline procedures, and enhance decision making at all levels of the organization. Information-driven technology includes aspects of information processing, communication, and process interaction. It can also be viewed as a component of product technology. Information technology can be used to enhance the adaptive capabilities of products and processes; facilitate communication and control, manipulate data and information, and facilitate external scanning of the manufacturing business environment. These definitions are consistent with the works of Edosomwan (1991; 1988), Frohman (1985), and Steele (1989). As shown in Fig. 9.1 advanced manufacturing technology can be utilized to achieve several outcome dimensions. Two of the outcome dimensions, productivity and quality, play a key role in improving manufacturing competitiveness.

Productivity

Productivity is a measure of how well resources are used to product output. It relates output to input and integrates performance aspects of quality, efficiency, and effectiveness. The four types of productivity measures recommended by Edosomwan (1987, 1989, 1991) for manufacturing companies are:

Comprehensive Productivity: The rates of tangible and intangible output to all tangible and intangible input factors.

Total Factor Productivity: The ratio of total measurable output to some of associated labor and capital (factor) outputs.

Partial Productivity: The ratio of total measurable output to one class of input.

Quality

Manufacturing companies have a great need to use advanced technologies to improve quality of design and quality of products and services. The quality of design that pertains to the quality obtained through changes or manipulation in design parameters should also be addressed by a market-driven enterprise. The differences in quality are a result of differences in size, materials used, reliability, equipment utilized, and temperature. Quality of conformance is a measure of how well the product conforms to the specifications and tolerances required by the design. Such factors as training, the production process, motivation levels, procedures, and quality assurance systems can have an effect on the quality of conformance. Quality characteristics typify the variety of uses of a given product. The quality characteristics may be of several types: behavioral or ethical (fairness, honesty, and courtesy), sensory (color, taste, beauty, appearance, etc.), time-oriented (serviceable, reliable, maintainable), and commercial (warranty).

Improvements in quality are therefore made by examining the design of the conformance phases with their associated characteristics through the application of appropriate techniques and methods including changes in design, timing, inspection procedures, and process control procedures. Quality when viewed as conformance to specification has great potential for being a very effective improvement strategy. However, this can result in a lack of commitment to quality by members of an organization that do not acknowledge the presented situation.

The quality of services pertain to the total satisfaction of the requirement and expectations of the internal and external customers receiving a service in a timely fashion, at minimum cost and price. The measure also pertains to both the tangible and intangible elements of service offered by individuals, operational units, and the total enterprise. Quality of service also involves physical factors, such as quality of products used for services, as well as the behavior and attitudes of the persons rendering services directly to the customer. Unlike product quality, which focuses on the level of relevance, uniformity, and dependability, service quality is a function of the production, delivery, and consumption processes. Most services have more tangible elements. It is very difficult to apply quantitative standards and measures to services. The quantitative standards that are usually available are applied in an indirect manner. Unspoken standards also exist in service quality management. The standards usually include the human attributes and characteristics that are unique. These include humility, empathy, courtesy, respect, care, and patience.

9.2 MANUFACTURING TECHNOLOGY MANAGEMENT MODEL

The key elements involved in managing manufacturing technology are presented in Fig. 9.3. Each element will now be discussed.

9.3 UNDERSTANDING TECHNOLOGY TRENDS AND CUSTOMER REQUIREMENTS

In order to effectively utilize advanced manufacturing technology for continuous improvement in quality and productivity, it is recommended that manufacturing companies begin with a thorough understanding of current and future generation technologies. The evaluation of existing and emerging advanced manufacturing technologies must focus on the use of competitive benchmarking approach to understand the capabilities of current and emerg-

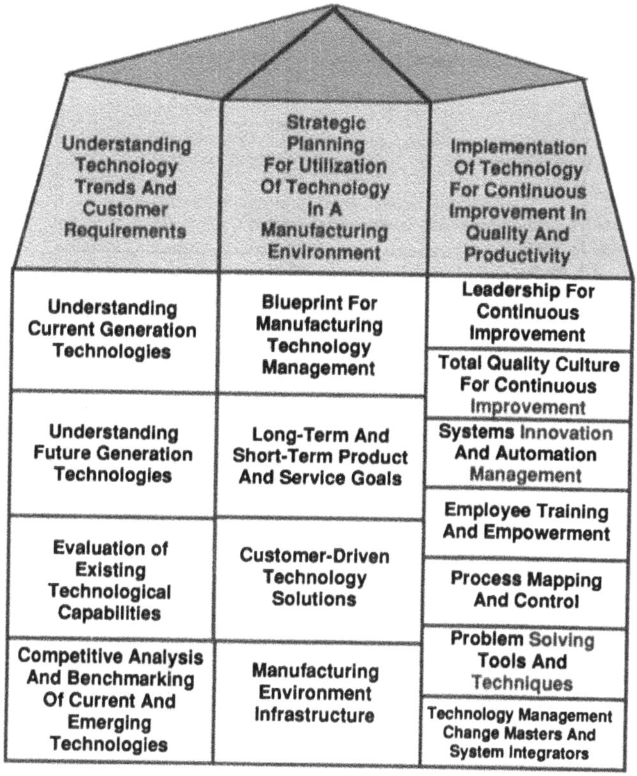

Fig. 9.3 Manufacturing technology management model.

ing technologies. Equally important is the understanding of customer requirements. The four-stage PEAD model shown in Fig. 9.4, is recommended by Edosomwan (1986 a & b, 1991) for understanding customer requirements. It can also be viewed as a system for capturing customer requirements at each stage of the product cycle. In identifying the needs of the customer, the use of the following alignment questions are recommended: What are the needs? What is the output required? What is the expected cost and quality of the output? How much is required where, and when? The process owner, who is responsible for the effectiveness and results of the operational work unit, now translates the needs and wants of the customer into specific product and service requirement. Marketing information and customer perception of existing and competing products and services are used to refine requirements. The process owner may elect to define the entire customer requirements through a matrix approach or by simply defining the roles and responsibilities of the supporting function or individuals that will help in the implementation of the final product and service specifications. One-on-one contact with customers with good channels of communication is highly recommended. Also,

PAY ATTENTION TO THE VOICE OF THE CUSTOMER AND SUPPLIER

- Identify their needs, wants, desires
- Obtain marketing information on customer and supplier perception for existing products and services
- Translate the needs, wants and desires into products and service requirements

EVALUATE THE EFFECTIVENESS OF MECHANISMS IN PLACE TO SATISFY THE CUSTOMER AND SUPPLIER REQUIREMENTS

- Focus on product, service, and process characteristics
- Specify requirements by operational needs

ANALYZE CUSTOMER AND SUPPLIER REQUIREMENTS AND MARKET INFORMATION AND SYNTHESIZE DATA FOR RELEVANCE

- Develop interrelationships between product and service characteristics
- Select the requirements that depict customer and supplier input

DEVELOP PRODUCT AND SERVICE CHARACTERISTICS AND TECHNICAL MEASUREMENTS THAT MEET THE CUSTOMER AND SUPPLIER REQUIREMENTS

- Review final specifications with development teams
- Implement the recommended product and service specifications

Fig. 9.4 The Edosomwan PEAD model for capturing the requirements of the customer (from Edosomwan, 1991).

customer surveys that are designed to measure perceptions of service and product quantity are recommended.

9.4 STRATEGIC PLANNING FOR UTILIZATION OF TECHNOLOGY IN MANUFACTURING ENVIRONMENT

A blueprint that describes the overall manufacturing product objectives, technology management strategies, and business strategy should be developed. As stated by Edosomwan (1991), long-term and short-term technology management objectives must be incorporated into overall business strategies. Such objectives must be included and not be limited to the following areas:

- Technology selection, specialization, or embodiment
- Sources of technological capability (internal vs. external)

- Research and development investment and staffing
- Competitive timing: initiative vs. responsiveness
- Product and service research development procedures and policies

The first step in formulating a manufacturing technology strategy involves the identification of all distinct technologies and subtechnologies in the value chain. The potential relevant technologies in other industries are then identified. The second stop involves the determination of which technologies are the most significant for competitive advantage and the likely path of change of key technologies. In the third step the firm's relative capabilities in the important technologies and the cost of making improvements are assessed. The fourth and final step call for the selection of a technology strategy. The strategic, tactical, and operational manufacturing plan should include all elements including current and future technology strategies, organizational change, people development, product design phases, logistics management, technical tools for just-in-time production, quality control and productivity improvement, and automation strategies. The operational manufacturing plan must complement the strategic plan. It should show the sequence of activities, including the development of a phased implementation plan, description of methods and procedures, and the total requirements for implementing new products and advanced technologies.

9.5 IMPLEMENTATION OF TECHNOLOGY FOR CONTINUOUS IMPROVEMENT IN QUALITY AND PRODUCTIVITY

Manufacturing companies interested in using advanced manufacturing technologies to achieve improvement in quality and productivity must adopt a philosophy of continuous optimization of products, services, technologies, systems, and processes. The philosophy involves the act of constantly and forever enhancing value and excellence. The overall aim of continuous optimization focus is to minimize the loss imparted by the product or service to the consumer from the time the product or service is developed to customer shipment. This also includes the acceptance of change. Positive changes in technology management strategies (product and service strategies) will occur through a regular, proactive focus on opportunities for improvement and action on the opportunities. In order for the activity to occur, employees must understand the product and service optimization process and techniques to resolve problems through identification and elimination of root causes. The improvement process deals with many technology management issues including:

- Integrating product and service development objectives with manufacturing and marketing capabilities, and finding ways to do this effectively.

- Institutionalizing total performance and improvement in the entire product and service life cycle from concepts to commercialization.
- Focusing on the use of tools and techniques that reduce total cycle time, contribute to improved product and service value, and are responsive to customer needs.
- Continuously meeting the challenges of the changing technologies, systems, processes, and demands of the marketplace.

Managing the product and service improvement process not only requires pushing the frontiers of technology but also involves the use of continuous improvement, evaluation, control, planning, improvement, and monitoring. The outcome of the improvement process must be benchmarked against that of competitors. This requires the knowledge not only of a specific business unit but to the industry and all competitors in the global marketplace.

The means for implementing advanced manufacturing technologies must be specified. Such means should address the following areas:

1. Management of product and service changes under conditions of rapid technological and social change
2. Management of the interface between the various organizations (external and internal) supporting the products and services
3. Methods and tools for improving product and service performance, project management, and the interrelationship between the requirements of the supplier, process owner, and customer
4. Adequate procedures, methodologies, and techniques for training the work force to the requirements of new products and services

The foundation for world-class manufacturing is a total quality culture. The tools, equipment, and training required in the quality improvement process must be provided to all employees. The formation and implementation of quality council and quality improvement teams must be encouraged. Management must take a leadership role to coordinate the establishment of quality goals, review progress against the goals, and participate in a reward and recognition program. Active management and employee involvement in quality improvement teams that investigate production problems and define quality projects must be encouraged. Ongoing analysis, review, and study of production methods must be performed with the sole purpose of making the process simpler, easier, more nearly perfect, and hence productive. Ongoing monitoring and evaluation of the manufacturing process is recommended. Review production measures periodically and identify failures and areas for improvement. The measurement system should be based on manufacturing objectives that will be tracked and reported with focus on progress reports.

9.6 CASE STUDIES

Manufacturing companies have realized significant benefits from the application of advanced technology. As reported by Moody (1991), NEXT Computer utilized computer-integrated manufacturing, materials requirement planning, fully automated assembly, optical scheduling, electronic Kanban to achieve improvement in productivity through scrap reduction, and improvement in quality with reject rates less than 4 to 6 parts per million. Hank (1991) reported that the Minneapolis valve plant of Dana Corporation utilized multidisciplinary teams, advanced cellular manufacturing technology, and process improvement techniques to improve manufacturing excellence. The plan won the 1991 Shingo Prize. Notable improvements included productivity up 32 percent, through time down 92 percent, inventory down by 50 percent, quality costs down by 50 percent, customer lead time down from 6 months to a matter of days, plant break-even points down 15 percent, and rate of return on investment up over 400 percent.

Mason (1991) reported that in 1988 Texas Instruments launched an ambitious program to improve performance in all of the manufacturing cycle. The major change was moving from a "push" system to a "pull" build to backlog system through improvements in scheduling and use of appropriate manufacturing technologies. The goal was to drive unsold inventory to zero, while creating ability to satisfy customer demand with short lead time. In two and a half years, cycle time has been cut in half, Texas Instruments maintains a 5-week build to backlog with needed flexibility and carriers less than 5 percent unsold inventory. On-time delivery has improved from 58 percent in 1987 to 92 percent in 1990.

Battles, (1990), reported that John Deere utilized a multifaceted approach to manufacturing improvement. Group technology cells, multifunction design teams in product planning, and focus on small job/quick change was utilized to improve the following parameters. Production design time is down 50 percent. As a result of innovative planning, a typical product now has 30 percent fewer parts and is produced in 50 percent less time. Overhead manufacturing cost is down 60 percent, quality is up 100 percent, and productivity is up 20 to 50 percent.

9.7 CONCLUSIONS

In order to effectively compete in a global marketplace, manufacturing companies must adopt a philosophy of using advanced technology for continuous improvement in quality and productivity. Advanced technology by itself is not a sufficient means for achieving competitiveness. When advanced technology is used in conjunction with appropriate business strategies, management practices, total quality culture, supplier and customer improvement strategies, employee development, and process control, manufacturing companies can produce competitive products at affordable prices and excellent quality. Managing advanced manufacturing technology encompasses certain boundaries. It involves setting the right manufacturing goals and objectives, leveraging all resources and assets by continuous optimization of all work processes and functions of the manufacturing organization. It also requires managing the totality of technology operation and product requirements from concept to commercialization. Utilization of advanced manufacturing technology is recommended to manufacturing companies seeking to increase market share, total customer satisfaction, reduce cycle time, improve quality and productivity, and improve competitive position.

REFERENCES

Abernathy, W. C., and Kantrow, A. (1981). The new industrial competition, *Harvard Business Review*, **59**(5).

Battles, C. S. (1990). World class manufacturing: A structured approach, *Productivity—The Best America Conference*, April 23–25, 1–16.

Edosomwan, Johnson A. (1986a). Managing technology in the workplace: A challenge for industrial engineers, *Industrial Engineering*, February, 14–18.

Edosomwan, Johnson A. (1986b). Technology Impact on the Quality of Working Life—A Challenge for Engineering Managers in the Year 2000, *Proceedings of the First International Conference on Engineering Management*, Arlington, VA.

Edosomwan, Johnson A. (1987). *Integrating Productivity and Quality Management*, Marcel Dekker, New York.

Edosomwan, Johnson, A. (1988). *People and Product Management in Manufacturing*, Elsevier, The Netherlands.

Edosomwan, Johnson A. (1989). *Integrating Innovation and Technology Management*, Wiley, New York.

Edosomwan, Johnson A. (1991a). The Connection Between Technology, Productivity and Quality, Technical Paper, Johnson & Johnson Associates, Inc.

Edosomwan, Johnson A. (1991b). *A Ten Step Approach for Justifying and Implementing New Technology*, Excellence Publications, Virginia.

Edosomwan, Johnson A. (1991c). Achieving Product Service Fast-Time to Market Objectives Through Technology Management, Proceedings for the International Conference on Engineering and Technology Management PICMET-91, Portland, Oregon.

Fortuna, Ronald M. (1988). Beyond quality: Taking SPC upstream, *Quality Progress* 23–28.

Frohman, A. L. (1985). Putting Technology into Strategic Planning, *California Management Review*, **37**(2) Winter.

Mason, J. (1991). Costs are reduced through build-to-backlog and autoreplenishment systems, February Issue, 40–42.

Moody, P. E. (1991). NEXT Computer, The Ultimate Computer Factory, *Target*, Spring Issue.

Porter, M. E. (1980). *Competitive Strategy*, Free Press, New York.

Rogers, H. (1991). What Dana learned from Japan, *Industry Week*, July, 36–40.

Skinner, W. (1984). Operations technology: Blind spot in strategic management, *Interfaces*, **14**(1),

Steele, L. W. (1989). *Managing Technology: The Strategic View*, McGraw-Hill, New York.

CHAPTER 10

REFLECTING ON WORK PRACTICE: THE ROLE OF PEOPLE AND ORGANIZATIONS

E. HAVN
Institute of Social Sciences
Technical University of Denmark
Lyngby, Denmark

10.1 INTRODUCTION

> Organisational problems are among the most important impediments to widespread AMT applications within the firms. Successful applications usually require re-organisation of work practice, skill use and skill development, re-organisation of production and in many cases of the entire enterprise. (STI Review, 1989)

This is one of the conspicuous conclusions in a recent study, published by OECD. It is not an exceptional conclusion. There are by now a vast number of empirical studies showing that it is the coherence of factory organization, horizontal and vertical division of labor, and skills that determine how new computer-supported tools are used, and therefore also whether or not the companies achieve the effect on rationalization and quality they had hoped

Organization and Management of Advanced Manufacturing. Edited by Waldemar Karwowski and Gavriel Salvendy.
ISBN 0-471-55508-8

for when they decided to invest in the, often, expensive new technology. In a Swedish study it is even stated that without organizational changes, including new skills, and new dispersion of responsibility in the jobs, advanced manufacturing technology may become the world's most expensive baby soother (Bengtsson, 1991).

These studies all demonstrate the systemic character of advanced manufactory technology (AMT), that is, that AMT influences the individual manufacturing processes as well as the functional coherence of the entire manufacturing system. The results of the surveys can be summed up in three principal findings:

1. Use of AMT increases the uncoupling of work and technology. Computers support a detailed and time-independent control of machines and production processes.
2. The qualitative importance of work activities for the continuous, uninterrupted, and optimum production process is growing. Production processes supported by AMT are often extremely complex and even minor mistakes and misunderstandings may result in major breakdowns. Preparing and controlling the processes require theoretical as well as empirical knowledge.
3. Use of AMT require reorganization of manufacturing processes to meet needs for augmented communication and cooperation between functions that are allocated to different departments in the traditional factory model.

Implementation of AMT is not, however, a question of merely "reinventing the factory," that is, developing new factory reference models that are tailor-made to suit the adaption of the organization to the technology. "Reinventing the tool" is just as important. Manufacturing technology has to be designed to meet the organization and people issues in its own implementation. Even though a configuration of computer-controlled tools does not prescribe a given organizational structure, it certainly is not neutral. Advanced manufacturing tools support some organizational forms better than others, and they are in themselves a materialization of designers views on organization and work.

In combination, therefore, the three principal findings exact new manufacturing models, and—just as important—new design processes of advanced manufacturing systems that sustain a joint design of technology, organization, and work.

Even though, however, the close relationship between technology and organization is well established in research, it is not common understanding, neither among industrialists nor among designers of the technology:

1. The surveys that demonstrates the firm weaving of technology and organization also point out that most companies implement advanced computer-supported technology in the existing organization with only

short training courses, often organized by the vendor company, and most vendor companies consider these preparations as sufficient. (STI Review, 1989; Schultz-Wild et al., 1989).

2. In a recent issue of *Communications of ACM* three editorials addressed human factor issues in design of AMT. One of them, by D. A. Norman, states that: "Many advances have been made in our understanding of the hardware and software of information processing systems, but one major gap remains: inclusion of the human operator into the system analysis. . . The designer must consider the properties of all system components—including the humans—as well as their interactions" (Norman, 1990). Two other editorials highlight the irony that the consequence of poor design of computer systems is not just that systems fail to perform properly but that "human error" is invariably cited as the cause.

The accumulation of research reports pointing out the mutual dependencies of organization, people, and technology in advanced manufacturing systems seem to have only limited influence on current praxis. One obvious reason for this is that factories are not designed. They are not products of a few people's analysis and description of production and its organization, functions, and flows. Factories are products of politics, visions, ideas, and people's perception of external and internal constraints, and available technology. Therefore, new factory models will only have an impact on history if they are concrete and accessible enough to penetrate many people's ideas and praxis. Ideas must be tangible to influence the evolvement of new manufacturing organization, and they can only be made tangible by good examples substantiated by experiments. In manufacturing processes using AMT, experiments require a design practice that allow for a tight coupling of organizational and technical development. Rethinking the rationality of systems design methodologies is a most important topic in development of new manufacturing organization, but methodologies are only useful if they are supported by relevant images of organization and work.

In the effort to promote a joint organizational and technical development of advanced manufacturing systems, it is important to discuss the images of people and organization that are depicted in the literature on human factors in manufacturing. It is not important because designers actually use textbooks on human factors in their work—we know they do not. It is important because what designers think about work practice profoundly affect what they do, and human factor studies contribute to professional designers' views on manufacturing.

Work practice is an individual and cooperative activity simultaneously. But it is too difficult to discuss both aspects at the same time. We will in this chapter discuss the individual worker at first and then take in the discussion of cooperative work.

10.2 WHAT ARE WE: "FACTORS" OR "ACTORS"?

The limitations of the current human factors discussion were accentuated by Norman, quoted above, in his description of the operator as a system component. The use of the term *component* subordinates people to a machine and reduces people to components with properties, rather than individuals with a formative role (Wynn, 1991). This may be an important cause of the fact that even though human factor studies have been on the academic scene for years, they have not led to any conclusive influence on development of computer artifacts. I think it is worthwhile to recapitulate the history of the concepts of *design* and *human factors* and to unravel some ambiguities in our everyday use of the terms.

The phrase "design of advanced manufacturing systems" is often used in two denotations:

1. Advanced manufacturing technology systems in the narrow sense of the word, that is, machines, computers, and applications, are, of course, the outcome of some kind of design activity. As a product AMT is designed in the sense that it has been mentally planned and delineated for later realization, although the design process may be very different from what we learn when reading design engineering textbooks.
2. The locution AMT system is, however, often used to describe the tools *as well as* the processes they are developed to support, and an AMT system is thus conceived of as a manufacturing system involving people and tools. Design of AMT systems involving people and tools is an activity quite unlike the design of mere tools, and by using the word *design* for both of the undertakings we may confuse the issue from the very start.

The last definition of AMT systems cannot be mentally planned in a more restricted sense. Thus we will discuss the role of design in development of AMT systems and the relationship between design in development of manufacturing systems and design in development of advanced manufacturing technology.

The term *human factor* is a well-known designation of the attempt to tailor the machines to the skills and limitations, both physical and mental, of the operators. In design of an artifact to be operated by people the constructor is bound to consider the prospective users' mental and physical capacity in work settings—effects of stresses, psychomotor ability, perceptual activity, mental processing workloads, and so forth. The utility of the tool depends on the designers' awareness of the user faculty, and their ability to integrate knowledge into the technical design.

In manufacturing systems, however, people are not simply affected by the technology, and tools are not necessarily good because design reflections embrace knowledge of users' capacity. We cannot describe people as merely "factors," we have to take as a starting point that people are "actors," who are operating a tool, for example, a computer application, and are using the application in work processes (Gerson and Star, 1986) or in situated actions (Suchman, 1987), and the quality of the application is the result of how well the user performs.

Human factor studies have a long tradition in design of manufacturing tools, dealing with questions of how to divide the work between the operator and the machine. The purpose of such studies is to establish a set of guidelines about general human and machine performance capacities. In the classical studies, intimately connected to assembly line work processes, the focus was on how to construct a machine to perform with the required functionality, and the human operator was seen as just another component, behaving, however, somewhat different from hardware components. Human factors studies were sprouting because companies were finding to their cost detriment that the "just another component" view of operators could lead to very inefficient systems, unless one knew a lot about this "component." It was gradually realized that controls and tasks designed to reduce both mental and physical strain of the operator allowed for improved performance of the human–machine system. Mental and physical strain were studied in the field of human factors engineering, but initially within the world of exceedingly well-described, well-planned and segregated work processes, and therefore without considerations for human problem solving and goal-directed behavior.

The fact that this view of manufacturing was seen for years as a suitable starting point for research in human factors was due to the immense impact that the successful introduction of mass production in the U.S. automobile production had on management thinking in the industrialized world. This vision of new manufacturing capabilities was based as a rigorously sequential process, with product development and manufacturing falling into a number of phases such as design, planning, production, and so forth. In this one-way, top-down model, the product design needs to be finished and proven, and that may take years before it enters production. The production methods and planning must be mapped out in least detail before the commencement of the processes on the very expensive and specialized machines. This production model had a strong impact in European manufacturing industry, in spite of the fact that for most products the European markets always have been too small and unstable to support the mass production (von Freyberg, 1989). The greater part of European industry consists of small- and medium-sized enterprises working in fluctuating markets, and without the financial capacity for mass production. These companies depend on skilled employees in most activities, and so do mass production, at least in all other activities but the most simple assembly work. As Shaiken (1984) notes:

> In the auto industry and elsewhere there proved to be two critical limits to Taylorism's effectiveness—one social and the other technical. The social limits stems from the distaste workers quickly develop for highly routinized and reg-

> imented jobs. The technical limits arises from the impossibility of fully eliminating skill and control from many occupations through the organization of work alone. (p. 26)

Working with machines in manufacturing usually involves task complexity, problem solving, and people's intentional activity. Therefore to be useful in the design of advanced manufacturing systems, human factors studies had to expand and take in problems of human action. Human factors engineering was supplemented by cognitive engineering, with the aim of understanding the fundamental principles behind human action and performance that are relevant for the development of engineering principles and design (Norman, 1986).

The goal of human factors and cognitive engineering studies is to give a theoretical grounding for design of advanced manufacturing technology, which gives an optimum setting for people to follow goals and solve their problems. The intention is to build better artifacts, as delineated by Bannon:

> They don't just want to know about systems after they have been built, they want to know how they should build them in the first place, and even what they should build. (Bannon, 1990, p. 266)

This approach has lead to several research traditions. First of all, to one working with formation of a theory of action, as described by Norman (1986):

> It seems clear that we need to develop theoretical tools to understand what the user is doing. We need to know more about how people actually do things, which means a theory of action. . . . We wish to attempt User-Centered System Design, to ask what the goals and the needs of the users are. (p. 37)

Norman suggest that we are starting the studies of structure, content, and dynamics of individual user cognition from a problem-solving model similar to the GOMS model (*G*oal, *O*perator, *M*ethods, *S*election) developed by Card, Moran, and Newell (1983). This is a stage model of user activity from establishing a goal, executing an action, and evaluating the system state with respect to the goals. Stage models, however, are simple models, allowing for all possible variations, but still simple. It is difficult to see how they could be developed to a level that is pervasive and yet precise and detailed enough to tell designers what to do in particular situations.

As a reader of the literature on human factors and cognitive engineering, you are not surprised that there has been serious criticism of the field for its lack of relevance to practitioners in systems design (Bannon, 1991).

The crucial difficulty in this approach is the idea that problem-solving activity is based on established goals, which is only the case in very simple actions, and the evidence referred to in the literature is usually elementary, one-person goal-seeking situations.

In an elusive introduction to cognitive engineering, Norman (1986) describes, as an instance of problem solving, a bather wanting to control the rate of total water flow and temperature. The problem-solving activity involved is surprisingly complex, and for most of us finding the right temperature in a trial-and-error process drags on, and we would welcome a device that could help us finding a satisfying temperature. The implicit assumption is that the results of studying a bather can be used in analyzing complex manufacturing situation, but this is only the situation if the goals are well defined and well described, and in manufacturing this is an assumption we cannot make.

Recent research in cognitive engineering realize the difficulties in defining goals and also that this complicates research considerably. Two new studies in cognitive task analysis are very suitable for illustration of these difficulties.

In problem solving we are all simultaneously a little butter-fingered and experts in what we are doing, and to understand what we do, and how we can use technology in our activities, Roth and Woods (1989) suggest an approach they call *cognitive task analysis*. During this analysis the cognitive engineer is

> attempting to define what makes the problem domain hard, what errors domain practitioners typically make and how an intelligent machine can be used to reduce those errors or performance bottlenecks. In this stage domain experts are not the sole subject of focus. Understanding the knowledge and strategies currently employed by average and low skilled practitioners is equally important. Understanding the limitations of the knowledge (e.g., incomplete or inaccurate *mental models of the domain*) and strategies (e.g., failures to consider multiple *interacting goals* in plan formulation) across the range of domain practitioners for whom the intelligent advisor is intended is a prerequisite for defining how machine intelligence can be deployed to enhance their performance. (Roth and Woods, 1989, p. 246, italics added)

In this elaborated description of cognitive engineering we do not just have goals or interacting goals, we also have practitioners' mental models of the domain. In his study in cognitive work analysis Rasmussen (1985) amplified the point by arguing that cognitive engineering cannot be useful for designing or planning without: "An intimate knowledge of the decision strategies applied by the operators, of the *subjective goals* adopted, and of the *mental representation* of plant properties used for the task." (p. 6)

All the simplicity of the early problem-solving studies is now gone. Our ability to design advanced manufacturing technology to support complex human work is dependent on the possibility of analyzing people's subjective goals and mental representation. Rasmussen has described what is involved in this kind of analysis and also the fundamental concepts of the discipline. Rasmussen and his colleagues are working on developing a *conceptual framework* for modeling work (Rasmussen, 1990). The aim is to find a conceptual framework for strategic planning of integrated information systems, and the ambition is to create a *rigorous and function oriented framework*, and *si-*

multaneously a framework that reflects the *subjective interpretation of the work environment* by the agents involved.

Rasmussen (1990) is not trying to establish a general model for manufacturing, but confines himself to work situations. The purpose of the project is to work out descriptions of tasks, activities, work domains, and so forth for developing a model framework, a framework for a description that can serve to compare results from analysis made in different contexts and domains, and that again can serve to *predict what kind of phenomena are to be expected in one work situation*, given the results from studies in other environments.

The representation of the work domain is based on interdisciplinary and ethnological studies of the actual work environment, and the project intends to combine several sciences spanning from engineering to basic human sciences. Concepts from these sciences are used to formulate several levels of analysis bridging from the technical description of work domains to the subjective values of the agents.

The design approach is based on the assumption that a trustworthy prediction of what could happen in a work situation is achievable by means of *two concurrent analyses* of work and work performances:

1. Identification of the functions or activities that an actor is faced with.
2. Identification of the role and characteristics of the individual agent.

By taking a firm *design approach* in constructing a methodology for development advanced manufacturing systems, Rasmussen (1990) has to *predict* the kind of *behavior of individuals and organizations* to be expected in response to changes in work conditions, such as those caused by introduction of new information. To do this, that is, to predict why a particular piece of behavior is chosen instead of another possible pattern, Rasmussen outlines a complex research project aiming at understanding how action alternatives in a particular situation is *eliminated* such that one unique sequence of behavior can manifest itself.

The basis of this project is empirical research looking for *constraints* and *subjective performance criteria*. The problem, as acknowledged by the project group, is that constraints are not all active at the time of behavior. History plays a role. It is necessary to identify the hidden constraints to predict and understand behavior. Rasmussen (1990) is trying to identify these hidden constraints by interfering from the work requirements, by analyzing the resource profile of the actor, and their subjective performance criteria.

This cannot simply be done by interviewing workers because of what Rasmussen calls the "operator logic." In stable work environments, a kind of operator logic, or myth about the goals and reasons, can evolve and replace the knowledge underlying the system rationale. Therefore, to base the design of decision support systems on operator logic will be a mistake. Reliable information must be based on analysis of the actual functionality of the domain. The analysis, therefore, has to include formulation of the intentions and reasons behind work domain structure, and also include inferences from field studies (Rasmussen, 1990; Rasmussen, et al., 1990).

Rasmussen (1990) is trying to use cognitive engineering to establish an analytical basis for prediction of behavior in work situations, and the result seems to be an analysis of the most extreme complexity. It is difficult to use such an analysis in design of advanced manufacturing systems because the analysis phase will be extremely long and yet never complete, even though it may yield valuable insights.

Human factors studies and cognitive engineering presuppose, as a basis for discussion of how people solve problems, a well-specified identification of the functions or activities with which an actor is faced. The advanced studies discussed above realize that problem solving is an individual activity dependent on culture and history not even known to the actor, and that our understanding of this aspect is essential for the development of manufacturing systems. It is, however, still solving well-defined and described problems, and the goals are basic as defined by an analysis of the functionality of the system. In that sense human factors studies and cognitive engineering, as we have seen them up to now, are children of Taylorism and scientific management, and their application in design of AMT systems depend on unambiguous descriptions of the functionality of manufacturing. The difficulty for this tradition is not so much whether it is useful to study human capacities and problem-solving activities; the problem is whether it is possible to combine these studies with a study of manufacturing function and use the result *to predict behavior of individuals and organizations as a basis of design* of a particular manufacturing system.

Gerson and Star (1986) have given expression to the problem for cognitive engineering as a basis for design in stating that:

> Every real-world system is an open system: It is impossible, both in practice and in theory, to anticipate and provide for every contingency which might arise in carrying out a series of tasks. No formal description of a system (or plan for its work) can thus be complete. Moreover, there is no way of guaranteeing that some contingency arising in the world will not be inconsistent with a formal description or plan for the system. (p. 266)

Gerson and Star (1986) conclude that the system for workplaces cannot be designed ahead of time without knowing the empirical workplace situation, and any representation of the workplace produced in order to help designers. The systems, therefore "must be *designed* and maintained *while they are in use* (p. 268, italics added).

In a similar connotation Suchman (1987) uses the term *situated actions* in her studies of problems of human–machine communication:

> Plans are resources for situated action, but do not in any strong sense determine its course. While plans presuppose the embodied practices and changing circumstances of situated action, the efficiency of plans as representatives comes

> precisely from the fact, that they do *not* represent those practices and circumstances in all of their concrete detail. (Suchman, 1987, p. 52)

What is at stake here is the idea that an objective reality can be usefully captured in a model and subsequently used as a sufficient basis on which to develop a computerized system. There is, as Bannon (1991) says, always an *interpretation* in the conduct of work, and therefore "development of computer-based applications requires the collaboration or involvement of a variety of distinct communities—workers with different skills, analysts, developers, programmers etc." (p. 40)

Winograd and Flores (1986) state that development of any computer-based system will have to proceed in a cycle from the design to experience, and back again. It is impossible to anticipate all of the relevant breakdowns and their domains. They emerge in practice, and the design is no longer an activity that creates a complete mental image or representation of a system before it is implemented. The resulting AMT systems are designed for supporting situated action, and not for controlling problem-solving activities.

10.3 DUE WORK PROCESS AND DEVELOPMENT OF COMPUTER ARTIFACTS

The tradition of designing computer artifact for human actors in situated actions is well know in Scandinavia. The classic Scandinavian research in participative design has created a strong basis for studies in and experimentation with user involvement in system development, or cooperative design as it is now called. The word *cooperative* refers to the user involvement but also to the interdisciplinary character of the research and the design process. The UTOPIA project is undoubtedly the best known Scandinavian research project from the eighties on system development and working life. It was a "project on trade union based development of, and training in, computer technology and work organization, especially text and image processing in the graphic industries" (Bødker et al., 1987). The research and development team consisted of graphical workers, social scientists, and computer scientists. The project, therefore, had to consider problems of interdisciplinary research as well as problems of bringing together of users and designers. In dealing with both these problems the UTOPIA project was blazing new trails. There was no tradition for interdisciplinary research, and design of systems for skilled work is not easily done, an essential reason being the existing tools and techniques of systems development. One of the Scandinavian pioneers in this field, M. Kyng, (1990) states that "in the quest to make workplace skill a more central aspect of computer system design, these projects ran into severe difficulties in trying to apply the tools and techniques of traditional system development. . . . In particular, they pointed out how the notion of tacit skills—those essential, yet not easily articulated qualities that we use in daily work—was difficult for computer system designers to grasp using formal system specifications." (p. 587)

These observations lead to an intensive research in methods and tools of *cooperative design*. Because designers and users in most cases have a completely different background, neither designer nor user groups can fully understand each other's practices or meanings. The *skilled work approach* therefore needs methods and tools to build a bridge that brings these experiences fully together. *Designing by doing* became a keyword in this striving toward integration of knowledge. The idea was to use mockups or rapid prototypes to simulate different kinds of computer-based tools and the surrounding organization, thus making it possible for the workers in the project to develop requirements and wishes on a concrete level by actually carrying out the work on simulation equipment.

The users were allowed to express their know-how in action, and the researchers could then contribute to the design process using their knowledge. Continuity of skilled work procedures is essential for this approach, and designing by doing is a work method suited for meeting the preconditions of the users. In the Scandinavian tradition the basic elements in this approach to system design are called *practice* and *family resemblance* (Greenbaum and Kyng, 1990). The word *practice* is used to insist on the *current* user practice as the starting point of systems design, albeit the design of a computer application and its use in a work setting will change the work practice. The word *family resemblance* is also used to stress that design should be based on situations having resemblance to prior work experiences of both the users and the designers.

The central issue in the UTOPIA project was to initiate a dialogue between employees, technicians, social scientists, and companies. It was a try-and-do-it kind of project, and the theoretical reflections came afterward (Ehn, 1988). One theoretical outcome was an amendment of the activity theory in research and development of information systems. This approach differs from the cognitive engineering activity theory by not accepting the key point that one can separate out the actual implementation of the system from its functional description. The emphasis on active subjects delineates the activity theory approach from traditional information systems, where predetermination of work sequences by the system is the normal case. On the other hand, the individual is not just acting in a vacuum, and the approach may suffer from a dichotomy between the individual and the social if these two aspects are not adequately mediated in the theory. The problem of *context*, which has become more and more recognized as a crucial issue for useful theory and empirical work, is therefore built into the very basis of the theory in terms of activities. As expressed by Kuutti (1991):

> The solution offered by Activity Theory is that there is a need for an intermediate concept—a minimal meaningful context for individual actions—which must form the basis unit of analysis. this unit . . . is called an activity. Because the

context is included in the unit of analysis, the object of our research is always essentially collective, even if our main interest lies in individual actions. (p. 254)

The activity approach is working at defining activity as a "fundamental type" of context. Context is essential because it is meaningless to study human qualities using a smaller object of research because "without that basic context one cannot grasp the essence of the phenomenon" (Kuutti, 1991, p. 255). The activities in which humans participate are seen as the basic unit of development and human life, and thus a foundation of the study of all contextuality. An activity, then, has the following properties (Kuutti):

- An activity has a *material object*, and activities can be distinguished according to their objects. The transformation of the object toward some desired state or in some direction motivated the existence of an activity.
- An activity is a *collective phenomenon*.
- An activity has an active *subject*, who understands the motive of the activity. This subject can be individual or collective. Not all participants involved in an activity necessarily understand the motive of an activity in which they are participating or even recognize the existence of it. In this case they are not active subjects of the activity.
- An activity exists in a *material environment* and *transforms* it.
- An activity is a *historically developing* phenomenon.
- Contradictions are the force behind the development of an activity.
- An activity is realized through conscious and purposeful *actions* by participants.
- The relationships within an activity are culturally mediated.

By taking this activity approach, a seminal change in the life cycle of AMT design is put on the agenda. The methodological cycle of design is moving from an analysis of the activity and the surrounding activities to the creation of instruments by which the users can transcend their own practice, thus creating a vision of the change, to an implementation of the final new instrument into the organization. The idea here is that inasmuch as we try to predict how praxis will change, artifacts are used differently from the original intentions, and new contradictions are always introduced, causing the need for new artifacts to arise. Methods that can be used are, for example, future workshops, reframing techniques, and prototypes (Bødker, 1990).

The role of the designer of advanced tools will consequently be widened considerably. Design will include activities that are not normally conceived of as part of the design process, for example, initial investigation of work activity and its historical development, following up and consolidation of the changed work activity. Moreover design involves a collective learning process among users and designers, and a change of the users praxis need to be dealt with at all levels, including both technical, educational, and work organizational conditions of the future activity.

In cases where introduction of AMT is the reason for changes in work praxis the designer must take the role of a facilitator, who makes sure that the process gets going in different ways. The systems designers initially get to choose the instruments to be applied, even though these may later be changes in the collective process with the users. In the wording of Bødker (1990, p. 226):

> The system designers get to set the stages whereas later moves must be done in cooperation with the users, and instruments must, in this sense be chosen depending on the material from the characterization of the use situation. Some of these tools may be traditional description techniques, or the like, and some may be some of the "new ones": mapping. . . , future workshops, role playing, mock-ups, prototyping. . . , setting up of limited test situations.

The UTOPIA project used some of these tools in the development of a computer application for text and image processing in graphic industries. Some very useful tools were developed in cooperation with typographers and journalists, but the project did not focus on the design of an integrated system for the newspaper industry, although in the beginning this was being considered. There are several reasons for not being able to take this momentous aspect of a new integration of newspaper production, and the majority of these have to do with politics and tradition. I think, however, that it also indicates the limitations of techniques like role playing, prototyping, and so forth: They are suitable, first of all, for design with small, well-defined teams.

All the good examples of activity approach tell of collective design with a fairly small group of people doing a job in close cooperation, albeit with some division of labor. A recent study in design by playing by Ehn and Sjögren (1990) can be seen as an instance of this tendency. The study describes a development process with 25 employees at the educational department of the national Swedish board for consumer policies. The employees were secretaries, case handlers, a technical support person, a marketing person, and the head of the department. The activity approach has no doubt proved to be useful for design of computer tools for teams like this one, but current changes in manufacturing praxis are not restricted to new work praxis in well-established teams, and design of advanced manufacturing systems are really wrapped up in problems of a new integration of complete manufacturing processes. What is missing in the activity approach in its present form is the involvement of the world outside the team. The context cannot be delimited to a small group of people with a relatively fixed definition of reality.

Companies implement AMT systems to obtain better productivity, flexibility, and reduce lead time. In doing this, new multiskilled groups may come into play, and interaction between groups may play a crucial role for the success of the system. This development results in innovations that cannot be dealt with in well-established groups. Or phrased differently: Delineation of the context of activities becomes a very difficult issue, dealing with new forms of cooperation, skills, and division of labor.

10.4 COOPERATIVE WORK IN ADVANCED INTEGRATED MANUFACTURING SYSTEMS

In a study of the use of computer systems in the German machine-producing industry, Schultz-Wild et al. (1989) characterize the individual computer systems as "*CIM components*." They use this expression because firms having some experience with individual computer systems now begin to experiment with the development of computer networks and shared databases. Only a small fraction of companies state that they now actively employ computer networks, but almost half of the companies plan to do so in the near future.

To make sure: Computer-integrated manufacturing (CIM) networks, covering all manufacturing activities, stand outside the strategies of the manufacturing industry, and the idea appears to be foremost an academic exercise. But even a moderate use of computer-controlled machines in a network results in changes in the areas of technology, work organization, and personnel policies.

Changes in computer-controlled technology caused by a new interfacing of machines covering various manufacturing processes bring about complex changes in the division of labor and in workers' use of machines. The overall work methods and procedures and skilled manufacturing work is undergoing a radical change. As a consequence, designing computer-based systems for cooperative work environments is to an extent like "writing in water." While the information system may be designed to match a current social structure of the labor process, the very change of technology induces a change in this social structure.

It is important, therefore, that the choice and development of systems allow a stepwise introduction of computer-aided methods, and that the evolution of a new kind of skilled work is systemized in the development of information and control technology, and in the design of work organization and personnel utilization, as well as in the training of workers. The organization should be seen as a learning institution, and one of its principal purposes should be the expansion of knowledge. When designing and introducing computer-based technology, we have to regard learning as the heart of productive activity. The behaviors that define learning and those that define being productive are one and the same (Zuboff, 1988).

In the Tayloristic design tradition the architecture of an integrated AMT systems is that of tightly coupled, sequential batch-processing system. Each step is described in detail and, therefore, easily controlled. The writing in water effect is repressed, but the price is a rigid system with limited learning capacities. A flexible or adaptable system needs more loosely coupled and decentralized processing structures, where the boundaries between different jobs are broadened. Group technology and autonomous island production are well-known examples of this kind of work organization. The horizontal integration of computer systems (e.g., PPS-CAM) and, especially the vertical integration [computer-aided design and manufacturing (CAD/CAM)], transcends, however, the boundaries of the production islands and points in the direction of *integration of design, planning*, and *production in the island*.

Product design and innovation are iterative activities. Strict division of labor, however, has turned iterations to a very difficult aspect of manufacturing. To put focus on skills and communication in the development of AMT leads to experimentation with a technical and organizational integration of the technical office and the shop floor. No single individual is able to overview all aspects of manufacturing and, therefore, a leap from the individual to the collective skill is always necessary. The utility of an AMT system using computer networks depends considerably on how this is done. To work on global systems includes development of a *new kind of collective skills and work* affecting design, planning, and shop floor production. (Havn, 1990)

In cooperative work there is a constant exchange of information. The actors inform each other about the state of the installation, requirements, decisions to be taken, or the most convenient moment to do so. This brings at least two things into perspective:

1. The importance of the way in which the information is presented, its accessibility, and its impact on the activity of the team.
2. The ability of the cooperating workers to create shared images of the work and its operations and, consequently, the individual members' ability to communicate and understand each other.

It is the exchange of information that enables cooperative workers, and, more particularly, certain of its members to account for their responsibilities, to develop global or partial syntheses, and these syntheses are then confronted with the images of the relevant operative processes available at both an individual and collective level. The actors do not have identical qualifications, but they must be able to create images of the processes going on, based on individual knowledge—but never limited to individual knowledge. The knowledge of the actors are always interdependent, but never identical.

The coordination of activities presupposes at least a compatible basis of orientation for the people involved. Integrated manufacturing work entails development of collective skills, a kind of group-oriented learning process for people with different backgrounds. On the historical background of a strong division of labor between design, planning, and shop floor production, the crucial problem for development of integrated AMT systems is to create *a shared basis of orientation* for the three groups.

10.5 DESIGN OF AMT SYSTEMS—WITH USERS

Development of integrated AMT systems entails innovations in work organization and division of labor, and thus *transcends the empirical knowledge of the people involved*. For skilled work AMT systems necessitate new individual skills and new forms of cooperation among workers at all levels in the factory. So even if innovations in integrated manufacturing build on and actively involve existing know-how within a multiskilled group of workers, it also reshapes the knowledge and communication structure in a way not known for the designers as well as the users. Methods and tools for analysis and description of the complete manufacturing process is thus necessary, not as the basis for design of the system but as an input to the user involving process.

Innovations in work organization and the division of labor are very sensitive domains for most workers, not least for skilled workers. The problem of power in system development is not restricted to the relations of top management and employees. Recent studies show, that

> opposition to various forms of participative redesign tends to come from parts of operational management, where there is growing awareness of the potential threat to the Taylorist hierarchy, coming from these new developments. (Dankbaar, et al., 1990, p. 127)

Involvement of users, therefore, is not a guarantee of successful work design, irrespective of how successful the cooperation itself goes off. Participative design depends on enough room for shared decision making. If most of the system choices are made in advance, user participation will only make people aware of their lack of real influence. But even if there is enough room the design alternatives and their consequences have to become clear and understandable—to the users and the designers. A new computer-supported work organization may disappoint the workers, and the outcome of the design process may result in collision of interests among the workers. People react, of course, to the concrete effects on the work situation. An attempt at designing participatively may, as den Hertog and Wielinga point out (den Hertog et al., 1990), stimulate workers to provide useful information about their work and suggestions for improvements. There may, however, be a bill to pay from the moment the system is in use, if the people involved realize that the designer did make use of the information but only for formulating solutions that, intentionally or not, imply a deterioration of the work.

System development process have to deal with power and interest within the organization, and have to deal with innovations that are not within the experiences of the actors in the design process. A three-day "future workshop" or "social simulation" may highlight some aspects of potential conflicts and innovation but cannot replace real-life experience, and the quality of the workshop depends on the representation of the complete system given in the workshop. So we are back then to the problem of representation or model.

We need models and representations in development of advanced manufacturing systems, but we don't have to take a traditional design approach, as Schmidt (1991) observes:

> The problem with incorporating of plans (established procedures, organizational structures, or conceptual schemes) in development of computer systems is not that plans are fictitious. Rather plans serve a heuristic function in action by identifying constraints, pitfalls, and strategic position in the fields of work. (p. 13)

Plans may serve different functions. As resources for competent development processes involving all relevant actors, they may convey information on the functional requirements to be met by the process and the product, and it may highlight decisional criteria of crucial importance and suggest strategies for dealing with specific type of problems. These plans are only available if we use all convenient methods and tools for establishing plans. These tools may be developed for traditional design purposes; that is not the problem. What we should concentrate on is to use the available tools and methods to create models accessible to actors and support actors (designers and users) in applying and adapting the model to the situation at hand.

REFERENCES

Bannon, L. (1990). A pilgrim's progress: From cognitive science to cooperative design, *AI & Society*, **4**, 259–275.

Bannon, L. (1991). From human factors to human actors: The role of psychology and human-computer interaction in system design, in *Design at Work*, J. Greenbaum and M. Kyng, Eds., Lawewnce Erlbaum, Hillsdale, N.J.

Bengtsson, L., and Berggren, C., Eds. (1991). *90-talets industriarbete i Sverige* (Industrial work in Sweden in the '90), Carlsson Bokförlag, Stockholm.

Bødker, S. (1990). Activity theory as a challenge to systems design, in *Proceedings of the IFIP TC 8 WG 8.2 Working Conference on the Information Systems Research Arena of the 90'S*, H-E. Nissen, H. K. Klein, and R. Hirschheim, Eds., Copenhagen, Denmark, December 14–16.

Bødker, S., Ehn, P., Kyng, M., Kammersgaard, J., and Sundblad, Y. (1987). A Utopian experience. On design of powerful computer-based tools for skilled graphic workers, in *Computers and Democracy*, G. Bjerknes, P. Ehn, M. Kyng, Eds., Gower, Aldershot, UK.

Dankbaar, B., and den Hertog, F. (1990). Labour process analysis and socio-technical design: living apart together? in *New Technology, Work and Employment*, **5**, 2.

Ehn, P. (1988). *Work-Oriented Design of Computer Artifacts*, Arbetslivcentrum, Stockholm.

Ehn, P., and Sjögren, D. (1990). From system descriptions to scripts for action, in *Design at Work*, J. Greenbaum and Kyng, M., Eds., Lawrence Erlbaum, Hillsdale, N.J.

Gasse, L., and Majchrzak, A. (1990). Development of a knowledge-based system prototype for predicting sociotechnical integration of FMSs, in *Ergonomics of Hybrid Automated Systems II. Proceedings of the Second International Conference on Human Aspects of Advanced Manufacturing and Hybrid Automation*, W. Karwowski and M. Rahimi, Eds., Elsevier, Amsterdam.

Gerson, E., and Star, L. (1986). Analyzing due process in the workplace, *ACM Transaction on Office Information Systems*, **4**, 3.

Havn, E. (1989). CIM and integration of work, in *Computer Integrated Manufacturing*, Halatsis and J. Torres, Eds., IFS Publications, UK/Springer-Verlag, Berlin.

Havn, E. (1990). Work organization, integrated manufacturing, and CIM system design, in *Computer-Integrated Manufacturing Systems*, (3)4 Butterworths, London.

Havn, E. (1991). In search of new structures: Issues in designing integrated manufacturing systems, in *International Journal of Human Factors in Manufacturing*, **1**, 4.

Kuutti, K. (1991). The concept of activity as a basic unit of analysis for CSCW research, in *Proceedings of the Second European Conference on Computer-Supported Cooperative Work*, Kluwer Academic Publishers, Dordrecht.

Kyng, M. (1990). Cooperative design: Bringing together the practices of users and designers, in *Proceedings of the IFIP TC 8 WG 8.2 Working Conference on the Information Systems Research Arena of the 90'S*, H-E. Nissen, H. K. Klein, and R. Hirschheim, Eds., Copenhagen, Denmark, December 14–16.

Norman, D. A. (1986). Cognitive engineering, in *User Centered System Design*, D. A. Norman and S. W. Draper, Eds., Lawrence Erlbaum, Hillsdale, N.J.

Norman, D. A. (1990). Viewpoint. *Communications of the ACM*, **33**(1), 4–7.

Parnas, D., and Clements, P. (1984). A rational design process: How and why to fake it, in *Formal Methods and Software Development*, H. Ehrig, Ed., Springer-Verlag, Berlin.

Rasmussen, J. (1985). A Framework for Cognitive Task Analysis in Systems Design, Invited tutorial presented at NATO Advanced Study Institute on Intelligent Decision Aids in Process Environments, September 1985, Pisa, Italy, p. 6. Cited from K. Schmidt and P. Carstensen, 1990. *Work Analysis. Theory and Praxis* (in danish), Risø, Roskilde.

Rasmussen, J. (1990). Use of field studies for design of work stations for integrated manufacturing systems, in *Proceedings of the International Conference on Human Factors in Design for Manufacturability and Process Planning*, Honolulu, HI.

Rasmussen, J., Pejtersen, A., and K. Schmidt (1990). *Taxonomy for Cognitive Work Analysis*. Riso National Laboratory, Roskilde (Esprit Basic Research, 3105: MOHAWC).

Robinson, M., and Bannon, L. (1991). Questioning representations, in *Proceedings of the Second European Conference on Computer-Supported Cooperative Work*, Kluwer Academic Press, Dordrecht.

Roth, E. M., and Woods, D. D. (1989). Cognitive task analysis: An approach to knowledge acquisition for intelligent systems design, in *Topics in Expert System Design. Methodologies and Tools*, G. Guida and C. Tasso, Eds., North-Holland, Amsterdam.

Schmidt, K. (1990). Analyzing cooperative work in advanced manufacturing, in *Ergonomics of Hybrid Automated Systems II*, W. Karwowski and M. Rahimi, eds., Elsevier, Amsterdam.

Schmidt, K. (1991). Riding a tiger, or computer supported cooperative work, in *Proceedings of the Second European Conference on Computer-Supported Cooperative Work*, Kluwer Academic Press, Dordrecht.

Schultz-Wild, R., Nuber, C., Rehberg, F., Schmierl, K. (1989). *An der Schwelle zu CIM. Stragegien, Verbreitung, Auswirkungen*, RKW-Verlag, Cologne.

Shaiken, H. (1984). *Work Transformed. Automation and Labor in the Computer Age*, Holt, Rinehart and Winston, New York.

Suchman, L. A. (1987). *Plans and Situated Actions*, Cambridge University Press, New York.

van der Meer, F-B., and Roodink, T. (1991). The dynamics of automation: A structural constructionist approach, in *Information and the Public Sector*, Vol. 1, Elsevier, Amsterdam, pp. 121–141.

von Freyberg, T. (1989). *Industrielle rationalisierung in der Weimarer Republik*, Campus Verlag, Frankfurt/New York.

Willcocks, L., and Manson, D. (1987). *Computerising Work*, Paradigm, Slough, UK.

Winograd, T., and Flores, F. (1986). *Understanding Computers and Cognition. A New Foundation for Design*, Ablex Publishing.

Wynn, E. (1991). Taking practice seriously, in *Design at Work*, J. Greenbaum and M. Kyng, Eds., Lawrence Erlbaum.

Zuboff, S. (1988). *In the Age of the Smart Machine*, Heinemann, Oxford, UK.

CHAPTER 11

PROJECT MANAGEMENT IN COMPUTER-INTEGRATED MANUFACTURING IMPLEMENTATION

A. B. BADIRU and R. E. SCHLEGEL
School of Industrial Engineering
University of Oklahoma
Norman, Oklahoma

11.1 INTRODUCTION

New manufacturing technologies have captured more attention in recent years. This is due to the high rate at which new productivity improvement technologies are being developed. The fast pace of the introduction of new technologies has created difficult implementation and management problems for many companies. New technology can be successfully implemented only if it is viewed as a system whose various components must be evaluated within an integrated managerial framework. Such a framework can be achieved through project management techniques. Thomas Murrin was quoted (*Industrial Engineering*, 1989) to have stressed the need to expedite the implementation of the advances in manufacturing technology. Such expedited implementation can be facilitated by a project management approach. The depressed state of technology implementation was addressed by a board of

Organization and Management of Advanced Manufacturing. Edited by Waldemar Karwowski and Gavriel Salvendy.
ISBN 0-471-55508-8

industrialists working under the auspices of the National Research Council (1984). The group stated:

> The most critical problem faced by too many U.S. industrial executives today is the steady decline in their companies' competitiveness at home and abroad, and the resultant loss in market share. A major reason for the decline has been the gradual emergence of a technology gap in manufacturing. It has not been a single identifiable event, but a slow erosion of the technological foundation of manufacturing. The keys to regaining competitiveness in most U.S. manufacturing industries are quality, productivity, and responsiveness in bringing new products to the market. A primary technology for attaining these attributes, across industries, is computer-integrated manufacturing. [p.1]

Due to the widely published concerns about computer-integrated manufacturing (CIM) technologies, it is imperative that a concerted effort be committed to implementation issues. This must be done both at the national and organizational levels. The CIM leadership role that the government of Switzerland has demonstrated (Rodriguez, 1991) is a good example for other nations and organizations. Weston (1991, p. 18) defines CIM as "physical, systems, organizational and management activity related to the design, production and distribution of goods produced by a manufacturing company." This definition captures the basic components of CIM as outlined by Aletan (1991). The management aspect of implementing CIM is an essential ingredient for success. The need to improve CIM management has been addressed by several authors (Avishai, 1989; King, 1991; Weston, 1989). This chapter presents a project management approach to improving CIM adoption and implementation. The behavioral (human factors) aspects of managing CIM projects are addressed. The triple C model of communication, cooperation, and coordination is recommended as an effective tool for ensuring the acceptance of new manufacturing technology. Computer-integrated manufacturing justification is also addressed as a function within the project management effort.

11.2 PROJECT MANAGEMENT APPROACH TO CIM IMPLEMENTATION

Project management is the process of managing, allocating, and timing resources to achieve a given goal in an efficient and expedient manner (Badiru, 1991). The objectives that constitute the specified goal may be defined in terms of time, costs, or technical results. A project may be quite simple or very complex. An example of a simple project is painting a vacant room. An example of a complex project is launching a space shuttle. Project management techniques are used widely in many enterprises, including construction, banking, marketing, health care services, transportation, and public service. The use of manufacturing, however, has been limited (Badiru, 1988). The adoption and implementation of CIM technologies constitute a suitable area for the use of project management techniques. The techniques of project management can help accomplish goals relating to better product quality, improved resource utilization, better customer relations, higher productivity, and fulfillment of due dates. These can generally be expressed in terms of the following project constraints:

- Performance specifications
- Schedule requirements
- Cost limitations

We define leadership as a project management function dealing with the *ability to integrate organizational requirements, personnel needs, and business limitations to generate decisions that create a conducive environment for achieving goals and objectives.* Project management can provide the leadership needed to successfully implement CIM. Doing the right thing the right way at the right time is a goal common to CIM and project management. Project management can address the managerial, technical, business, and human factors aspects of implementing CIM.

Steps of Project Management

Figure 11.1 presents the major steps in project management. The steps are typically performed strategically in accordance with the specified project goal. The steps are discussed briefly below.

Problem Identification This is the stage where a need for a proposed technology is identified, defined, and delineated. For example, a prevailing low level of productivity may trigger a need for a new CIM technology.

Project Definition This is the phase at which the purpose of the project is defined and clarified. A *mission statement* is the major output of this stage. The mission statement should specify what should be achieved in terms of cost, time, or performance.

Project Planning Project planning determines the nature of actions and responsibilities needed to achieve the project goal. It entails the development of alternate courses of action and the selection of the best action to achieve objectives making up the goal. Planning determines what needs to be done, by whom, for whom, and when. Planning should be done for both long-range strategic purposes (Cleland, 1990) as well as short-range operational purposes.

Organizing Project organization is a crucial component of project planning. Organizing a project involves determining the interfaces needed among personnel in order to assure a dedicated pursuit of the project goal. It involves

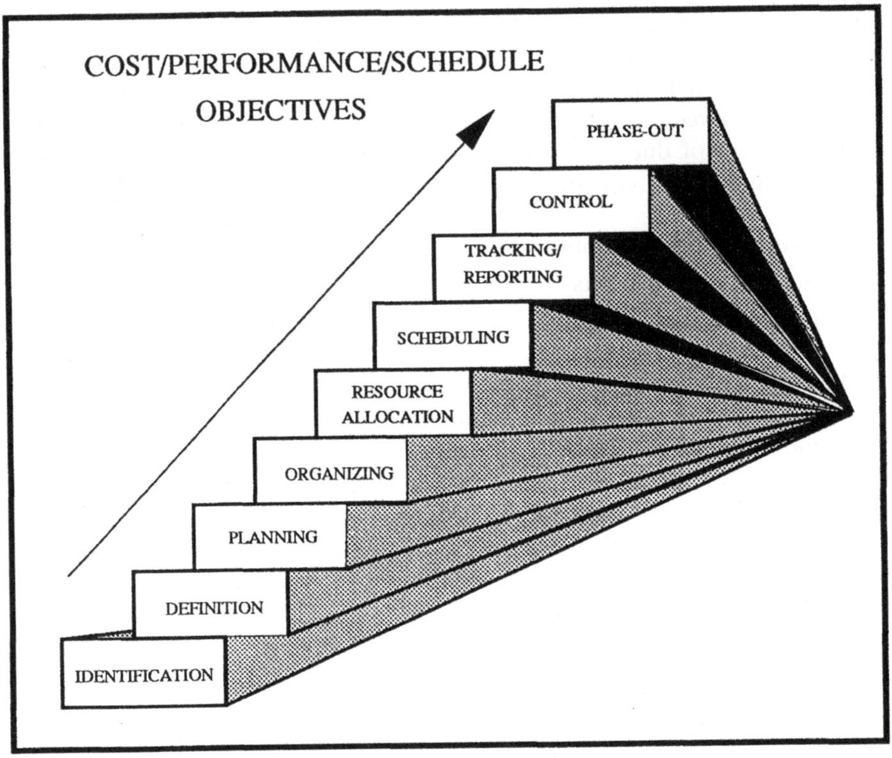

Fig. 11.1. CIM project management steps.

the determination of what activities should be performed in what functional areas and under what supervisory hierarchy. Staffing is a component of organizing. Staffing is the process of defining the required personnel and the assignment of the personnel to appropriate functions with respect to the project goal. The staffing process may include hiring new people, reassignment of existing personnel, training or retraining, and realignment of responsibilities.

Resource Allocation Project goals and objectives are accomplished by applying resources to functional requirements. Resources, in the context of project mangement, are generally made up of people and equipment, which are typically in short supply. The people needed for a particular task may be committed to other ongoing projects. A crucial piece of equipment may be under the control of an uncooperative department.

Project Scheduling Project scheduling is a major function in project management. The project schedule shows the timing of the efforts and resources committed to the project. A master schedule should be developed and made known to all project personnel. Scheduling involves the following functions:

TABLE 11.1 Format for Resource Availability Database

Resource ID	Brief Description	Special Skills	When Available	Duration of Availability	Number Available
Type 1	Technicians	Wiring	1/1/93	Two Months	20
Type 2	Programmers	FORTRAN	6/25/95	Indefinite	5
Type 3	Engineers	Design	Current	Five Years	35
•	•	•	•	•	•
•	•	•	•	•	•
•	•	•	•	•	•
•	•	•	•	•	•
Type *n*	Tech Aide	Manuals	9/2/93	Six Months	1

- *Basic Scheduling* The basic scheduling computations are performed by using forward-pass and backward-pass computational procedures. These computations give the earliest and latest start and finish times for each activity. Schedule management often involves performing time–cost trade-off analysis. The scheduling procedure makes use of tools such as Critical Path Method (CPM), Program Evaluation and Review Technique (PERT), and Gantt charts (Wiest and Levy, 1977; Moder et al., 1983).
- *Resource-based Scheduling* Constrained resource allocation refers to the process of allocating limited resources to competing activities in a project. Scheduling heuristics or mathematical allocation models are used in allocating the limited resources during the scheduling phase. A database of resource availability, similar to the one presented in Table 11.1, should be developed for each CIM project prior to starting the scheduling process.
- *Resource Leveling* Resource leveling refers to the process of scheduling activities in a way to reduce the period-to-period fluctuations in resource requirements.

Tracking and Reporting This phase involves the diagnostic process of checking whether or not project results conform to plans and specifications. Tracking and reporting are essential for the control phase of a project.

Control In this function, necessary actions are taken to correct unacceptable deviations from expected performance. Control is effected by measurement, evaluation, and corrective action. Measurement is the process of measuring the relationship between planned performance and actual performance with respect to project objectives. The variable to be measured, the measurement scale, and the measuring approach should not be clearly specified during the planning stage.

Project Termination This is the phase-out stage of a project. The termination of a project should be implemented forthrightly. A project should not be allowed to drag on needlessly after the expected completion time. A terminal activity should be defined for a project during the project planning phase. An example of a terminal activity may be the submission of a final report, the "power-on" of new equipment or the signing of a release order. The conclusion of such an activity should be viewed as the completion of the project. Follow-up activities may be necessary in certain cases, but those should constitute a different function outside the project itself. If a project is not terminated when appropriate, the motivation for it will wane and subsequent activities may become counterproductive. This is particularly true for technology-based projects where the "fear of the unknown" and "resistance to change" are already major obstacles.

Obviously, CIM project management has some unique characteristics that must be addressed when implementing the project steps presented. For this reason we devote the next two sections to two aspects of CIM project management: *CIM technology planning* and *CIM technology control.* These two aspects fall, respectively, under the general categories of planning and control in Fig. 11.1.

CIM Technology Planning Computer-integrated management planning is the process of establishing courses of action within the prevailing operating environment to achieve established goals. Thus CIM planning is needed to:

- Minimize the effects of uncertainties
- Clarify technology goals and objectives
- Provide a basis for evaluating project progress
- Establish measures of technology performance
- Determine required personnel responsibilities

In addition to the usual contents of general project planning, the following items are essential when planning CIM projects.

Overview and Assessment This specifies the goals and scope of new CIM technology as well as its relevance to the organization's mission. The major milestones, with a description of the significance of each, should be documented. A specification of the operating characteristics of the technology should be developed. Questions should include: What inputs will be required by the manufacturing technology? What outputs are expected? What is the scope of the manufacturing technology implementation? How will its performance be measured and evaluated? Is the operating environment suitable for the configuration of the technology hardware? What maintenance is needed and how will the maintenance be performed? What infrastructure is required to support the manufacturing technology?

There is a multitude of new technologies that have emerged in recent years.

Technologies such as group technology, cellular manufacturing, and artificial intelligence have received much attention in the literature. However, much more remains to be done in actual implementation. It is important to consider the peculiar characteristics of a new CIM technology before establishing adoption and implementation strategies. The justification for the adoption of a new technology is usually a combination of several factors rather than a single characteristic of the technology (Badiru et al., 1991, Falkner and Benhajla, 1990). The important benefits of CIM technology may include productivity improvement, improved product quality, production cost savings, flexibility, reliability, and safety. An integrated evaluation must be performed to ensure that CIM technology is justified both economically and technically. The scope and goals of the proposed technology must be established from the beginning of the project. This entails the comparison of departmental objectives with overall organizational goals in the following areas:

1. *Market Position*: This should cover items such as market cost of proposed product, assessment of competition, and market share.
2. *Growth Potential*: This should address short-range expectations, long-range expectations, future competitiveness, future capability, and prevailing size and strength of the competition.
3. *Contributions to Customers*: Any prospective technology must be evaluated in terms of direct and indirect benefits provided to the customer. These may include product price versus value, product quality, product reliability, and customer service.
4. *Profitability*: An analysis of how the technology will contribute to profitability should consider past performance of the technology, incremental benefits of the new technology versus conventional technology, and value added by the new technology.
5. *Cash Flow and Cost Analysis*: The conventional economic analysis with proper modifications should still play a significant role in the technology assessment process. This may cover a comprehensive analysis of fixed and sunk costs, cost of obsolescence, maintenance requirements, recurring costs, installation cost, space requirement cost, capital investment requirements, return on investment, tax implications, cost of capital, and other concurrent projects.
6. *Resource Utilization*: The utilization of resources (staffing and equipment) in the pre-technology and post-technology phases should be assessed. This may be based on material input–output flows, high value of equipment versus productivity improvement, required inputs for the technology, expected output of the technology, and utilization of technical and nontechnical personnel.
7. *Risk Exposure*: Uncertainty is a reality in technology adoption efforts. Uncertainty will need to be assessed for the initial investment, return on investment, payback period, and volatility of the technology.

8. *Effectiveness of Product*: The product of the new technology should be evaluated for functionality, durability, and aesthetics.
9. *Organizational Productivity*: An analysis of how the technology contributes to organizational productivity may be verified by studying the shop throughput, delivery times, scrap rate, personnel idle time, learning rate, and design-to-production cycle.

CIM Technology Goal This consists of a detailed description of the overall goal of the proposed technology. A goal is the combination of a series of objectives. Each objective should be detailed with respect to its impact on the project goal. The major actions that will be taken to ensure the achievement of the objectives will also need to be identified. The overall long-range purposes of the technology should be defined. Its feasible useful life should be defined. A frequent problem with manufacturing technology is the extension of its useful life well beyond the time of obsolescence. If a technologically feasible life is defined during the planning stage, it will be less traumatic to replace the technology at the appropriate time.

CIM Technology Policy The technology policy refers to the general guideline for personnel actions and managerial decision-making relating to the adoption and implementation of manufacturing technology. The project policy indicates how the project plan will be executed. The chain of command and the network of information flow are governed by the established policy for the project. A lack of policy creates a fertile ground for incoherence in technology implementation due to conflicting interpretations of the project plan.

CIM Technology Procedures Technology procedures are the detailed methods of complying with established technology policies. A policy, for example, may stipulate that CIM must be justified before approaching management for funding. A procedure will then outline the specific approaches for the required justification (e.g., economical, technical, etc.).

CIM Technology Resources The resources (staffing and equipment) required for the adoption and implementation of CIM technology should be defined. Currently available technology resources should be identified along with future resource needs. The timeframe of availability of each resource should also be specified. Issues such as personnel recruiting and technical training should be addressed early in the project.

CIM Technology Budget Technology cannot be acquired without an adequate budget. Some of the cost aspects that will influence technology budgeting are first cost, fixed cost, operating cost, maintenance cost, direct/indirect costs, overhead cost, and salvage value. The bottom-up approach to developing a technology budget is recommended for CIM.

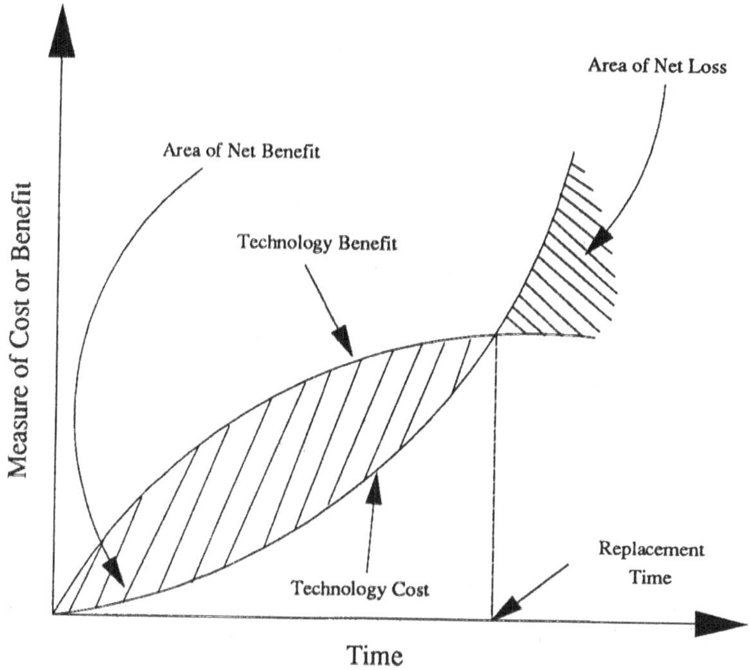

Fig. 11.2. CIM benefit versus cost (adapted from Badiru, 1991).

Technology Performance Measures Performance standards should be established for any new CIM technology. The standards constitute the criteria for evaluating the progress of the implementation process. In addition, the method by which the performance will be analyzed should also be defined to avoid ambiguities in tracking and reporting. The bottom line in any technology implementation is composed of profit, benefit, and/or performance. An analysis of the expenditure required for implementing CIM versus its benefits should be conducted to ascertain that it is worthwhile. Can an existing technology satisfy the needs more economically? Even if future needs dictate the acquisition of the technology, economic decisions should still consider prevailing circumstances. Figure 11.2 (Badiru, 1991) shows an example of a plot useful for evaluating performance measures for CIM technology.

Potential CIM Technology Problems New manufacturing technologies are prone to new and unknown problems. Contingency plans must be established. Preparation must be made for unexpected problems such as technical failure, software bugs, personnel problems, technological changes, equipment failures, human errors, data deficiency, decision uncertainties, and so on.

CIM Project Control Time, cost, and performance form the basis for the operating characteristics of a project. These same factors help to determine

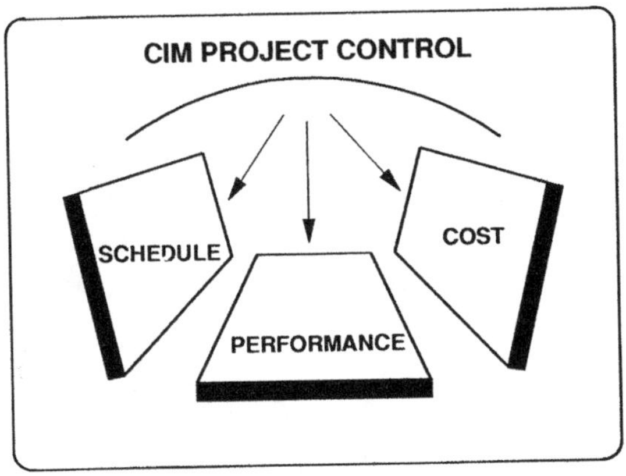

Fig. 11.3. Elements of CIM project control.

the basis for project control. Project control is the process of reducing the deviation between actual performance and planned performance. To be able to control a project, we must be able to measure performance. Measurements are taken on each of the three project constraints: time (schedule), performance, and cost. Figure 11.3 presents a model of the interaction between these control elements in a CIM project management environment.

The traditional procedures for measuring project progress, evaluating performance, and taking control actions are not adequate for technology management where events are more dynamic. Some of the causes of control problems in project schedule, performance, and cost are summarized as follows:

Schedule Problems	Performance Problems	Cost Problems
Delay of critical activities	Poor quality	Inadequate budget
Unreliable time estimates	Poor functionality	Effects of inflation
Technical problems	Maintenance problems	Poor cost reporting
Poor precedence relationships	Defective equipment	Increase in scope of work
Change of due dates	Lack of training	High overhead cost
Improper priorities	Lack of clear objectives	High labor cost

Schedule Control The Gantt charts developed in the scheduling phase of a project can serve as the benchmark for measuring project progress. Project status should be monitored frequently. A record should be maintained of the difference between task expectation and task realization. This information is then conveyed to the appropriate personnel. The more milestones or control points there are in the project, the better the monitoring function. The larger number allows for more frequent and distinct avenues for monitoring the schedule. Thus, problems can be identified and controlled before they become unmanageable. Some corrective actions that may be taken for schedule problems are:

- Expediting
- Eliminating marginal activities
- Combining related activities
- Revising milestones
- Adjusting time estimates

Performance Control Most project performance problems will not surface until after the project is completed. This makes performance control very difficult. However, every effort should be made to measure all the intermediate aspects of the project. After-the-fact measurements are generally not useful for project control. Some of the performance problems may be indicated by time and cost deviations. So, when schedule and cost problems exist, an analysis of how the problems may affect performance should be made. Since project performance requirements usually relate to the performance of the end products, controlling performance problems may necessitate altering project specifications.

Cost Control Numerous accounting and reporting systems have been developed over the years for cost monitoring and control. The level of interest in this topic is indicative of its importance in operations management. The process of controlling cost covers several key issues that must be addressed. These include:

- Better estimation of time, resources, and cost
- Clear communication of project requirements, constraints, and available resources
- Better coordination of project functions
- Timely reporting of project requirements: time, materials, labor, budget, and so forth.

Human Side of CIM Project Management

CIM requires synthesizing technical, business, quantitative, and qualitative considerations in the production environment. The role of the human is a significant aspect of achieving success with CIM. No matter how technically capable a technology is, the tangible and intangible contributions of personnel will play a major role in its successful adoption and implementation.

Operators are required to interact with CIM hardware. The design and installation of the hardware must consider the human interactions needed for successful operation. Both the behavioral and performance aspects of human interaction with technology should be evaluated. Even in highly automated installations, the human role remains a significant factor.

With proliferation of sophisticated CIM hardware, there is a danger of complacency in the design and installation processes. Such complacency must be overcome by incorporating human interaction consideration in the early stages of CIM planning. Issues involving worker safety, labor–management relations, job enrichment, and personnel training are important in CIM implementation.

There are several project management approaches that are particularly useful for addressing the human side of managing complex projects. The next two sections discuss two popular approaches: *the triple C model and the matrix organization*. These can be implemented within the framework of the project management steps presented earlier in Figure 11.1.

Using The Triple C Model for CIM Implementation

The successful implementation of CIM requires a coordinated approach that should utilize conventional project planning and control techniques as well as innovative management strategies. Intricate organizational and human factors considerations come into play in the implementation of today's complex technologies. The success of a project depends on good levels of communication, cooperation, and coordination. The triple C model (Badiru, 1987) facilitates a systematic approach to the planning, organizing, scheduling, and control of a project. It addresses the following functions of project management.

- Communication
- Cooperation
- Coordination

Project Communication Many new manufacturing technologies are just reaching the market. There are still apprehensions and controversies regarding their potential impacts. Implementing new technology projects may generate concerns both within and outside the organization. A frequent concern is the loss of jobs. Sometimes, there may be uncertainties about the impacts of the proposed technology. Proper communication can help alleviate the fears of those to be affected by the implementation effort. CIM project communication should address the following:

- Nature, scope, and impact of CIM
- What resources are available to implement CIM
- Who is in charge of CIM implementation

- Why CIM is needed
- Direct and indirect benefits of CIM
- What personnel contribution to CIM implementation will be needed
- Who will be affected by the failure of the CIM effort
- Scope of the implementation effort

Wide communication is a vital factor in securing support for a project. A concerted effort should be made to inform those who should know. The communication channel must be kept open throughout the project life cycle. In addition to in-house communication, external sources should also be consulted as appropriate.

Project Cooperation Not only must the personnel be informed, but their cooperation must also be explicitly pursued. Merely signing off on a CIM project is not enough assurance of full cooperation. In effect, CIM technology must be sold to management and employees. A structured approach to seeking cooperation should help identify and explain the following items to the project personnel:

- What cooperative functions are needed to implement CIM
- What timeframe is involved in the implementation
- Criticality of cooperation to the project
- Organizational benefits of cooperation
- Implication of lack of cooperation

Staffing, time, and equipment needs should also be addressed during the pursuit of project cooperation.

Project Coordination The efforts of the CIM implementation team must be coordinated after successfully initiating the communication and cooperation functions. Coordination facilitates conducive organization of project efforts. The development of a responsibility chart can be very helpful at this stage. A responsibility chart is a matrix consisting of columns of individual or functional departments and rows of required actions. Cells within the matrix are filled with relationship codes that indicate who is responsible for what. The responsibility chart, shown in Fig. 11.4, helps to avoid overlooking critical functional requirements and responsibilities. The matrix should indicate the following:

- Who is to do what
- Who is responsible for which results
- What personnel interfaces are involved
- Who is to inform whom of what

Codes:

R = Responsible
I = Inform
A = Approve
S = Support
C = Consult

Responsibilities	Management	Engineer	Technician	Project Manager	Technical Writer	Plant Manager	Training Dept.
1. Problem Definition						R	
2. Personnel Assignment	C			C		R	
3. Project Initiation				R		A	
4. Technology Prototype	C	R	R	I		S	
5. Full Technology	C	R	R	I		S	
6. Technology Verification	R	R	R	C	I	C	
7. Technology Validation	R	R	R	R	I	C	
8. Technology Integration	R	R	R	R	I	A	
9. Technology Maintenance	R	R	R	C	I	A	
10. Documentation	C	C	C	C	R	A	I

Fig. 11.4. Example of CIM implementation responsibility chart.

- Whose approval is needed for what
- What support is needed from whom for what functions

Matrix Organization for CIM Implementation

Matrix organization is a popular project organizational structure that facilitates resource sharing and system integration. It is particularly suitable for technology implementation projects. It permits both vertical and horizontal flows of information. The matrix model is sometimes called a multiple-boss organization. It is a model that is becoming increasingly appealing as the need for resource sharing increases. Technology projects, for example, require the integration of specialized skills from different functional areas. Under matrix organization, projects are permitted to share physical resources as well as managerial skills. An example of a matrix organization model is shown in Fig. 11.5.

A matrix organization is suitable where there are multiple managerial accountability and responsibility for CIM implementation. There are typically two chains of command. The horizontal and vertical chains of command deal with functional and project lines. The project line in the matrix is usually of a temporary nature while the functional line is more permanent. The operation

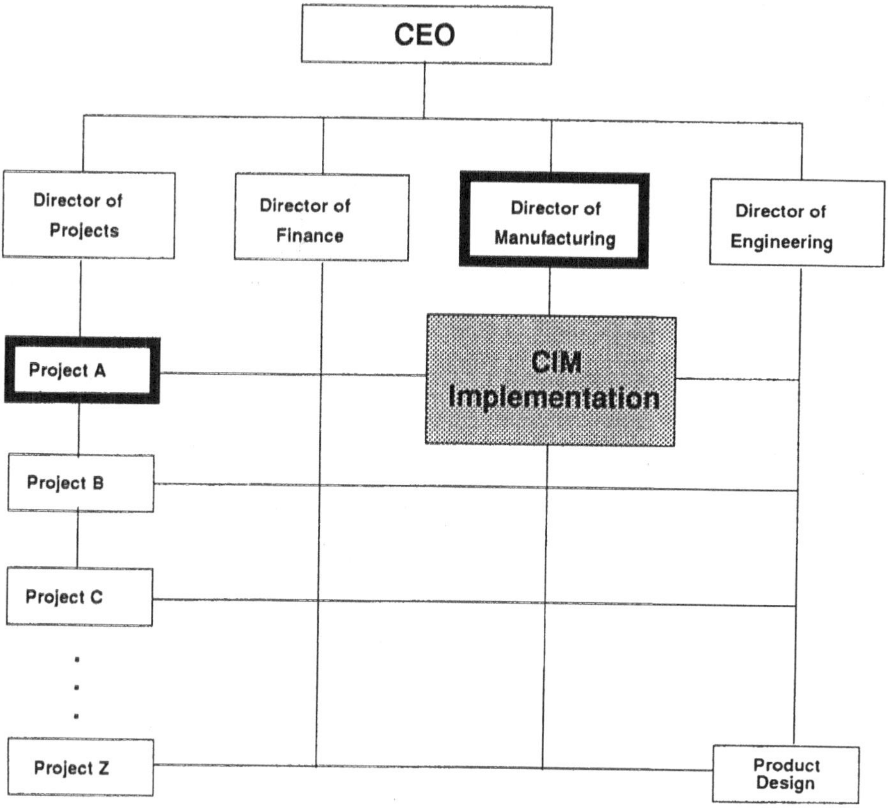

Fig. 11.5. Matrix organization for CIM implementation.

of the matrix structure is very dynamic depending on the prevailing project scenarios. The advantages of matrix organization include the following:

1. Consolidation of objectives: The objectives of the task at hand are jointly shared and pursued by multiple functional departments.
2. Efficient utilization of resources: The allocation of resources is more streamlined. Staffing and equipment can be allocated jointly among departments to improve utilization levels.
3. Better flow of information: Since departments are cooperating rather than competing, there is a better flow of information.
4. Interfunctional contacts: The joint responsibility for projects creates an atmosphere of functional compatibility and lateral functional interactions.

Despite its several advantages, the matrix organization does have some disadvantages. The major disadvantages include potential lack of sponsorship for shared projects, the multiple-bosses problem, lack of a unified project priority, higher overhead costs, and power struggle between managers. However, these disadvantages can be avoided or minimized with prior awareness and prudent implementation.

11.3 CIM TECHNOLOGY CONVERSION STRATEGIES

The implementation of CIM can be effected through one of several strategies. Some strategies are more suitable than others for certain types of technology. The most commonly used technology conversion strategies include:

Parallel Conversion: The existing technology and the new technology operate concurrently until there is confidence that the new technology is satisfactory.

Direct Conversion: The old technology is removed totally and the new technology takes over. This method is recommended only when there is no existing technology or when both technologies cannot be kept operational due to incompatibility or cost considerations.

Phased Conversion: Modules of the new technology are gradually introduced one at a time using either direct or parallel conversion.

Pilot Conversion: The new technology is fully implemented on a pilot basis in a selected department within the organization.

Figure 11.6 shows a graphical representation of the technology conversion options.

Computer-integrated manufacturing technology integration can facilitate effective sharing of resources to boost utilization levels. Technology resource sharing can involve physical equipment, facilities, technical information, ideas, and technical personnel. Technology integration is a major effort in technology adoption and implementation. Technology integration is required for proper product coordination. Integration facilitates the coordination of diverse technical and managerial efforts to enhance organizational functions, reduce cost, improve productivity, and increase the utilization of resources. Technology integration ensures that all performance goals are satisfied with a minimum of expenditure of time and resources. It may require the adjustment of functions to permit sharing of resources, development of new policies to accommodate product integration, or realignment of managerial responsibilities. It can affect both hardware and software components of an organization. Important factors to consider in CIM technology integration include:

- Unique characteristics of each component in the integrated technologies
- Relative priorities of each component in the integrated technologies
- How the components complement one another
- Physical and data interfaces between the components

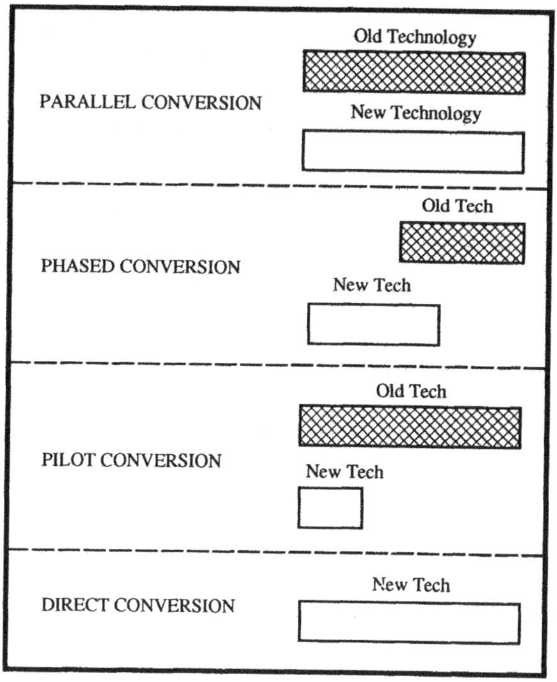

Fig. 11.6. Technology conversion strategies.

- Internal and external factors that may influence the integrated technologies
- How the performance of the integrated system will be measured
- Human interface required for the technology implementation

A formal justification process will facilitate giving due consideration to the various factors (qualitative and quantitative) involved in CIM implementation. The next section discusses this important aspect of CIM project management.

11.4 CIM TECHNOLOGY JUSTIFICATION

Computer-integrated manufacturing justification must be performed as a part of the project management effort; it must be justified on the basis of both technical and economic considerations in order to ensure success of the implementation process. Without proper justification, it may be difficult to achieve acceptance of the technology. Several analytical and computer tools (Badiru, 1992) are available for CIM justification. Presented below are two computer approaches that are easy to implement. The expert system approach is needed to address the inexact nature of making technology implementation decisions.

JUSTEX: An Expert System for Technology Justification

Sundaram and Badiru (1991) presented the *JUSTEX* (justification expert) expert system. The main objective of JUSTEX is to serve as a decision aid to managers and engineers in the justification process. The expert system provides a synergistic approach by combining tactical, economic, and strategic methods in the justification process. After evaluating all the factors specified, JUSTEX presents a recommendation of GO, DEFER, or NOGO. The knowledge base is structured in the form of production rules. Four different knowledge bases are linked to form JUSTEX. The sources of knowledge included experts in industry, books, and journals. For the purpose of JUSTEX, advanced manufacturing technology is classified as one of the following:

- Design venture
- Specialized manufacturing venture
- Miscellaneous ventures

Under each of these, there are further subheadings that narrow the types of advanced manufacturing technology. For example, under design venture, the subheadings are CAD, CAE, SIMULATION, and NC/CAM/CAD. The categories under specialized manufacturing venture are MATERIAL HANDLING, FMS, and ROBOTICS. The categories under miscellaneous ventures are TOTAL QUALITY MANAGEMENT and GROUP TECHNOLOGY. The categories are illustrated in the decision flow diagram shown in Fig. 11.7.

JUSTEX adopts a comprehensive search approach to the justification problem. It takes into account the interactions between the strategic components that play a vital role in the justification of any type of technology. JUSTEX considers a tactical approach first. This enables the user to decide the specific areas of technology to explore. The tactical approach is short term, but the decisions are based on the match between the company's objectives and the exact nature of the technology venture. If these two do not match, then JUSTEX displays an appropriate decision on whether to implement the technology or not. The strategic approach is more long term in nature and takes into account several organizational aspects. The economic approach provides the quantitative economic aspects of the technology investment. The most important role of the expert system is to combine the three different approaches so that a comprehensive decision can be reached.

JUSTEX uses a backward chaining search strategy. It is developed with the VP-Expert shell. In addition to the four linked knowledge bases used by JUSTEX, there are two externally linked BASIC programs. These are called from within the VP-Expert shell. The tactical module of JUSTEX has ap-

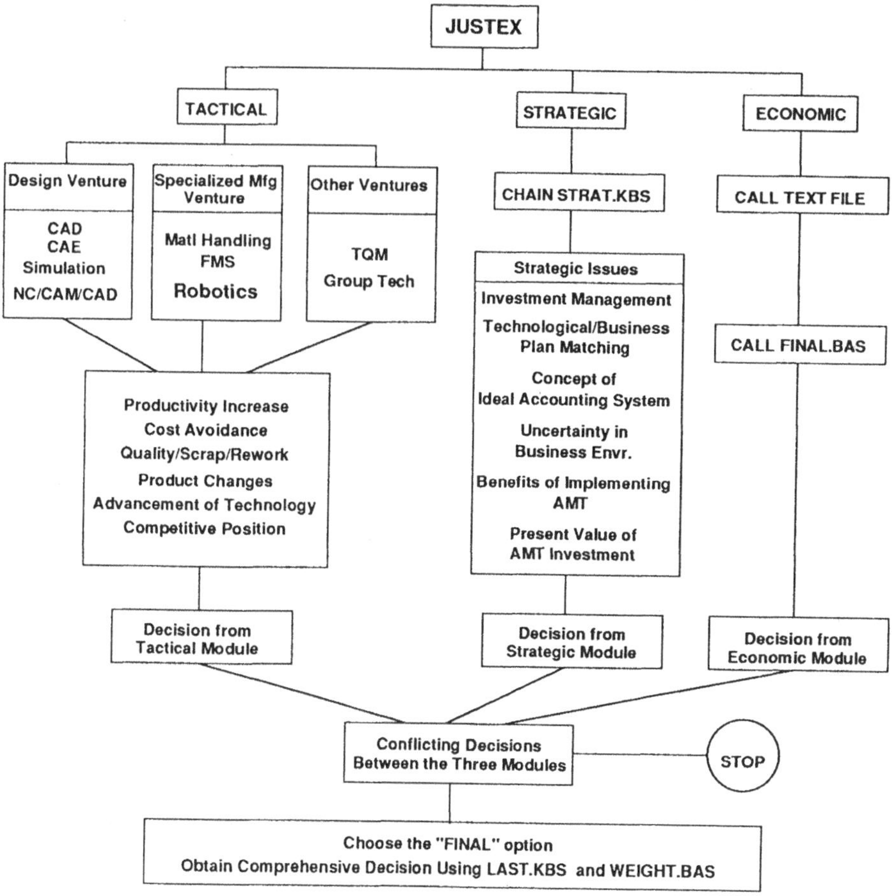

Fig. 11.7. Decision flow diagram for JUSTEX.

proximately 200 rules. It is linked to the strategic module by using the *chain* clause. The strategic module consists of six issues that are vital to manufacturing technology decisions:

1. Effective management of high-technology investment options
2. Match between the technological and business plans
3. Ideal accounting system
4. Uncertainty involved in the industrial environment
5. Benefits arising as a result of the implementation
6. Present worth of the investment

The first issue is the effective management of high-technology investment options. Many strategic investment decisions can be temporarily postponed to give the decision maker enough time to resolve the uncertainty before making a commitment. The second issue addressed in JUSTEX is the match between the technological and business plans. The long-range business plan must be consistent with the firm's technological resources. To have good investment alternatives, the firm's technological and business portfolio must be integrated into the company's business policies. The third issue that is considered in the strategic module is the concept of an ideal accounting system. Companies seeking to compete effectively must develop cost accounting systems that reflect their investment decisions and cost structures. Internal accounting practices should be guided by corporate strategies. The fourth issue that is addressed in the strategic module is the uncertainty involved in the industrial environment. The fifth issue is the benefit arising as a result of the implementation. This issue attempts to capture the key benefits of advanced manufacturing technology (AMT). The sixth issue addressed in the strategic module is the present worth (PW) of the investment. This helps to combine monetary and nonmonetary aspects of the justification process. The DCF (discounted cash flow) method is used since it is the most commonly used economic justification technique. It takes into account the time value of money and other factors such as inflation. The DCF model used is:

$$
\begin{aligned}
\mathrm{PW} &= \sum_{j=0}^{j=N} P_j \\
&= \sum_{j=0}^{j=N} C_j\left(\frac{1}{1+i}\right)^j \\
&= \sum_{j=0}^{j=N} C_j'\left(\frac{1}{1+\theta}\right)^j\left(\frac{1}{1+i}\right)^j \\
&= I + \sum_{j=1}^{j=N} S(1+k_1)^j + M(1+k_2)^j\left(\frac{1}{1+\theta}\right)^j\left(\frac{1}{1+i}\right)^j
\end{aligned}
$$

where I = initial cash flow, S = annual savings in present dollars, M = annual cash flow in present dollars, k_1 = annual rate of increase in the relative value of annual savings due to price differentials, k_2 = annual rate of increase in the relative cost of operating and maintenance due to price differentials, θ = annual rate of general inflation, i = inflation free discount rate, C_j' = cash flow in period j in current (then) dollars, and C_j = equivalent cash flow in period j in present dollars. In the above method, the savings were classified as direct and indirect. The program starts with a display of selected text files, which give the user an introduction to the system and its capabilities. The economic module was written in BASIC and is run from within the VP-Expert shell. The user is asked to input numerical values for the capital, amount of savings, the inflation rate, and so on. The final knowledge base used in JUSTEX enables the user to arrive at a comprehensive decision when there

```
Which justification methodology would you like to try first?
The ideal path recommended by the system is, TACTICAL, ECONOMIC
and STRATEGIC.
 TACTICAL                    *                        *
 STRATEGIC                   *                        *
 ECONOMIC                    *                        *
 FINAL                       *                        *
```

Fig. 11.8. Sample JUSTEX consultation screen.

are conflicting decisions from the three modules. This knowledge base also invokes a BASIC program from within the VP-Expert environment. Selected screen displays from a JUSTEX consultation are presented in Figs. 11.8 to 11.10.

COST-JUST: A Spreadsheet Model for Cost Justification Badiru et al. (1991) presented a multiattribute spreadsheet model for manufacturing technology cost justification. The model, named COST-JUST (cost-justification), contains modules for the analytic hierarchy process (AHP), utility models, and the system value model. The multidimensional structure of the spreadsheet model is useful in representing the multitude of attributes that are encountered in the cost justification of CIM technology. Spreadsheet models have the following advantages:

- They offer an environment for performing sensitivity analysis
- They are familiar to most management analysts and decision makers
- They have built-in routines for making graphical plots for visual assessment

```
The following list shows some of the areas which have scope for
improvement. Please identify the areas where you seek improvement
or change. Limit your choices to TWO (2). Press <End> after confirming
selection.
 Production Increase         *                        *
 Cost Avoidance              *                        *
 Quality Scrap Rework        *                        *
 React to Prod Change        *                        *
 Advancement of Tech         *                        *
 Competitive Position        *                        *
```

Fig. 11.9. Sample JUSTEX consultation.

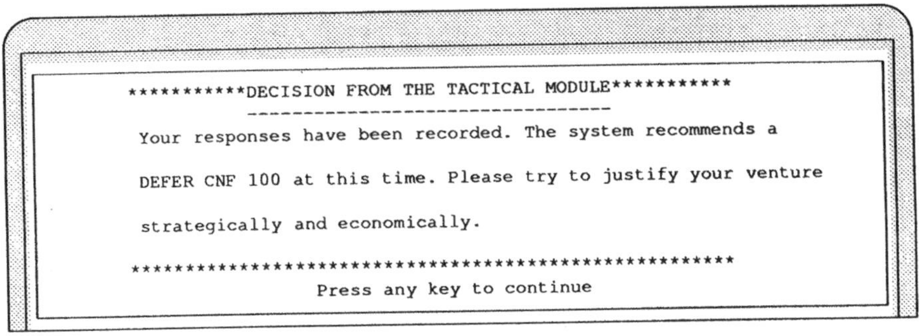

Fig. 11.10. Sample JUSTEX conclusion screen.

The spreadsheet model was developed with the Lotus 1-2-3 macro language and can handle up to five manufacturing technology alternatives and five attributes at the same time. It permits the inclusion of both monetary and nonmonetary factors in the analysis. Figure 11.11 shows the design layout of the system.

The following examples illustrate the use of the spreadsheet model. Suppose we have the objective of selecting the best overall manufacturing technology out of three feasible alternatives. The alternatives are to be compared on the basis of factors or attributes that the organization considers to be very important. Such factors may be determined based on goals relating to productivity, quality, customer service, and so on (Badiru, 1990). For the purpose of this illustration, five attributes are used. The set of data follows:

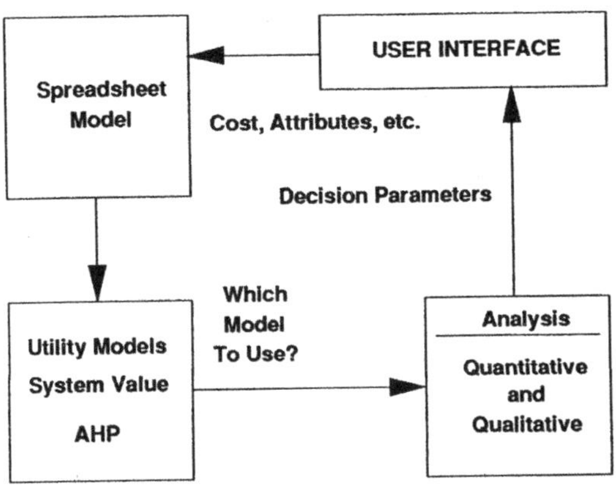

Fig. 11.11. Operational interface for COST-JUST.

TABLE 11.2 Summary of AHP Results for Technology Alternatives

	Attributes					
	A $i=1$	B $i=2$	C $i=3$	D $i=4$	E $i=5$	
$w_i \Rightarrow$	0.288	0.489	0.086	0.041	0.096	
Technology j			k_{ij}			α_j
Technology 1	0.093	0.118	0.500	0.633	0.562	0.208
Technology 2	0.211	0.243	0.250	0.106	0.375	0.244
Technology 3	0.685	0.639	0.250	0.260	0.063	0.548
Column Sum	1.000	1.000	1.000	1.000	1.000	1.000

CIM Technology Alternatives	CIM Technology Attributes	
1. Technology type I	A: Training requirements	D: Quality
2. Technology type II	B: Flexibility	E: Productivity
3. Technology type III	C: Serviceability	

Using the AHP module of COST-JUST, the results presented in Table 11.2 were obtained for a hypothetical example involving the three alternatives and five attributes specified. The three technology alternatives are rated on the basis of all five attributes. It is seen that technology type 3 is the best choice. This section presents only a brief introduction to the need to include formal justification in CIM project management. More detailed justification approaches should be adopted when dealing with CIM implementation projects.

11.5 CONCLUSION

Computer-integrated manufacturing technology can provide significant benefits if properly implemented. Past failures can be traced to improper adoption and implementation approaches. New CIM technologies that are not implemented at appropriate times will quickly become obsolete and may never provide their highly acclaimed benefits. A project management approach can provide the mechanism through which CIM objectives can be achieved within organizational goals.

This chapter has presented guidelines for applying project management techniques to CIM implementation. The major steps in a project management process are presented. The chapter suggests the triple C model of communication, cooperation, and coordination as an important tool for ensuring the acceptance of new CIM technology. The role of the human in CIM implementations is also discussed. Two computer approaches for justifying CIM within a project management framework are also presented. Detailed guidelines are provided for CIM technology planning and control.

REFERENCES

Aletan, Samuel (1991). The components of a successful CIM implementation, *Industrial Engineering*, **23** (11), 20–22.

Avishai, B. (1989). A CEO's common sense of CIM: An interview with J. Tracy O'Rourke, *Harvard Business Review*, (Jan/Feb), 110.

Badiru, Adedeji B. (1987). Communication, cooperation, coordination: The triple C of project management, in *Proceedings of 1987 IIE Spring Conference*, Washington, D.C., May 1987, pp. 401–404.

Badiru, Adedeji B. (1988). *Project Management in Manufacturing and High Technology Operations*, J Wiley, New York.

Badiru, Adedeji B. (1990). Strategic Planning for Automated Manufacturing, in *Justification Methods for Computer Integrated Manufacturing Systems*, H. R. Parsaei, T. L. Ward, and W. Karwowski, Eds., Elsevier Science Publishers, Amsterdam, pp. 17–39.

Badiru, Adedeji B. (1991). *Project Management Tools for Engineering and Management Professionals*, Industrial Engineering & Management Press, Norcross, GA.

Badiru, Adedeji B. (1992). *Expert Systems Applications in Engineering and Manufacturing*, Prentice-Hall, Englewood Cliffs, NJ.

Badiru, Adedeji B., Foote, B. L., and Chetupuzha, J. (1991). A multiattribute spreadsheet model for manufacturing technology justification, *Computers & Industrial Engineering (Proceedings of 13th Annual Conference on Computers and Industrial Engineering, Orlando, Florida, March 1991), 21* (1–4), 35–43.

Cleland, David I. (1990). *Project Management: Strategic Design and Implementation*, TAB Professional & Reference Books, New York.

Falkner, Charles H., and Benhajla, Saida (1990). Multi-attribute decision models in the justification of CIM systems, *Engineering Economist*, **35** (2), 91–114.

Industrial Engineering (October 1989). Commerce deputy: U.S. isn't automating quick enough, *Industrial Engineering*, **21** (10), 7.

King, Leslie (1991). Today and tomorrow: Speaking the language of the factory, *Industrial Engineering*, **23** (11), 18–19.

Moder, Joseph J., Phillips, Cecil R., and Davis, Edward W. (1983). *Project Management with CPM, PERT and Precedence Diagramming*, 3rd ed., Van Nostrand Reinhold, New York.

National Research Council (1984). Committee on the CAD/CAM Interface, Manufacturing Studies Board, *Computer Integration of Engineering Design and Production: A National Opportunity*, National Academy Press, Washington, D.C., p. 1.

Rodriquez, Moises-Enrique (1991). Switzerland establishes CIM Centers to study technology, *Industrial Engineering*, **23** (11), 23–24.

Sundaram, Deepak, and Badiru, Adedeji B. (1991). JUSTEX: An expert system for the justification of advanced manufacturing technology, in *Knowledge-Based Sys-*

tems and Neural Networks: Techniques and Applications, Sharda, Ramesh, et al., Eds., Elsevier Science Publishing, New York, pp. 89–98.

Weston, F. C. (1989). CIM from a top management perspective, *CIM Review*, **5** (2), 18–22.

Weston, F. C. (1991). Functional goals are often in conflict with each other, *Industrial Engineering*, **23** (11), 25–29.

Wiest, Jerome D., and Levy, F. K. (1977). *A Management Guide to PERT/CPM with GERT/PDM/DCPM and Other Networks*, 2nd ed., Prentice-Hall, Englewood Cliffs, NJ.

CHAPTER 12

MANAGING THE CHANGE TO AUTOMATED MANUFACTURING

J. LAMARSH
LAMARSH & ASSOCIATES
MORTON GROVE, ILLINOIS

12.1 INTRODUCTION

There is no single change that achieves automated manufacturing. Each change—dynamic shop floor control scheduling systems, robotics, work cells, bar coding—is by itself difficult to implement. But when all the changes are occurring at once, they compete for limited resources. Sometimes they even cancel each other out. It is no wonder timetables are destroyed, budgets get eaten up, and the end result, an automated manufacturing operation, does not get implemented. If the company does achieve an automated environment, it often does not yield the payback it originally promised.

The cost of failure when attempting to implement automated manufacturing is very high. The immediate cost is the lost dollars and time spent in the various projects. But the long-term cost is much greater. People in the organization learn that the company does not know how to implement change, that it does not know what it is doing. Therefore, the next change attempted has even less potential for success because people have lost confidence and do not sign up. This chapter will examine the potential risk of a change failing

Organization and Management of Advanced Manufacturing. Edited by Waldemar Karwowski and Gavriel Salvendy.
ISBN 0-471-55508-8

because the human factors are not understood or managed. It will explore the critical change variables and present a systematic way of dealing with each of those variables in order to reduce the risk factor.

12.2 THE RISK OF FAILURE

The risk of failure is determined by the ability of the company to address the human variables that will impact success or failure. The risk of failure is high when people in the company do not have a clear, well-defined picture of what the changes are. It is impacted strongly by the ability and willingness of senior management to sponsor the changes. Employees, management, customers, and vendors will be resistant to changes. The amount and type of their resistance will contribute to the potential for failure. When changes have little impact on the current culture of the company, the risk is dramatically reduced. However, when the current beliefs, behaviors, and unwritten rules are challenged by the changes, the risk goes up.

The men and women on the various project teams and task forces must figure out how to overcome these serious obstacles to reduce the risks. They need to build a central focus and a strategy to integrate all the changes successfully. They must use a structured, systematic methodology of identifying all the changes, assessing them on critical variables such as cost, time requirements, amount of potential resistance, commitment needed by senior management, and the impact on the company's culture. From this assessment the company can build a matrix that will lead to an overall plan for prioritization and integration of each change in relation to the desired end result.

To successfully manage the change process requires management, both sponsors of the change and the primary change agents, to understand the change process and develop the skills and expertise to use this information and available tools and resources to architecturally design a project plan (Harrison, 1990).

12.3 UNDERSTANDING THE CHANGE

Before applying a change management methodology it is imperative to understand the process. There are five critical factors that affect success in changing to an automated environment:

1. **Manage the Three Stages of Change**

   Many companies have made a commitment to automate manufacturing. They have spent a lot of money on seminars, education,and charts on the wall. But the payback just is not coming. Six months to a year later, people still say they are commited to improved manufacturing processes.

But they keep making the same mistakes, the same customers are still angry, rework keeps piling up, and the overtime is growing. Management starts to turn these failures around when it understands the elements of change.

2. **Hold Sponsors Accountable for Their Role**

   Everyone knows that senior management must be behind a change for it to be successful. But what does that mean? Senior managers must learn how to *sponsor* change. They cannot just be commited to automated manufacturing, they must be obsessed by it. They cannot tolerate anyone in the organization who does not support it. They must communicate their support constantly, both publicly and privately. And they must change the reward mechanisms to be sure that new behaviors are rewarded.

3. **Analyze Resistance Factors and Decrease Them**

   People do not resist the concept of automated manufacturing. But they do resist the impact of the resulting changes on the way they do their work and the way they think about themselves in relation to that work. Successfully introducing automated manufacturing means assessing what those specific changes are, who needs to change, and what the sources of resistance will be. Then the company can build a plan to address those resistance factors and reduce them.

4. **Build an Environment That Supports an Automated Manufacturing Environment**

   The environment or culture builds up over the life span of a company. The change to automated manufacturing cannot be successful unless old behaviors, beliefs, and unwritten rules that guide today's actions are analyzed to determine which ones to keep and which ones to replace. The current environment is strong because it has been around for a long time. Getting rid of undesirable behaviors, beliefs, and unwritten rules will take careful study and even more careful planning.

5. **Develop Skilled Change Agents**

   To implement automated manufacturing, people need to learn what it is and develop procedures to introduce it. People also need to build the skills of change management. Those responsible for automated manufacturing must understand the change process and know how to build a comprehensive implementation. Eventually, lots of people in the company become change agents and the need for this knowledge spreads throughout the company.

These change agents have to integrate the technical and the human elements of change into a single implementation plan. Companies cannot settle for a high degree of technical competence only. They must demand that

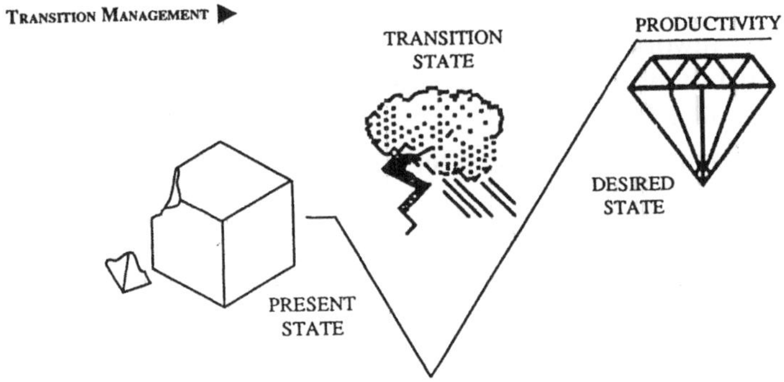

Fig. 12.1. The unfreezing–transition–refreezing approach. (Copyright IMA, 1990.)

aligned with the technical know-how is the understanding of what happens to people when they must change.

There is always a price to pay for change. Smart companies chose to pay that price up front, at the beginning of the process of introducing automated manufacturing by spending the necessary time and effort figuring out how to manage the change factors. Companies that do not, pay at the back end of the process by never achieving their goals, spending a lot of time, effort, and education and then having to do it all over again. Worse yet, management may end up putting the change on the shelf along with all the other aborted efforts at improvement like qualities circles and team building.

Each of these factors requires intense effort and understanding.

Manage the Three Stages of Change

Everyone working toward automated manufacturing knows that there is a gap between the way we work today and the way we want to operate. If there were not, we would not be making the commitment to automated manufacturing. But understanding what automation is, learning to use statistical process control or figuring out ways to do faster changeovers is only part of the process of achieving automated manufacturing.

Companies that have been successful know that there is a very critical step in the process that is often overlooked or taken for granted. They recognize that achieving automated manufacturing is going to be a major change from the way things are today. They know that in order to get there from here, the change *itself* must be managed.

Change is a process not a discrete event. Managing that process requiring "unfreezing" the present state, managing the transition state, and "refreezing" in the desired state (see Fig. 12.1) (Lewin,1947).

The Desired State Clearly defining where the company needs to be is the critical first step. The strategic vision defines the desired state. The vision must be clearly understood by everyone in the company. What does automated manufacturing mean? Equally important, everyone must be able to translate that vision into a picture of their own area and what it will look like when automated manufacturing is in place.

The Present State Once the desired state is defined, we can go back and begin to systematically unfreeze the old way. Even when the end result of our change is highly desirable, it is not easy to give up the old way. It is comfortable, it is predictable, and it is easier. We cannot expect that agreeing to a change is an automatic guarantee that it will happen.

The Transition State The transition state is where the changes occur, where new definitions and new behaviors are defined, where people learn how to do things the new way. The transition state is very uncomfortable for people. The ambiguity and uncertainty create a high degree of stress. Productivity drops, stress increases. Many people welcome the changes automation brings. They may be very glad the company is making this comittment. But that does not mean they change easily.

In this stage of the change process things cannot be done the old way, but the new way is not well established yet. For instance, a company introduces a policy that says it will never ship products out of spec. But there was a long period when the company could not meet its own standards. When a customer screamed, supervisors would make the same decision they had made in the past: "Ship it this way and maybe they won't notice." That conflicting message told workers the company was not really serious about the change; it was just talking about it. The company had to find those mixed messages and get rid of them.

Hold Sponsors Accountable for Their Role

Who Needs to Do What? Everyone talks about the need for management support if change is going to be effective. But few companies have successfully defined the specific actions that management must understand and do if a change is to be successful.

Therefore, change agents try to do it all. That will not work. On the organization chart the project manager may look powerful because he or she reports to the vice president of manufacturing. But most of the people who have to change do not report to the project manager. Their greatest source of influence is the person they *do* report to. That is where they get the information, direct and indirect, on what will make them successful in the company and what will not.

That is why a lot of people come to training and get excited about what *could* be. But the next day, back at their desk or in front of their machine, nothing changes. Their management, on up the ladder, has not sponsored the changes. So Charlie comes back from training and says "Hey, boss, I think we'd be able to improve the quality of this operation if we'd change the angle of the feeder line." If Charlie is told, "Forget it! That would never work. Besides, I'm the boss. It's my job to think, not yours," Charlies gets the message: The company really does not want automated manufacturing; it is just a lot of fancy talk.

Everyone in management must take a close look at themselves and how they sponsor change. Senior management has to be sure that *all* managers and supervisors are doing what it takes. No one can be exempted from sponsoring automated manufacturing if the company is serious about change. So, as anyone in management looks at the organization chart, he or she needs to see a cascade of sponsors from the president all the way down to the managers and first-line supervisors in the office.

There cannot be any black holes in the sponsor cascade; no division or department that is not committed to change. Everyone must be a good sponsor (see Fig. 12.2).

What Needs to Be Done Even the most committed sponsors need to do a lot more than just say they are committed. What do sponsors have to do to cause a *real* change in the company? The sponsor's role breaks down into three parts:

1. Understand the change.
2. Manage the change resources.
3. Exhibit appropriate management skills.

Understand the Change In order to sponsor change successfully, management throughout the organizational cascade must understand change, the change process, the change variables, and their role in the change. This learning must occur in a shared environment and be presented in an organized, structured learning situation such as a seminar or workshop. Sponsors throughout a company have a range of understanding of the change process gained from experience and observations. This understanding must be organized into a change management methodology that has as its outcomes:

- A common, agreed-upon overview of the change and its impact
- Acceptance of the importance of managing the critical change variables
- Identification of the roles of each of the parties to the change process: sponsors, change agents, and the targets of the change

In this learning environment all sponsors in the cascade must start from a point of dissatisfication with the way things are today and a picture of where the company should be heading. They must also think through the impact the change is going to have on the company, both organizationally and in

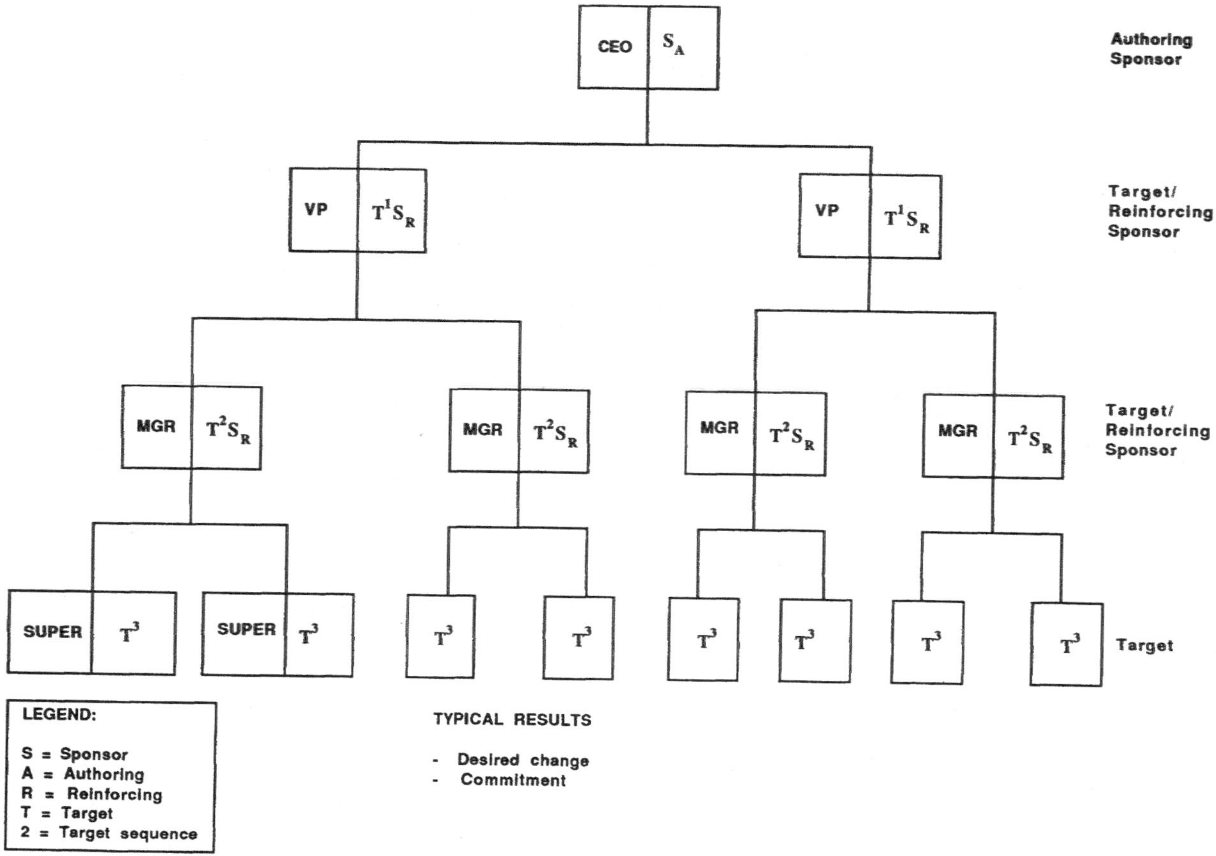

Fig. 12.2. Key roles. (Copyright IMA, 1990.)

human terms. A major change such as building an automated manufacturing focus means a lot of changes going on at once. Sponsors must decide how these changes are to be prioritized and what the relationship is between energy focused on the change process and energy focused on business as usual. The most important thing that sponsors must do is to show that they want this change, that they believe in it.

Manage the Change Resources One of the things that defines a sponsor is that he or she is in a position to determine how the company's resources are allocated. Sponsors need a clear picture of what resources will be required. They must be prepared to allocate these resources or, to the degree that the required resources are not available, they must alter the change. Sponsors *cannot* ask change agents to bring change in on time and within budget if they are not able to provide those agents with the resources to do the job.

But the sponsor's role does not stop at resource allocation. Continuous monitoring and support are required to ensure that resources are properly managed.

One resource sponsors frequently overlook is the reward and recognition system in the company. Rewards are a vital component in the change process and need to be systematically used as a resource that supports the change.

Exhibit Appropriate Management Skills To be effective, every manager/supervisor in the sponsor cascade must be able to communicate his or her sponsorship. That means both publicly and privately they must convey the message that they are serious about this change. It is critical that they do this using the language and the frame of reference of their audience; the accounts payable department, the East Coast plant, the shipping crew on third shift.

Equally important, sponsors must know how to solicit and value feedback from all the target groups affected by the change. Because automated manufacturing requires many technological changes, sponsors must learn what these are and how they will impact the company.

At all times, management must have good working relationships with their direct reports. But these relationships become especially important during times of change. The basic skills of leadership, motivation, communication, and mutual problem solving are minimum requirements of effective sponsors.

Analyze Resistance Factors and Decrease Them

Why do people resist change? Why would they fight improvements and efficiencies that keep their company competitive, indeed, keep it alive? During the 1980s American manufacturing watched foreign countries become more successful by making changes that improved quality, reduced costs, reduced manufacturing cycle time, and brought products to market faster. If we are going to be able to compete, we will have to make a lot of changes in the way we do business.

Do American workers and management care? Would they rather keep sitting there, doing things the same old way while the ship slowly sinks below the surface? Do they see what the cost of not changing is? Are they just too lazy or too stupid to change?

Why Do People Resist Change? When change agents bump into resistance, it is often hard to understand. They can so clearly see the logic of installing visioning inspection equipment, consolidating the financial functions using a common database, and integrating the various islands of automation. It makes no sense for people to keep coming up with excuses as to why it will not work. But that is the first mistake of agents of change. They cannot assume that resistance to change is logical. People who resist change have serious reasons for doing so, but those reasons may seem to be very illogical (Coch and French, 1948).

Targets of change do their own analysis of the changes. They have a strong need to:

1. Maintain the Status Quo

Often the targets of change understand perfectly well the need for the change and will be the first to agree that this specific change is long overdue and very necessary. But there is an illogical side to everyone that is frequently the source of resistance to change. People are all creatures of habit. They get so used to doing things a particular way it becomes second nature. Change takes time, energy, and effort. That is hard. Think about morning routines such as showering, dressing, eating breakfast, driving to work. After a while, it becomes so routine people do not think about it; it is the status quo.

2. Protect the way in Which We See and Explain the World

Everyone in the company looks at the proposed changes from their own perspective, their own frame of reference. And each individual tries to determine whether the new way is compatible with what was important to him or her in the old way:

Will the new production scheduling system require new behaviors?
Yes, now the production scheduling will be done on the shop floor in real time.

Does it challenge old values?
Yes, we have always felt that such important work must be done by guys in white shirts in a special room. People on the line could not figure this stuff out.

Will it require people to throw out old ways of doing work and thinking about the work?

Yes, the line workers will know when and what to schedule before the department heads do. We have never done things that way before.

The old ways made us successful for a long time. It is hard to give them up because they defined people and how they saw and interpreted the world. The logic of a change does not automatically mean people sign up for it.

When Does Resistance Increase? When the change is announced, each individual affected by the change evaluates it from his or her own perspective. Consciously and unconsciously people ask themselves a series of questions. The answers will determine whether or not each person is supportive of the change or resistant to it. Understanding this process and knowing these questions is a critical first step for the change agent charged with the responsibility to diminish and manage resistance.

Do we really need this change?
What are the strategic business issues that led the company to make this decision? What would it mean if we did not change? What are our competitors doing? What do our customers expect us to do? How does this particular change solve the problems we have?

Does this mean they think I have been doing a poor job in the past?
Quality inspection has been a dedicated and loyal group of people for 20 years. They gave their all to the company. Now they think the company is telling them it has all been worthless. Their whole work group is going to be disbanded. Their impression is that nobody values what they have been doing.

How much disruption of the day-to-day routine will this cause?
It is fine to make changes, but companies still have to get product out the door. Putting these changes in place is going to really throw things into confusion. People become concerned. "How can I concentrate on the change and still do my job? How long do I have to live with all this disruption?"

What's the payback for doing this? What's the cost?
Because everyone judges the change from their own frame of reference, people look at what the impact of the change is on them individually and on the company. One critical analysis is the relationship between the cost of changing (individually and as a company) and the reward for changing. Some changes will have a high cost for specific individuals and groups of people, with very little reward. The jobs of quality inspectors may be eliminated. The production scheduling department may be reorganized and the job functions altered.

What if management's wrong? Can we go back to the old way?
Every change is a gamble. Those who made the decision worry about

whether or not they made the right decision. Those who were not in the decision loop also worry. They were not there for the discussions. They did not see how the decision was made. One of the critical questions is how difficult it would be to restore the status quo if the decision was not right. If the old way is gone forever and the new way does not work, then what will the company do?

What's the probability that this change can be implemented successfully? The history of a company is powerful. Part of that history is the changes that have been implemented in the past. Does the company have a good record of making changes? How difficult will this change be? Does it have the resources to do it? Is senior management prepared to do what it takes to get the changes in place?

What will it look like when the change is in place? What will it look like when the company reaches the desired state? How will work get done? How will success be judged? How will people behave? Asking people to change without a clear idea of what they need to change to makes it harder for them to get up and follow the leader.

What part will various groups play in the implementation of the change? Companies that have a history of successful change know that the more people are involved in the change process the less resistance there is. Everyone needs to be a part of the process in order to have some control.

How Do People Resist? Individuals and groups go through this series of questions, analyzing the change from their own frame of reference. If the outcome of this analysis is resistance to change, people then make a choice. They either decide to express that resistance or they decide to keep it to themselves.

Change agents need to understand the resistance analysis process, know what questions targets of the change are asking of themselves and what those answers are so they can address these issues. Systematically, they need to go through each issue and determine how they can diminish its impact. Change agents cannot do that, however, if they do not know what people are thinking.

Therefore, it is critical that change agents set up an environment where people are encouraged to express their resistance. It must be safe to question things, to seek answers to the questions that arise, to express doubt and fear.

Change agents manage the change process by understanding that resistance is inevitable. Then they can begin the process of safely surfacing that resistance and working to diminish it.

A planned approach to managing resistance includes a variety of action steps based on a structured process of education, communication, participation, and involvement (Kotter and Schlesinger, 1979). When people understand the change process itself and can fit what is happening to them into a model that explains their reactions and validates them, they are better prepared to move through the change stages of unfreezing, transition, and refreezing.

The degree and type of involvement in making decisions about the change itself and about the process of changing have a direct affect on people's acceptance of the change and their commitment to the end result (Lawrie, 1990). The technological aspects of the changes being introduced to support advanced manufacturing require an active participation by the targets of the change in the planning, implementation, and integration phases of the project. Simultaneously, many companies are changing the culture of the organization to create an environment that asks the shop floor worker to commit to a quality product and a continuous improvement orientation to their work. Achieving these cultural changes and the smooth integration of the technological changes requires that companies treat targets of the change as partners in the process, not adversaries who must be conquered.

Build an Environment that Supports Automated Manufacturing Environment

Achieving automated manufacturing means changing a lot of old behaviors. Many things people always believed will not be true any more. There are many old unwritten rules that kept companies successful for a long time but now are going to get in their way. These beliefs, behaviors, and unwritten rules are the company's culture. Committing to automated manufacturing is really committing to a new culture The core of this change is the struggle between old beliefs and behaviors and the beliefs and behaviors needed for automated manufacturing (see Fig. 12.3). Many companies have found out the hard way that in that kind of struggle the old culture usually wins. The old culture took a long time to evolve and it is hard to fight it.

An early step in this process is to systematically analyze the current culture. Then the company can decide what to keep and what beliefs, behaviors, and unwritten rules it wants to change. That includes looking at performance measurements and informal reward and recognition systems—two critical aspects of the culture.

Again, the first step is to define the type of organization the company will be when it achieves automated manufacturing. Without this step, it is like putting a high-performance racing engine into an old 1950s pick up truck. Old beliefs, behaviors, and unwritten rules are like the old tires and transmission of the truck. They will not support the engine. The old beliefs, behaviors, and unwritten rules will not support the move to automated manufacturing. A good mechanic would take a long, hard look at that truck, analyze all its parts, and decide which ones to keep and which to throw out or replace.

Define the Desired Culture First we need to define the culture we want; the culture that will support automated manufacturing.

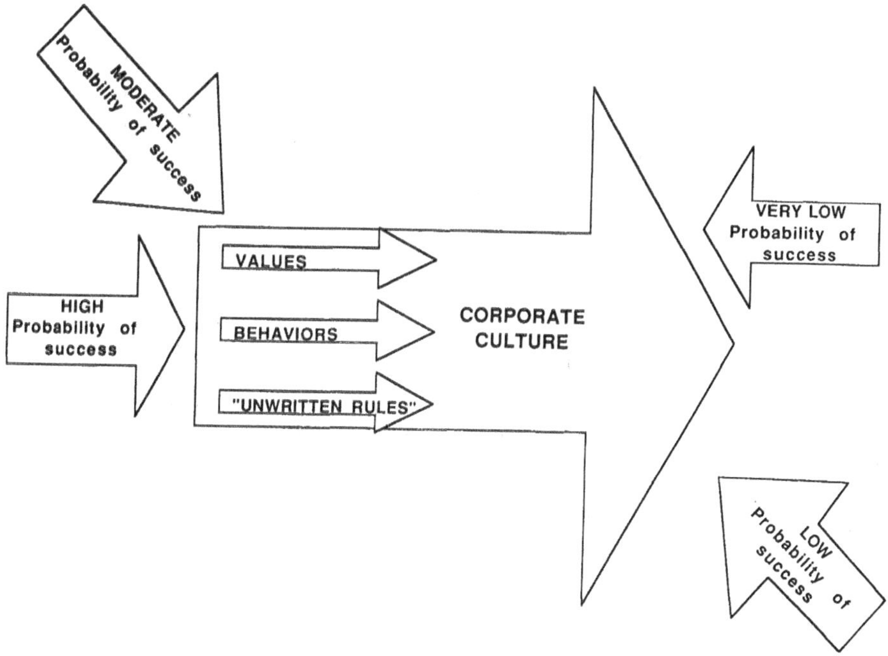

Fig. 12.3. Managing resistance, change, and corporate culture. (Copyright IMA, 1990.)

What will the company be like when it achieves automated manufacturing?

Within departments
Across department lines
Across the line to customers/vendors

How will it institutionalize automated manufacturing?

The process
The ownership

What changes will management make in the way in which it manages?

Communication
Reinforcement/consequence management
Decision making
How much responsibility/authority will be passed down through the chain of command?

What will the management structure look like?

More/less managerial positions
Skill set required

What role will technology play in creating this desired state?

What could prevent this future picture?

At management's level
At the worker level
Because of technology
Because of customers

Analyze the Current Culture Next, look at the current culture. Determine what exists today that will fit the desired state and what to get rid of. Start with the unwritten rules.

This list of unwritten rules fits many manufacturing companies. These are the type of rules that must be identified. Then determine those that will stay and those that will go.

Manufacturing's Unwritten Rules

- Fourth quarter will always be our busiest, it is the nature of our business.
- The end of the month is the best time to ship product.
- If you need 20, ask for 30.
- Sales and marketing will always look out for themselves first.
- Nobody on the shop floor should ever be standing around with nothing to do.
- We need fire fighters, they get the job done.
- The customer is *always* right.
- If it isn't broke, don't fix it.
- The bigger the office, the more important the occupant.
- The bigger the batch, the lower the unit cost.
- The faster the machine, the more economical the product.

Next, look at the way people behave and the beliefs those behaviors reflect. Look, for example, at the communication process, the way the company makes decisions, and the things it measures and rewards.

Develop the Skills of the Change Agent

Many companies make the commitment to change to automated manufacturing, but few have achieved it. Those companies that are successful do not have extraordinary resources, especially brilliant people, or deeper pockets than others. The overriding element that differentiates the successful companies is the willingness to take the time to systematically manage the required changes.

Throughout this discussion there has been a key role in the change process that keeps surfacing: the change agent. Change agents are the men and women that senior management turns to and says, "Make it happen!" They are the people who must have the greatest understanding of the change process and the way to manage it.

Change Agent Criteria There are seven critical elements to consider when selecting and/or developing a change agent.

1. *Experience*: Change agents must know the company, its people, its cuture, and its operations.
2. *Trust*: Change agents must be viewed by senior management and the targets of the change as individuals who do not have any hidden agendas, carry no political baggage around, or play games with people or departments.
3. *Understanding*: Change agents must have a clear understanding of what needs to change, why it must change, what the desired state will look like, what impediments stand in the way of success, and what sponsors are willing and unwilling to do to achieve success.
4. *Knowledge*: Change agents must know what automated manufacturing is and how to achieve it. Equally important, they must know what change is, and how to manage it.
5. *Commitment*: Change agents must believe in the change to automated manufacturing, be convinced of its benefits, and be prepared to make the sacrifices in time, energy, and pain it will take to achieve it. Frequently, automated manufacturing efforts fail because a change agent took the job because his or her boss believed in the change, but the change agent did not.
6. *Tolerance*: Knowing what to do and how to do it are critical to the skill set of the change agent. But change has to do with people, and people do not follow nice simple charts and steps in an implementation plan. They are illogical, often confusing, and frequently stubborn. Therefore, the change agent must have a high degree of tolerance for the vagaries of human behavior and the difficulties of making changes happen. Without it, the length of the change process can eat up their energy and their commitment.
7. *Ability to Communicate*: Change agents must be particularly good at communicating, sharing information, and developing a picture of the desired state and asking for feedback from sponsors and targets. To accomplish this, change agents must understand the communication process, be able to package their messages in the frame of reference of

their various audiences, and be especially skilled at listening. Listening will be a key ingredient in understanding what will impede the change process and how to overcome the obstacles: whether they are weakening sponsorship or resistance on the shop floor.

Our current knowledge of engineering processes, design elements, machine tooling, and inventory management will, most likely, be obsolete in a few years. The world of knowledge and the tools to use that knowledge is changing so rapidly there is only one thing we can be sure will remain constant: the amount and degree of change. Therefore, the ability to manage change successfully, the ability to be a change agent, becomes a highly prized skill in business.

12.4 PUTTING IT ALL TOGETHER

This structured process of managing the changes required for automated manufacturing needs to be put into a context. When and how is it used in the implementation process?

Steps to Implement Automated Manufacturing

Define Automated Manufacturing for the Company Senior management must start the process. Before ordering up an automated manufacturing organization, they need to sit down and define what that will look like in their company.

- Exactly what do we mean by automated manufacturing?
- What do our customers expect?
- How will we measure automated manufacturing?
- What will be the role of workers? Of managers?
- How will we reward people when they achieve automated manufacturing?

Build an Implementation Team The more clearly defined the automated manufacturing picture is, the easier it will be to determine the resources it will take to get there. The first resource is the change agent. A move to automated manufacturing requires an implementation team headed by a highly skilled change agent. The implementation team is not responsible for achieving automated manufacturing, however. They *are* responsible for ensuring that all parties to the change process do their job to achieve automated manufacturing. That means they help sponsors to sponsor, targets to manage their own resistance, and line change agents throughout the company to manage the changes in their areas.

Cascade the Change Down Through the Company The automated manufacturing definition built at the senior management level is, by definition, global and all-encompassing. One of the first steps in the process is to cascade that definition down through each functional line, geographic region, and department/division. At each level the targets must take that definition and refine it to apply to their specific area. If the definition of automated manufacturing includes reducing the price of nonconformance, for example, senior management will determine companywide measurements. The Dallas plant, the Seattle plant, and the Charleston plant will use that measurement as a parameter to determine their own measurements.

Everyone in the company must understand automated manufacturing at two levels: the global level and their own specific level. These two levels must always be in alignment. Change agents drive this process. Equally important, they monitor it on an ongoing basis.

Assess the Present State Now that the desired state of automated manufacturing is defined from the board room to the stock room, the implementation team begins a process of assessing the present state of all aspects of the company that will be affected by the change. Whether they use the Malcolm Baldridge Award criteria or any other assessment instrument, it is time to determine where they are today. Used well, this assessment is the foundation of an implementation plan.

Determine the Degree of Change Required and the Specific Changes to Be Made When the company has the information it needs, it can begin the process of change. It knows what it wants to look like (the desired state) in a cascade down through the organization. It knows what it looks like today (the present state) in each of the critical areas affected by an automated manufacturing environment. Comparison of the two will determine the type and degree of changes required. Analysis of these changes will lead to an understanding of the resource requirements to make the changes and a prioritization of the various changes necessary. Too many companies, because they do not do this step systematically, launch an automated manufacturing process in a blizgkrieg of changes that vie for scarce resources and eventually consume one another in a battle of politics and infighting.

Build an Implementation Plan Armed with this information, the implementation team can now build a plan to achieve automated manufacturing. The plan will include all changes required, resource allocations, and the steps in managing the change itself.

Action Steps in the Implementation Plan

- Define the change.
- Assess the amount of disruption the change will cause.
- Determine the impact of the history of previous changes.

- Measure the type and degree of cultural impact of the change.
- Identity and assess the sponsorship.
- Identify all targets of the change and measure their resistance.
- Select and educate change agents.
- Build the education, communication, and reinforcement steps into the plan.

Measure, Monitor, and Revise Plan During Implementation Companies that achieve automated manufacturing know that it is made up of a series of small victories getting closer and closer to the ultimate definition set by management. They also know what these small victories must be measured and monitored and, most important, celebrated along the way. Successful change requires an obsessive focus on the end result, with incremental steps that show that the company is still on the right path and is getting there. The achievement of an automated manufacturing environmment takes years. People must see that the movement is forward. And they must be rewarded for the effort and the accomplishment of each step along the way.

12.5 WHEN THE CHANGE MANAGEMENT METHODOLOGY IS APPLICABLE

The application of the change management process occurs at three stages of the project:

Stage 1: Risk Assessment

Determining what the cost of moving to an automated manufacturing environment will be. This assessment must include the cost of managing the changes, the degree of resistance, the sponsor commitment, and other critical change variables.

Stage 2: Planning

When determining the sequence of activities and building the budgets and critical paths for the various projects required in this change, the change process must also factor in and plan for the human elements.

Stage 3: Implementation

Change management methodology provides a series of specific action steps, strategies, and tactics to introduce and apply at various stages of the implementation process.

12.6 SUMMARY

Just as the paradigms regarding what our factories should look like are changing, so to are we forced to examine our ways of implementing change. The old methods of slamming changes across the organization are no longer valid. A commitment to automated manufacturing processes and a budget for hardware, software, and formats will not be sufficient. Managing the critical change variables is also required.

REFERENCES

Coch, L., French, Jr., J. R. P. (1948). Overcoming resistance to change, *Human Relations* **1**, 512–532.

Harrison, D. (1990). *A Structured Approach to Managing Change*, Implementation Management Associates, Denver.

Lawrie, J. (1990). The ABCs of change management, *Training & Development Journal*, March, 87–89.

Lewin, K. (1947). Frontiers in group dynamics, *Human Relations*, January, 5–41.

Kotter, J., and Schlesinger, L. (1979). Choosing strategies for change, *Havard Business Review*, March.

FURTHER READINGS

Bouldin, B. (1989). *Agents of Change: Managing the Introduction of Automated Tools*, Prentice-Hall/Yourdon Press, Englewood Cliffs, NJ.

Carlisle, H. M. (1976). *Management Concepts and Situations*, Scientific Research Associates, Palo Alto, CA.

Conner, D., and Kelley, D. (1979). The emotional cycle of change, in *The 1979 Annual Handbook for Group Facilitators*, J. E. Jones and J. W. Pfeiffer, Eds., University Associates, San Diego.

Dalziel, M., and Schoonover, S. (1988). *Changing Ways: A Practical Tool for Implementing Change within Organizations*, American Management Association, New York.

Fulmer, R. M. (1974). *The New Management*, Macmillan, New York.

Hellriegel, D., and Slocum, J. W. (1987). *Management: A Contingency Approach*, Addison-Wesley, Reading, MA (4th ed., 1986).

Hellriegel, D., and Slocum, J. W. (1976). *Organizational Behavior—Contingency Views*, Addison-Wesley, St. Paul (4th ed., 1986).

Kanter, R. M. (1983). *Change Masters: Innovation for Productivity in the American Corporation*, Simon & Schuster, New York.

Miles, R. E. (1965). Human relations or human resources? *HarvardBusiness Review*, **4**, 148–163.

Seashore, S. E., Ed. (1983). *Assessing Organizational Change: A Guide to Methods, Measures, and Practices*, Wiley, New York.

Senge, P. (1990). *The Fifth Discipline*, Doubleday, New York.

Tichy, N. M., and Hornstien, H. A. (1976). Stand when your number is called: An empirical attempt to classify types of social change agents, *Human Relations*, October, 945–967.

Woodward, H. (1987). *Aftershock: Helping People through Corporation Change*, Wiley, New York.

Zeithaml, V., Parasuraman, A., and Barry, L. (1990). *Balancing Customer Perceptions and Expectations*, Free Press, New York.

CHAPTER 13

INTERNATIONAL PERSPECTIVES ON COMPUTER-INTEGRATED MANUFACTURING

R. J. BADHAM
Department of Science and Technology Studies
University of Wollongong
Wollongong, Australia

13.1 INTRODUCTION

In the nineteenth century, Max Weber, one of the founders of organizational studies, made the classic statement that it is possible to "rationalize life from fundamentally different basic points of view and in very different directions" (Weber, 1974, p. 26). Unfortunately, since that time many students of technological change in the workplace have forgotten this insight and acted as if there is one path to technical change and associated work organization. This view has become so predominant that even where different forms of technology or work organziation are observed, it is often assumed that these are either trivial local variations (in the case of technology) or, in the case of work organization, are caused by nontechnical factors influencing the way technology is *used* and that the "logic of technology" is of much lesser significance than was previously thought. The problem is, however, far deeper

Organization and Management of Advanced Manufacturing. Edited by Waldemar Karwowski and Gavriel Salvendy.
ISBN 0-471-55508-8

than such analyses suggest, and its resolution profoundly affects how we analyze and assess international experiences in computer-integrated manufacturing (CIM).

Weber's insight is important in two senses. First, it points to the fact that the direction of technical development may take a number of different forms depending on the interests and requirements of the groups and organizations involves (Badham, 1991). With regard to CIM, it is only to be expected that, to paraphrase Weber, it will be designed and implemented from fundamentally different points of view and in very different directions. If this is so, only a partial insight into the nature and diffusion of CIM will be gained if a search is made for indicators of "levels" of CIM and attention focused on the implementation techniques and impacts of these levels (or their "requirements" if they are to be employed productively). At the level of the firm, the sector, and the nation, CIM systems may be far from similar, and important variations exist that both reflect and impact upon skills, jobs, and work organization.

Second, Weber's observation also highlights the fact that rationalization can occur in both a nontechnological or a technological form—if by technology we mean here the hardware and software of production systems. As the dramatically effective introduction of Japanese *organizational* techniques has shown us in the last two decades, changes in human skills, motivation, and organization are at least as important as developments in hardware and software systems in achieving productivity gains. Whether and in what way different organizations develop the technological and organizational components of CIM may be undertaken, again, from very different points of view and in very different directions.

In combination these two observations raise fundamental questions over how we define and think about CIM: First, to what extent are organizational as well as hardware/software components included in our definition. Second, how do we classify and rank CIM components and systems as *similar* or *different* for the purposes of examining their requirements and impacts?

In regard to the first issue, such questions are far from academic for they determine the basic approach that is taken into analyzing and explaining CIM systems. As commented in Zylstra (1989, p. 10):

> Is the system an accumulation of the hardware and software, or does "system" include the associated people, procedures and operating environment? Without the human element of the operating condition, the system performance can only be determined from the fully automatic mode (if it can operate with "lights out"). This automation view of CIM misses the effects of system problems, manually tended operations, new product changeovers, maintenance programs and other human intervention that will not be understood or included in the CIM plans.

If hardware and software components of CIM are identified as "computer aids for integrated manufacturing," as suggested strongly by the European *human centered*, *skill based*, or *anthropocentric* approach to CIM (Badham

TABLE 13.1 National Conditions and Alternative Approaches to Production Management and Control.

UNITED STATES	JAPAN
Standardised product/Customised Options	Product Variety/Packaged Options
High Volume/Slow Market Change	Lower Volumes/Rapid Product Innovation
'Supply Push' - MRP & MRP II Systems	'Demand Pull' - Just-in-Time
'Closed Loop' Automatic Factory Visions	'Open Loop'/Customer Oriented/ Organisation Based Systems
Complex Shop Floor Control Systems/Firefighting Approach	Simpler Shop Floor Control Systems
High Inventory - Just-in-Case	Low Inventory
NATIONAL CONDITIONS AND ALTERNATIVE APPROACHES TO PRODUCTION MANAGEMENT AND CONTROL	

Source: Compiled from P.Clark and S.Newell, (1992), 'Societal Embedding of Production and Inventory Control systems: American and Japanese Influences on Adaptive Implementation in Britain, **International Journal of Human Factors in Manufacturing** (in press)

and Schallock, 1991a), the key research focus is upon *computer- and human-integrated manufacturing* (CHIM) or *task-integrated production* (TIP) (Haywood and Bessant, 1990; Koetter et al., 1989; Martin, 1990). In this case CIM development at all the stages of invention, innovation, or diffusion refers centrally to components and combinations of technical/organizational change rather than merely hardware/software systems. From such a perspective, for example, the attempt by American computer-aided production management (CAPM) systems to automate all production planning and scheduling within a centralized computer command and control system is an important technical, organizational, and cultural phenomenon that contrasts strongly with the Japanese tendency to focus on just-in-time organizational techniques as a basis for planning and scheduling (see Table 13.1). The study of the source and impact of these different technological *and organizational* solutions is equally important in examining the diffusion of CIM.

In regard to the second definitional issue, key questions are raised over how to define CIM and operationalize this concept for the purposes of examining its origins, diffusion, requirements, and effects. If the components and architecture of CIM in different organizations reflect the different requirements and interests of developers and users, then the definitions of CIM put forward should be capable of picking up these differences as a basis for analysis. For these purposes, taking, for example, the diffusion of flexible manufacturing system (FMS) as an indicator of CIM development, and examining international experiences in FMS as a basis for generalizing about CIM, would be totally inadequate. In this regard, not only are the different levels of diffusion of different components of CIM important [e.g., computer-aided engineering or drafting, computer-aided design or manufacture (CAD/CAM), etc.] but also their development and configuration within specific system architectures (e.g., physical location of workstations, distribution and access to computer-based systems and databases, etc.). These variations may reflect not the trivial customization of generic systems for specific production circumstances but major and systematic variations in CIM trajectories. Important questions are raised, for example, by the contrast between the attention paid to the United States to the development and diffusion of CAD systems, and the focus in Japan upon flexible and integrated computer-aided manufacturing (CAM) systems (Schallock 1991). Similarly, there has been a focus within Germany upon the development of workshop-oriented programming (WOP) systems for computer numerical control (CNC) machines and decentralized production scheduling or electronic control systems in contrast to U.S.-developed centralized CAD/CAM and material requirements planning (MRP) systems (Bandemer et al., 1991).

Any international overview of CIM that ignored these variations and their consequences would provide a seriously restricted and misleading perspective on contemporary experiences with CIM. This is even more the case as there is such a dominance of rhetoric and analysis that adopts the alternative position, that is, that stresses the universal evolution of CIM as the inevitable onset of the "factory of the future" and offers quick generalizations about *the* character of *best practice* CIM implementation and production.

In reviewing international perspectives on CIM, this chapter will provide a critical analysis of the dominance and impact of traditional models of CIM development and diffusion that ignore or underplay this variation in the vision that they put forward. It will stress the importance of variations in CIM and outline a number of the approaches and case studies developed to understand and utilize these variations. In conclusion, it is argued that more attention should be paid in the development of CIM models to *systematic* variations in the context and purposes for which CIM is employed, and the means for integrating international perspectives in CIM to addressing this issue.

13.2 TRADITIONAL PERSPECTIVE ON CIM AND ITS CRITICS

Traditional View of CIM as the "Automatic" Factory of the Future

Underlying much of the traditional thinking on the nature and international diffusion of CIM is a distinctive yet inadequate and outdated view of technological change. This image is supported by perspectives from a number of intellectural disciplines (such as economics, innovation studies, and development studies), the rhetoric and ideology of the "socio-technical constituencies" (Molina, 1989) promoting CIM, and deeply embedded cultural assumptions about technology, work, and society. Because of this pervasiveness it is difficult to place a convenient label on the view but many of its key features have been variously labeled "technocentric" or "technocratic," "Taylorist," "linear," "machine oriented," and so forth. Although many of these terms refer to varying elements of the perspective, and different analysis of CIM may adopt more or less of these elements, in its most extreme or "ideal typical" form it can be identified as a relatively clear and generally distinctive vision of CIM diffusion and the nature and meaning of international experiences in CIM. Because of the power of this vision, much of the contemporary debate over CIM can be best seen as different reactions to this view and the selective adoption or rejection of its key assumptions. In Norman Mailer's terms, even those not aware of it remain "in the belly of the beast."

In its ideal typical form, the traditional view of CIM has at its heart a vision of the *automatic factory* in which all information and material flow is performed automatically under the command of computer-based systems that control the increasingly *process*-like (rather than discrete product or batch) production system. In its most extreme technocratic form this is embodied in visions of the unmanned factory that represent the culmination of Western utopian aspirations and fascination for "automata"–the creation of what Boguslaw has documented and described as the "new utopian" breed of system designers (Boguslaw, 1969). In its popular form this view is represented in such imagery as the following:

> Imagine an engineer sitting at a computer terminal punching in data for the design of a new product and sketching freely with a lightpen on the screen before him. Happy with the design, he presses a button and the details are passed electronically to another computer running software that checks to see whether the design's stresses and strains are within prescribed limits. The information then zips along to a third computer which generates instructions that command the tools in the workshop to machine, assemble and store the engineer's product ready for distribution—all done automatically, without hassle, delay or hefty manhandling, and all before the morning's coffee break. One more satisfied customer. Welcome to the "factory of the future." (*Economist*, 1987)

This image of the CIM factory of the future has been exemplified in model factories owned by such firms as General Motors and Fujitsu Fanuc. Its principles of operation are laid out in engineering and management journals and textbooks, as well as guidelines for its introduction. In addition, it has been eulogized in more popular journals and magazines such as *Fortune*, *The Economist*, and *Business Week*; it has been promoted by vendors of new computer-integrated manufacturing equipment such as General Electric, IBM, and McDonnell Douglas; and it has been endorsed and encouraged by government programs and agencies such as the American Air Force's Integrated Computer Aided Manufacturing (ICAM) program, the Japanese Ministry for

International Trade and Industry's Unmanned Factory program, the European Community's ESPRIT program, and so forth. Within the United States, as Nicholas Valey observed in *The Economist* (1987, May 30), this vision of "CIM has become a kind of holy grail. With all the trappings of an industrial crusade, CIM (or something much like it) is the big stick America is going to be wielded in the coming battle for global competitiveness."

The claim is made that we are entering a new age of manufacturing, that we are witnessing the arrival of the high-tech, computer-integrated, flexible and reprogrammable factory of the future. It is, in the words of one pundit, "highly technological and anti-industrial," a cybernetic era "beyond mechanisation." Some indication of its flavor is given in the more popular press mentioned earlier. Thus, as a 1983 *Fortune* article on "the Race to the Automatic Factory" described the Fanuc Factory, "automatic machining centres and robots typically toil unattended through the night, with only subdued blue warning lights flashing as unmanned delivery carts move like ghostly messengers through the eerie semi-darkness."

In Detroit, Michigan, in 1986:

> A row of silver car bodies glides along the production line and halts. Six welders on either side immediately spring into action, moving swiftly around, over and inside the metal skeleton, sparking the many spot welds needed to hold the incipient automobile together. Within 23 seconds flat, they have knocked off 250 welds between them; the convoy moves on, and the next half-dozen cars are ready to be born. The welders never take a break, they don't drink coffee and they've never felt the need for conversation. No graffiti, no half-eaten sandwiches or soiled copies of the *National Enquirer* clutter their pristine work-space. That's because these ideal workers are robots, not people . . . The "Factory of the Future" is on the way. (Forrester, 1987, pp. 170–171)

These images of the unmanned factory of the future appear high-tech, contemporary, even futuristic—yet in an important sense they are strongly traditional. Automation is identified with the long established image of a mechanized utopia. Often this vision is explicitly related to a profound distruct of human beings and a desire to eliminate their role in production. As commented by Teicholz and Orr (1987, p. 13) in their "CIM Industry Overview": "Why is CIM desirable? Because it reduces the human component of manufacturing and thereby relieves the process of its most expensive and error prone ingredient." Where the emphasis is not so strongly based on the "unmanned" character of CIM, there is a stress upon the removal of traditional craft, manual, and information transfer skills and personnel, and the rise of a relatively small elite of highly technically trained personnel organized into flexible teams that program and maintain the "autonomous" CIM systems. In the words of one CIM proponent: "German arts and crafts and the well known German craftsmanship of skilled workers will no longer be required in the future but will be replaced by flexible systems and intelligent control" (Martin, 1990, p. 206).

Accompanying this vision is a diffusion model that, in the words of Ayres and Ranta, "seems to imply a single center where CIM technology is created, and from which it spreads out spatially in all directions" (Ayres and Ranta, 1989, p. 407). In this view the process of technological development proceeds in a linear fashion from invention, through innovation, to diffusion. Computer-integrated manufacturing is thus perceived as emerging and being developed within a heartland of CIM innovation located primarily within large electronics and computer companies and major developers/users of CIM located primarily in the United States, and Japan. From this center increasingly more sophisticated "generations" of CIM systems emerge with enhanced capacity, decreasing costs, increasing reliability, and greater standardization—all essential for the effective takeup and use of CIM.

The dynamics of the process are seen as governed by the twin forces of technology and markets. Technological advances in electronics and computer-based command and control systems are perceived as driving CIM innovation in a direction guided by the dual market demands for reduced costs and design/production complexity and increased product innovation and variety. Firms and nations are consequently classified as *leaders* or *followers* in the introduction of CIM, with larger more sophisticated manufacturing firms and nations leading in a race in which the followers lag behind. To the extent that other economic, political, and cultural factors have a part to play in this process, it is in the role of either facilitating or obstructing the inevitable march of best practice technology driven by the requirements of world competitiveness. In the words of General Electric, the alternative for manufacturers in the advanced industrialised world is only to "Automate, Emigrate or Evaporate."

This vision is supported by views on technology prevalent within neoclassical visions of the market economy. As outlined in a recent report to the European Commission Monitor–FAST Program, neoclassical economics assumes stylized economic agents (firms, households), allows for diversity only in the unexamined presence of comparative advantages in opportunity costs or factors of production, and views the local context of technological change purely in terms of the speed of local adaptation and adoption of best technology. There is, consequently, no systematic investigation of the source of factors of production, the origins of differences, and the active influence of local contexts in generating and perpetuating differences (Bruno et al., 1991).

In the area of innovation studies, despite a currently more complex model of the linear process, there has been a predominant macro-concern with the diffusion of the "one best way" of technology and organization and an emphasis on the central role of "long waves" of technological change and the market in determining outcomes. In the arena of development studies, the classic model of deindustrialization emphasized the export of simple mass production to third world countries and the development within the first world of the more complex forms of manufacturing—with ideas of CIM development following this model. Underlying it all, however, is the technological determinist viewpoint of the autonomous development of technology, in a process the social character of which is not examined, and the adaptation of the economy, organization, and society to the *imperatives* of this advance when combined with market *requirements*.

Critique of Linear Models and Generic Systems

Over the last decade this image has been increasingly challenged at a number of different levels. In the area of innovation studies, the linear theory of innovation has been widely discredited for its one-dimensional and unidirectional view of the process of technological change. A more flexible and interactionist view is now more accepted, with stronger emphasis on the influence of later "stages" of the innovation process on the nature, degree, and direction of technological change. This has been particularly noticeable in the work of economic historians of technology and evolutionary theories of innovation, in which the interaction between technological *trajectories* and *selection environments* is strongly influenced by the piecemeal and incremental nature of "learning by doing" within the innovation process (Badham, 1993).

This view is also noticeably present among contemporary commentators on CIM. As Ayres and Ranta observe about more linear views of CIM that assume diffusion from a central origin: "This image is clearly wrong, inasmuch as there are many such centers, in scores of institutions located in several continents." Moreover, "leadership in the underlying electronics technology does not necessarily coincide, or co-locate, with leadership in downstream applications technology" and, furthermore, "that CIM should be seen as a case of the type of technology, like the telephone or the electric light, that is introduced in relatively primitive form and continuously improved thereafter. Here progress and diffusion occur in parallel, and interact closely" (Ayres and Ranta, 1989, p. 407) This view is also well examplified by Warnecke, who comments that "CIM is . . . not an organisational and technological advance which can be achieved overnight. Once initial developments have been completed they must be integrated and partly updated. For a successful CIM installation, therefore, a modular concept, specific to the firm in question must be developed. This individuation must apply to the functional models themselves as much as to the problem solving within the models" (Warnecke, 1989, p. 3).

Generalizing on what has increasingly become conventional wisdom as presented in engineering and management journals, Warnecke stresses the importance of in-house expertise, demonstration, training production personnel to develop CIM solutions for their area, and production personnel skilled to operate CIM. As illustrated by titles such as "CIM: Much More Than Adding Computers" (Teresko, 1987, p. 47) and statements to the effect that "CIMs is not a software package or set of machine tools that can be purchased and installed using the latest project-management techniques" (Westong, 1988, p. 64), there is an increasing emphasis on the firm-specific, management, business strategy, and human/cultural aspects of CIM. Computer-integrated management systems have to be selected, customized, developed, and implemented in organizations sufficiently committed, financed, and staffed to make this possible. "In the real world of CIM technology," Ayres and Ranta comment, "there is an active, conscious, *feedback* between the primary technology developer and the market," (Ayres and Ranta, 1989, p. 418). It is the local, context-specific, development of CIM that lies at the heart of understanding its nature and consequences in specific firms—not the passive introduction and adaptation to the requirements of generic systems.

This last point is a key focus of research currently underway into the *configurational* nature of complex technologies such as CIM. These technologies are not driven by a central system dynamic nor purchased as a self-sufficient technological entity but, rather, have to be *configured* to fit the specific context within which they must operate. Any examination of the forces influencing CIM structures and the consequences of CIM for skills and work organization within specific firms must therefore pay central attention to local organizational structures, interests, and strategies (Badham, 1992; Badham, 1993). There can be no simple reading off of CIM *impacts* from a linear model of innovation or a deterministic analysis of the requirements of generic systems.

Critique of One-Dimensional Diffusion Models: Variability in Diffusion Patterns

The critique of the linear model is not, however, solely restricted to emphasizing the active contribution of the *diffusion* process in effective CIM implementation. Nor, even, to the active contribution that these developments make to the final character of CIM systems. Priority is also attached to two phenomena: the systematic differences in diffusion of the different components of CIM *and* the sectoral or firm-type differences in innovation styles and types of CIM developed. In the latter case, as observed by Pavitt, there is "enormous variety between firms in sources of technological opportunities and in the rate and direction of their development" (Pavitt, 1990, p. 19). In arguing for a systematic relationship between firm size and core business, on the one hand, and technology strategy on the other, Pavitt formulates the typology outlined in Table 13.2.

What is most important about this typology for our present purposes is not the details but, rather, the possibility of characterizing different types of technology/business strategy relationships, and the clear consequences that these have for the type of CIM systems and the priorities that the different types of firm will have. Latest research in the economics of innovation clearly points, therefore, to the need for analyses of CIM characteristics and requirements to pay specific attention to the differing requirements of different types of innovating firms. From within the CIM community, Warnecke explicitly recognizes this, stressing the significance of the fact that "computer

TABLE 13.2 Basic Technological Trajectories.

	Definition			
	Science-Based	Scale Intensive	Information Intensive	Specialized Suppliers
Source of Technology	R&D Laboratory	Production Engineering And Specialized Suppliers	Software/ Systems Dept. Specialized Suppliers	Small-Firm Design and Large-Scale Users
Trajectory	Synergetic New products Applications Engineering	Efficent and Complex Production and Related Products	Efficient and Complex Information 'Processing, and Related Products	Improved Specialized Producers Goods (Reliability and Performance)
Typical Product Groups	. Electronics . Chemicals	. Basic Materials . Durable Consumer Goods	. Financial Services . Retailing	. Machinery . Instruments . Specialty Chemicals . Software
Strategic Problems for Management	. Complementary Assets . Integration to Exploit Synergies . Patient Money	. Balance and Choice in Production Technology among Appropriation (Secrecy and Patents) Vertical Distintegration (Cooperation with Supplier) and Profit Center . 'Fusion' with Fast-Moving Technologies . Diffusion of Production Technology among Divisions . Exploiting Product Opportunities . Patient Money		. Matching Technological Opportunity With User . Absorbing User Experience . Finding Stable or New Product 'Niches'

Source: K.Pavitt, 1990, 'What we Know about the Strategic Management of Technology', California Management Review, Spring, p.20

manufacturers have no company-specific solutions to offer, only computer-specific concepts, which must be 'company-neutral,' " yet users need "practical and concrete demonstrations of the problems, advantages and possible solutions, and to obtain objective technology and branch specific assessments of solutions." (Warnecke, 1989, p. 4).

Critique of the Unmanned Factory

In reaction against CIM images of the unmanned factory, a number of points are now commonly stressed. First, at least 70 percent of all manufacturing is conducted in small manufacturing firms involved in the production of one-off or short batches of products—not the large high-tech firms described earlier. Second, the spread of the various technological components of high-tech manufacturing has not been as rapid as is often claimed. Robots are still limited in their range of application, interlinking CAD and CAM systems has proved extremely complex and difficult. There are still only a relatively few systems of flexible computer-interlinked machine-transfer lines (commonly known as FMS or flexible manufacturing systems), and even the distribution of numerical and computer numerically controlled machine tools is less than is often proclaimed (Schutz-Wild, 1990). Third, the productivity successes of many large firms who have embarked on this CIM path has often fallen well below expectations (Forrester, 1987, p. 188). General Electric's $600 million pioneering CIM plant in St. Louis making dishwashers was believed by many GE managers to have been too expensive for the results. The six flexible appliance factories reviewed by *The Economist* in the late 1980s were developed using a far lower level of automation. Similarly, the degree of robotization employed by Fiat in its famous heavily robotized car production lines is not being adopted by the company as the model for future factories. In a survey by Ingersoll Engineers of 40 companies introducing CIM, it was found that the degree of introduction was varied and unimpressive, and the profitability of sophisticated systems was low (Ingersoll, 1984). Increasingly management and engineering consultants are emphasizing the faulty thinking that lies behind simplistic high-tech solutions to manufacturing productivity. Emphasis is increasingly placed on the detailed examination of companies' manufacturing processes before, and possibly instead of, computerization. Moreover, the "people factor," the crucial significance of extensive retraining and difficult organizational restructuring, is now commonly regarded as a crucial yet exceedingly difficult component of any transition toward CIM.

If we take statements from management journals, consultants recommendations, governmental reports, and trade union or industrial association publications as an indication of industry's conventional wisdom, it would appear that the importance of the human factor in successful innovation is now fully recognized, and the "automatic factory" utopia is now far from fashionable (ACTU, 1987; BMFT, 1990; Hayes and Jaikumar, 1988; OECD, 1988a; Walton and Susman, 1987; Weiner and Brown, 1989). Similarly to Margaret Thatcher's claim that "the Cold War has not been won," there seems to have been victory in the human factors battle, at least at the level of public displays of conventional wisdom. As detailed by a leading U.S. information technology expert, U.S. firms have an annual 12 percent increase in their data processing budget, but only 2 percent productivity increases—they are therefore being directed to spend more attention on the "human element" (Daniel, 1987; McLoughlin and Clark, 1988; OECD, 1988b).

15.3 ALTERNATIVE PERSPECTIVES

Most critics of the traditional model of CIM have directed their attention to the general need for attention to specific user requirements, the presence of different sectoral or national variations in CIM implementation, and the importance of the people factor in implementing CIM and ensuring that the systems are effectively maintained and upgraded. Many of these critics have, however, concentrated on the demands of technology and changing markets, identified human factors issues in terms of post hoc implementation or the manner in which CIM systems are *used*, or continued to search for an alternative one best way of international best practice. For others, however, *central* importance is attached to variations in CIM technology as well as organizational systems, the influences upon these variations are sought in a *wide range* of institutional, organizational, and social as well as technical and market factors, and attention focused on the significant value-laden *choices* to be made between more or less appropriate or desirable forms of CIM. These perspectives on CIM are, however, often unknown or neglected within the technical writings on CIM or the more traditional human factors literature. In the interests of fostering interdisciplinary understanding and debate on CIM variations, this chapter continues by outlining three of the main international perspectives.

Societal Effects and Industrial Culture Perspectives

From within the disciplines of industrial relations and industrial sociology there has been a sustained resistance against technologically determinist theories that offer linear models of the international convergence of industrial societies toward similar and universal production systems. In contrast, it is emphasized that there are different paths or routes of industrialization and that different types of skills, work organization, and management–labor relations may characterize firms, sectors, and nations with similar levels or types of technology (Badham, 1986). In the ensuing debate over the nature and effects of new CIM technologies, the *societal effects* approach makes two assumptions: first, the culturalist assumption that interprets "socio-technical events as *strategically* influenced by tentative, piecemeal experimentation and innovation, under a focus of action which brings *technical, organizational and labour* considerations to bear on each other." (Sorge et al., 1983, p. 9) (emphasis added); and, second, the societal effect assumption that "the causes that underlie the continuous adaptation, improvement and modification of technology may be looked for, in central social and economic tendencies, and not only in the internal dynamics of science and technology as reasonably self-contained areas" (Sorge et al., 1983, p. 16).

On the one hand it is observed that within the constraints imposed by their environment, organizations possess a range of *strategic choice* and *design space* in the selection and development of CIM systems (Bessant and Grunt, 1985; Child, 1987). In particular, as emphasized by McLoughlin and Clark, there can be great variation in the "tightness of coupling" between senior management objectives and lower level management influences on system structure through the decisions they made on the implementation strategy and operation of new technology (Buchanan and Boddy, 1983; McLoughlin and Clark, 1988). Within the adopting organization, CIM configurations are directly influenced by what Child has called the "inherited characteristics" of a firm, that is, the size of the companies and plants, the stock of competencies and skills, traditional modes of operation and labor control, and the internal power structure. Continuing this theme, Eckart Hildebrandt (1989), a proponent of the CAPIRN network *industrial culture* approach, has emphasized the central role of the *social constitution* of firms in determining the design and introduction of CIM systems.[1]

It is a key component of this approach, however, that its focus transcends that of the internal organization of individual firms and, even, the influence on CIM systems of management–labor relations within different industrial relations systems. In explaining the dynamics of change in computer-based technologies and associated forms of work organization, an argument is made for an "extended" industrial relations perspective that extends into the broader areas of national patterns of skill formation and sets of governmental and nongovernmental institutional arrangements that form a broader industrial culture or the "sociological" components of production inputs. Particular emphasis is placed upon product markets and labor markets. The key theme of both Arndt Sorge and Wolfgang Streeck's research on the West German system of skill formation, for example, is that the crucial determinant of sociotechnical change is the strategic approach of firms to the opportunities created by changes in technology and product markets, but that this strategic approach is heavily influenced by national institutional contexts. In their words, "managerial product strategies—manufacturing policies—respond to the market but are not determined by it" (Sorge and Streeck, 1988).

For Sorge and Streeck, increasing uncertainty and indeterminacy brought about by changing product markets and flexible technology increasingly opens up for exploitation a product strategy arena of *diversified quality production* involving the high-volume production of customized high-quality competitive goods, *but it does not determine which firms will enter such an arena.* As Streeck argues on the basis of research into the West German auto industry and industrial relations system, the same conditions in West Germany that influence the more decentralized and flexible use of computer-based technologies are also influential in deciding which product strategy will be adopted by management (i.e., in the case of the car industry diversified quality production). The high level of shop floor skills that encourages a decentralized system of production organization, in combination with extensive system of worker's rights and job guarantees in West Germany, has influenced management to search for the kind of markets that would enable it to utilize comparative skill advantage and avoid the potential problems created by external labor market rigidities. The strategy to move toward diversified quality production reflects these established institutional structures rather

[1]Defining social constitutions as the "overall ensemble of the most important norms and rules permitted or effective in the plant that influence the employee's attitude toward work and behaviour at work," Hildebrandt emphasizes the central impact on production planning systems in German machine building firms of plant size and status, relationship between planning and execution, the independence of plant areas, means of personel coordination, nature of parallel structures and contradictory principles, the position of the works council in the structure of plant pressure groups, and local culture and its influence on the integration of the factory.

than a product market/technology determinism. The broader implications of this argument are that only particular types of "institutionally rich" societies (i.e., those involving extensive institutional support for education and training and developed internal labor markets) are capable of moving into the arena of diversified quality production, and the associated sociotechnical CIM trajectory (Matzner and Streeck, 1991). Moreover, as those large West German firms that have entered into diversified quality production have tended to create systems that involve "more organic organization with more enriched skills and overlapping work roles" (Sorge and Streeck, 1988), these conditions act back to reinforce this particular sociotechnical CIM trajectory.

Case Study 1: CNC Machine Tools in Great Britain and West Germany

In a series of books and articles throughout the 1980s based on research comparing manufacturing plants in West Germany, Great Britain, and France, Arndt Sorge and his colleagues investigated the influence of national socio-economic conditions on the development and use of computer-based process technologies. Their general orientation was employed in an investigation of how comparable engineering firms in West Germany and Great Britain responded differently to the technological trajectory of computer-based machine tools.

As outlined by Sorge et al. (1983), a uniformity has been imposed on the development of CNC equipment through trends in information and machine tools combined with universal changes in production and economic conditions. Machine tools have become mechanically more sophisticated with improvements in tools, materials, and cutting mechanisms, while in comparison control system price increases have declined in significance compared to overall machine costs, and their reliability and capacity has increased. In combination, these trends have enhanced the potential for small and medium-sized firms to introduce CNC equipment, and for the equipment to be used in a wider range of products and batch sizes. Simultaneously, there has developed an increasing demand for machines capable of producing smaller batches of parts with increased variability while also providing increased accuracy and ability to cope with complex machined parts in order to assist assembly. In the face of increased competitiveness from newly industrialized countries and Japanese manufacturing techniques, these conditions represent "more flexible production, geared to more exacting requirements with regard to design and precision, and subject to tighter control. CNC development and application has responded to such challenges, and has helped to bring them forth" (Sorge et al., 1983, p. 112).

Yet in the face of these similar developments in information technology, machine tools, and production conditions, the research of Sorge et al. shows that the reaction of firms differ in different national contexts, shaped by "unchanging socio-technical traditions." The factors discovered to have a key influence were company and plant size, batch size, type of cutting and machinery, national institutions and habits of technical work, management and training, and socioeconomic conditions relating to natural resources, size of product markets, and rate of growth. These result in a national patterning of firms' responses to the selection and development of CIM systems and associated forms of work organization.

A key feature of the differences between sociotechnical trajectories of CNC use and development in West German and Great Britain proved to be the degree to which "indirect" production functions (including work planning, programming, machine setting, quality control, maintenance, technical design and development, etc.) were allocated to shop floor personnel involved in "direct" production. Particular emphasis was placed upon the location of programming between the planning/programming department, foremen/supervisors, setters, and operators, with West German firms having a substantially greater degree of devolution of programming responsibilities. This phenomena was related by Sorge et al. to the different organizational and labor conditions in West Germany and Great Britain. Labor conditions refer to the recruitment, training, retraining, remuneration, and general treatment of labor; but Sorge et al. emphasize the central role of the education and training system broadly defined to include less formal ongoing workplace learning or "acquisition of competence." In West Germany this is characterized by a "pervasive practical professionalism."

These labor conditions interact closely with and strongly influence organizational structures and processes, and for Sorge et al. are primarily determined by them. In West Germany the organizational hierarchy from the foremen or Meister upward is more highly technically based, with less functional division between line management and specialists; the ratio of technical support staff to shop floor workers is relatively low, with the largest proportion of works personnel and the lowest percentage of supervisors as a percentage of works employees; in general there is only one level of staff supervision (Meister) compared to two in Great Britain and France; there is less departmental differentiation or fragmentation, and formal organizational charts or structures are more personalized and in general less formalized; and plant control by industrial firms is more technical and production oriented rather than finance based. It is not surprising, therefore, that operators tend to operate more flexibly across jobs, and tend to have greater autonomy. For example, work plans and time allowances are challenged more often in Germany by supervisors; work groups and foremen tend to be given a greater role in planning; and foremen in Germany are freer to detail workers to particular jobs during a shift as specialist departments tend to detail work less precisely.

The result of this analysis provided by Sorge et al. was to show the manner in which the West German national systems of education and training influenced the development of organizational structures and processes which, in turn, influenced the development and use of CNC technologies. However, despite their statement that they are examining the forces influencing "the continuous adaptation, improvement and modification of technology," their work stops short of showing how the incremental modification and development of CNC technologies is influenced by and combines with organizational and labor conditions to form significantly different sociotechnical trajectories. They discuss, for example, the increasing simplicity of computer–user interfaces, and the use of dialog-based programming languages, but this is not extended into an analysis of the form and extent to which a specific software/hardware trajectory may be developed and diffused in a German context that gives greater programming to skilled craft workers and involves an innovation system that includes close relationships between Fraunhofer and other engineering institutes, machine tool producers, and technologically sophisticated user firms. This matter has been addressed, for example, in the development, with the assistance of the German Research and Technology Ministry's Program Fertigungstechik, of workshop-oriented programming methods. Moreover, many of the key issues are addressed more directly in the human centered or anthropocentric CIM perspectives. The societal effects case study gives us an insight into key variations in the organizational components of CIM systems rather than the effect of variable organizational characteristics on CIM "technologies."

Anti-Taylorist or Antitechnocratic CIM Design Approach

This approach incorporates the proponents of *human centered*, *anthropocentric*, or *skill-based* CIM as well as Scandinavian *work-oriented* and *computer-supported cooperative work* perspectives (Ehn, 1988; Warner et al., 1990; Rosenbrock, 1990). Anti-Taylorist or nontechnocratic forms of CIM are those technical systems that support rather than undermine the utilization and development of both "tacit" and "formal" production skills in both the *use of systems* and the *understanding and development of systems*. The design ideal is opposed to Taylorist or traditional engineering/technocratic approaches that (i) identify efficiency with the maximum extraction, codification, and automation of all production knowledge and skills, (ii) view the remaining human activities within engineering systems as temporary sources of error or disturbance, and (iii) define human factors issues as training in the requirements of system use and measures necessary to "sell" systems to users and prevent opposition. In contrast to this view a design ideal is upheld that is committed to the creation of human–machine systems that *subordinate* the hardware and software of machine systems to the *control* of human operators.

In computer system design the emphasis is on computer *support* of worker initiative and group work, and the creation of *transparent tools* for use by users skilled in the production area in which they are to be deployed. In European initiatives to promote human-centered or anthropocentric technologies, the "human" is placed at the "center" of such systems in a vision of production that both draws upon and develops traditional images of *craft* work and *organic* forms of production. The image of the skill-based factory of the future is centred on computer-aided craftspeople rather than monitoring and error recovery personnel maintaining an independent integrated manufacturing system. The informating potential of new technologies is utilized to the maximum to assist skilled operators to make crucial judgments concerning system operation and development. The solution to the problems of manufacture are sought in creating technology to develop and support human creativity, interaction, communication, and skills in computer and *human*-integrated manufacturing or computer-*aided* production systems—not removing system dependence on human skills through maximum automation, computer-*guided* systems or *computer*-integrated manufacture (Badham and Shallock, 1991a).

These approaches to CIM continue and develop upon the traditional sociotechnical and organization theory contrasts between *mechanical* and *organic* systems, low-trust and high-trust cultures, Theory X and Theory Y management philosophies, and so forth, with a strong commitment to the establishment of more organic systems and high-trust production cultures supported by Theory Y management philosophies (Badham and Mathews, 1989). In this sense there are a number of similarities with the image of the skill-based sociotechnical *integrated factory* provided by writers such as Susman and Chase in the United States (Susman and Chase, 1986). According to this view, the contemporary introduction of flexible and integrated technologies in a manufacturing environment that emphasises product variety, reduced lead time, and rapid product innovation results in a general increase in the need for worker's skills, responsibility, and teamwork. Key emphasis by supporters of these approaches are: maximum decentralization of direct and indirect production functions to direct production workers, the creation of semiautonomous groups in production cells, and continuous incremental innovation in skill enhancement, organizational effectiveness, and technical system capabilities.

The nontechnocratic or anti-Taylorist approach to CIM further develops this approach in three main ways. First, detailed attention is paid to the character and shaping of the technical system. in a manner that goes beyond the design of human–machine user interfaces, and incorporates the embeddedness (and consequences) of wider organizational interests and assumptions in system architecture (Badham and Shallock, 1991a; McLoughlin and Clark, 1988). As Charles Perrow has aptly observed in his discussion of the relevance of organizational theory to human factors engineering, it is of crucial importance to pay attention to "the way (things)—equipment, its layout, its ease of operation and maintenance—are shaped by organizational structure and top management interests, and in turn shape operator behaviour . . . a largely unquestioned social construction of reality—one that should be questioned" (Perrow, 1983, p. 540). Many of the German and European projects, for example, have successfully addressed these issues in the process of CIM system design by providing practical suggestions in areas such as the formulation of utilizable human–machine interface criteria (Corbett, 1987) and methods for dealing with issues such as the allocation of functions between individuals and machines in the process of system development (Clegg et al., 1989), and developed approaches to address the overall data management requirements of computer- and human-integrated CIM models (Corbett et al., 1990).

Second, the development of principles and working manuals for assessing the adequacy of new system designs incorporates not only the range of job functions for employees but also the depth of skill involved in the performance of these functions, suitable stress levels as well as creative demands, and the requirements for meaningful definitions of group work and control over working conditions. The utilization of these principles and manuals assists in overcoming many of the traditional problems and ambiguities in the discussion of group work, that is, intensification masquerading as enrichment, decentralization of functions accompanied by the standardization and deskilling of the performance of these functions, and so forth.[2] Many of the projects carried out by adherents to this perspective have been undertaken in conscious recognition of the previous limitations of sociotechnical, user-oriented, and participatory system design techniques, and the legitimatory and antitrade union purposes that such projects can embody (Corbett et al., 1990).

Third, in adopting a broader view of CIM system characteristics subjected to anti-Taylorist or antitechnocratic design principles greater attention is paid to the broader economic and organizational factors that have to be considered in appropriate and effective system design. Peter Brödner, author of *Factory 2000*, has provided a classic "onion model" for informing a design process that takes into consideration such factors. This specifies system design steps as proceeding from an analysis of the strategic requirements of the firm to the specification of the work functions and tasks that have to be fulfilled, and then the sequencing of work, and only subsequently the division of functions between human and machine, human–machine interaction, and hardware and software requirements (Brödner, 1988). In considering strategic issues, attention is paid to corporate strategic aims and managerial philosophies, interdepartmental structures and interests, the structural system of authority and control in the organization, job security and payment systems, and even the set of informal cultural expectations and practices of the employees whose jobs are to be redesigned (Kelly, 1982). The anti-Taylorist CIM approaches stress, in particular, the significance of addressing broader considerations of

[2]In Germany the work of work psychologists Ulich, Hacker, and Volpert has been developed, particularly at the Institute for Human Sciences in Work and Training at the Technical University, Berlin, into a precise set of principles and working manuals (RHIA, VERA, and micro-VERA job analysis techniques) for assessing the humane character of new work systems and assisting in system redesign. Among a number of successful projects utilizing these principles, there is presently an important study being conducted of the varying conditions of CNC use and production island organization, as well as a proactive system design project in Berlin to establish production islands utilizing the full range of these principles (Badham and Schallock, 1991a).

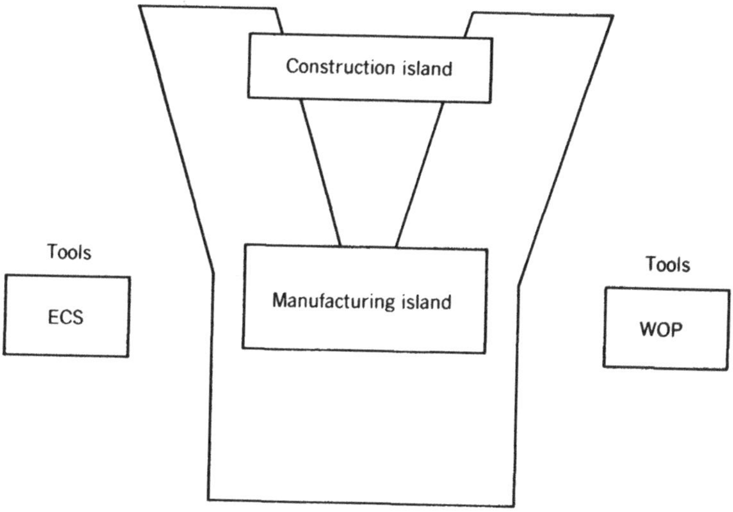

Fig. 13.1. Decentralized CIM philosophy [Source: Bandemer, S., Henning, J., and Hilbert, J. (1991), p. 19].

labor market conditions and product market characteristics and strategies as a central influence on CIM design and operation.

The promotion of human-centered approaches in Europe is explicitly based on the assumption that European product markets (high quality, customized, smaller batch size) and labor markets (high skill, unionized) support the search for a non-Taylorist antitechnocratic form of CIM development. This view is promoted, among others, by the European Commission Forecasting and Assessment of Science and Technology (FAST) program, the West German Research and Technology Ministry's Manufacturing Technology Program and Work and Technology Program, and projects within the European Commission ESPRIT and BRITE-EURAM programs (Warner et al., 1990).

Case Study 2: Decentralized CIM in West Germany

Utilizing a view of CIM outlined in Fig. 13.1, the decentralized CIM philosophy within West Germany has been schematically represented in Fig. 13.2. Workshop-oriented programming (WOP) is based on the increasingly extensive use of workshop programming devices on CNC machines in mechanical and electrical engineering (60 to 70 percent by the second half of the 1980s). However, most of these systems, based on DIN 66025 (German industrial standard), despite increase use of graphical interfaces and tools for calculation, use abstract codes, and it takes about one year to become familiar with the programming system. Contemporary WOP systems, developed and promoted with the assistance of the German Manufacturing Technologies Program, have been created to overcome this problem and make these systems

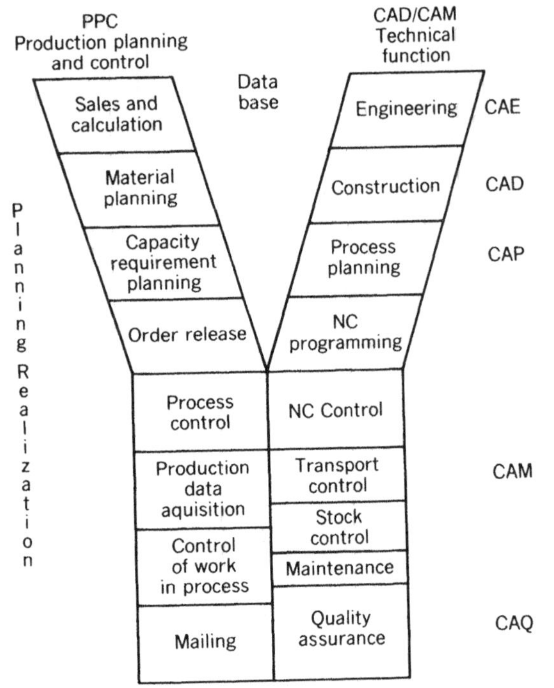

Fig. 13.2. CIM philosophy [Source: Scheer, A-W. (1988), CIM, Der Computergesteuerte Industriebtried, Springer, Berlin/Heidelberg, p. 3].

more accessible for use by traditionally skilled shop floor workers. As developed by the Traub Machine Tool Company: "These do not require the knowledge of abstract codes but ask the workers to fill in data in plain words or graphical symbols. . . . Inputs can be immediately controlled and corrected on the screen. The procedure is supported by the possibility to take recourse to similar programmes, which can be adjusted. The administration and transfer of data for tools and work pieces are provided. Functioning and reliability of programs may be tested by simulation of procedures" (Bandemer et al., 1991, p. 21).

Computer-aided design/workshop-oriented programming strategies are a more decentralized version of CAD/CAM strategies in which design data are passed onto the NC programming office for production programming rather than being centralized in the design office. Programming is carried out by operators or supervisors, where geometric data are taken over from the CAD-system by the shop floor programming system where programming and adjusting tasks can be dealt with directly by the shop floor personnel. There are, however, a variety of technical barriers to the diffusion of these systems created by the incompatibility and lack of integration of existing systems.

In the earlier production planning and control (PPC) systems, shop floor scheduling and particularly detailed production scheduling was identified in Germany as a major weak spot in the systems. This applied particularly to industries in which a high number of unscheduled disruptions occur and where rescheduling at short notice is often necessary, as in the field of mechanical engineering. The earlier PPC modules had two main problems: The immense quantities of data could only be processed in weekly runs yet lists of detailed job sequences often became outdated in hours; and highly automated scheduling processes focusing on capacity balancing and sequencing in line with alogarithns did not allow ad hoc alternatives resulting from dialog. Over the last few years, however, new decentralized scheduling systems have emerged, described in Germany as *elektronische Leitstände* (electronic control stations or ECS). Systems developed and utilized in the European Commission ESPRIT and German Research and Technology Manufacturing Technologies Program have been influential in this development. These systems run on autonomous small computers (PCs or workstations) with their own data bank and PPC interface; concentrate on the core function of detailed scheduling with their functional range built up successively; and use a graphic display modeled on the Gant chart, comprising workstations on one axis and a time scale on the other. Decentralized production scheduling still only occurs in a minority of firms, however, despite evidence of the widespread need for frequent production rescheduling. This situation is, however, predicted to change (Bandemer et al., 1991, p. 28).

In combination, the WOP and ECS systems only provide some technical tools that facilitate non-Taylorist forms of CIM. The technical/organizational model most favored in Germany is the manufacturing island (*Fertigunginsel*) concept. This concept has found its most popular expression in the reorganization of assembly line production in the car industry and in the elaboration of the group technology concept for discrete parts manufacturing. The latter incorporates the idea of group technology, combined with semiautonomous work groups and shop floor computer support, and reemerged in the 1980s as a major initiative in West Germany to meet the demands of small-batch production, rapid product innovation, highly costly and risky centralized scheduling systems, and the effective utilization of craft skills and less dependance on scarce technical skills. In the islands, operators' tasks (both direct and indirect) are widened as far as possible; computer-aided planning facilities are located at shop floor rather than planning department level; and, as far as possible, planning and scheduling functions are supported at production island rather than supervisor/area control level. Since the establishment by the Research and Technology Ministry of its production island project in 1980, there have been four major conferences held by the AWF Association in Eschborn to communicate experiences of manufacturing island developers, elaborating the principles and case study experiences of production islands. These proceedings represent the most extensive coverage of production island experiments. Some of the most significant projects are those established at Sulzer-Weise (Bruchsal), Felten & Guilleaume (Nordenham), and Kodak,

(Berlin), as well as others carried out by such firms as MAN Roland, Heidelberger Druck, Mercedes Benz, Krauss Maffei, Mannesman Demag, and others.

National Systems of Innovation Approach

This approach to CIM derives from analyses of the emergence of *techno-economic* paradigms within revolutionary periods of technological change. In the work of Freeman and Perez (Perez, 1983) the analysis of the new era of *systemation*[3] emphasizes the important role played by new technology systems in both promoting the technical side of CIM and creating a new "technical common sense" among managers and engineers. The heart of the contemporary technological revolution is identified as the set of radical and incremental innovations driven by the declining cost, increasing capabilities, and pervasive application of the new key factor of production: microelectronics.

This approach also recognizes the importance of the *institutional framework* of change and recognizes the existence of variations in both levels of CIM and sectoral or national institutional frameworks. As observed by John Child, the general concept of integrated manufacturing theoretically creates a technical situation in which decision making could be made at any point within the system. The types of decisions that personnel are called upon to make, and which personnel make these decisions, may therefore be variable. In Perez's words, "the new paradigm combines trends towards centralization and decentralization, towards more control and more autonomy, so the variety of combinations is likely to be quite wide, not only in the present transitional period, but probably into the future upswing" (Perez, 1983, p. 456). The variety of different CIM components provide firms with a number of options. "Firms certainly can draw upon a variety of new technological tools, from NC machines to CAD/CAM systems including telecom networks. But what do they do with them and how do they use them? The range of new manufacturing arrangements (e.g. the various implementation of 'just-in-time'), new logistics know-how (e.g. autonomous production islands linked through new parts circulation schemes, such as Group Technology), and stock management methods (e.g. Materials Requirement Planning) together provide

[3]New production systems are appropriately characterized as primarily sociotechnical rather than technical systems, embodying a "new trend towards merging all activities—managerial and productive, white and blue collar, design and marketing, economic and technological—into one single interactive system." The explicit intention is to emphasize "the systemic, feedback nature" of the organizational software as well as the hardware as the "essential distinguishing feature" of the new firm. In the development of this production model, they recognize the important autonomous role to Japanese organizational techniques in the "parallel and even antecedent change" of Japanese organization to the microelectronics revolution, and the origins of the present system as lying in both microelectronics and the Japanese flexible organizational model (see Perez, 1983, p. 453, and Freeman, 1984, p. 503).

the firm with a substantially rejuvenated and enriched reservoir of methods to organize work and production" (Zysman et al., 1991, p. 38).

This is clearly observable in sharp contrasts between the strategies of General Motors and Toyota or Levi Strauss and Benetton. In the case of GM, the attempt to compete with Japan and reduce lead times has been a high-tech strategy to use the "4Cs" (CAD, CAM, CIM, CAE) to achieve "fast to market" benefits. Toyota, in contrast, focuses on carefully crafted interaction within and between work teams and with business partners (it is significant that one Toyota director recently remarked that the company is not a parent company to subsidiaries but a sister). Little attempt has been made to automate kanban paper slip based systems with electronic mail networks. As Zysman et al. observe: "While Toyota resists introducing information technology within its adaptable process, GM hopes investment in adaptable information technology can substitute for process changes, or even drive them" (Zysman et al., 1991, p. 41).

In Benneton's case, a *flexible specialization* response to fast changing product markets has been to detect and produce for fashion changes fast, through an electronically linked network of subcontractors and retail stores. Levi's *modified mass production* strategy, in contrast, has focused on the continued production of variations on a standard product and the use of a vertical production and distribution system that has attempted to implement a proprietary communications system with retailers in order to foster an industry-wide electronic data interchange (EDI) standard.

In contemporary ongoing research on national systems of innovation and national conditions of enterprise calculation, the *institutional frameworks* that are seen as directing technological innovation include financial institutions, government activity and policy, national systems of skill formation, market structure, capital goods user/producer relations, dominant national economic blocks, and the composition of demand, as well as the industrial relations system.[4] Interest is focused on such factors as the key national industrial complexes, the quality of domestic demand, factor endowments, the system of structure and competition, and the R&D and learning system (Dalum, 1990).

In the work of Edquist and Lundvall (1989), this is systematized into a distinction between the influence of a country's *institutional setup* and *economic structure*. The former provides the set of formalized rules and non-formalized norms and habits that govern general "learning" values and practices, and the latter refers to the production structures that provide the integrated set of user–producer innovation trajectories in a country's leading economic sector or "development block." As this approach stresses the importance of

[4]See, for example, the discussion in R. Badham and B. Schallock (1990), "The Human Shaping of CIM," *IEE Proceedings IT and People Conference*, Bournemouth; and K. Williams, J. Williams, and J. Thomas (1984), *Why are the British Bad at Manufacturing?* Routledge and Kegan Paul, London.

sectoral as well as national influences on technological trajectories, the type and organization of industrial sectors in the economy is clearly of great significance. Central importance is attached, for example, to the kinds of alternative innovation trajectories outlined by Pavitt earlier, and the different dynamics of semiconducter, automobile, production equipment/component, and process sectors. The institutional structure is regarded by Edquist and Lundvall as the "software" for the "hardware" of these production structures that together form "a framework defining the space for interactive learning and affecting its rate and direction" (Edquist and Lundvall, 1989, p. 16).[5]

As this indicates, a key emphasis of this work on national systems of innovation is the evolutionary and incremental nature of most learning and innovation patterns. Central importance is attached to routine learning within production, marketing, and R&D as a process that both generates problems to be solved by "searching" organizations (e.g., R&D laboratories) and contributes to their solution. The incremental pattern of such changes within national systems, influenced by the difficulty of establishing transnational learning systems, leads to the relative stability over time of dominant paths or trajectories within countries, even in the face of conscious national policies to change their direction.

This approach is based on extensive studies of the process and history of technical change that reveal that "technology is not a set of blueprints given by scientific advances that occur independently of the production process. It is, rather, a path dependent process of learning that reflects the community and market context within which it is developed." This view is succinctly summarized in Zysman et al., 1991, p. 17/8):

> Technological knowledge is not simply information that can be bought or sold, but rather often a subtler set of insights that develop only in conjunction with both design and production. The process is step by step, an iterative interaction between opportunities and knowledge in specific settings . . . . technological knowledge is often local in nature and advances incrementally. Such knowledge accumulated in local institutions in the form of learned know-how is intimately connected to locales in which it is generated and does not necessarily move between regions within countries, let alone nations. It accumulates in firms in the form of skilled workers and proprietary technology and difficult-to-copy know-how. It accumulates in communities in such diverse forms as suppliers, repair services and networks of know-how. It accumulates in nations in the skills and experiences of the work force and in the institutions that train workers and diffuse technology. This kind of local or non-traded knowledge is crucial during the initial development phase of new products and processes. Such accumulated knowledge, implicit in some cases and institutionally rooted in others, does not flow easily or quickly across national borders.

[5]Dalum (1990) illustrates this most effectively with evidence of the evolutionary trajectories of high-tech, medium-tech, and low-tech sectors within individual OECD countries, trajectories which, as measured by their relative position in the nation's economy, are relatively little affected by the varied governmental attempts to promote high-tech industry.

The resulting *technological trajectories* are consequently centrally influenced by knowledge strongly linked to context-specific production problems, linkages between production companies, suppliers, distributors, and supporting institutions mediated by both market and nonmarket ties, and the spillover of knowledge between these entities fostered by these linkages.

Case Study 3: Swedish National System of Innovation

One important example of this approach is the analysis carried out by Edquist and Lundvall of the Swedish and Danish national systems of innovation and their influence on the nature and direction of CIM innovation. In the examination of the Swedish system they focus on two well-known phenomena of the Swedish experience of CIM: first, the contrast between a rapid diffusion of CIM technologies accompanied by a relatively low level of product innovation; and, second, the internationally high proportion of group work and job enrichment accompanying new CIM systems.

The central point of Edquist and Lundvall is that the small country status of Sweden (ruling out huge investment and sales efforts in new products), its collective consumption patterns, plus its traditional successes in process diffusion, have led to a very low level of product innovation and diffusion. Yet, it has probably the highest rate of process diffusion as a result of its facilitative institutional framework and economic structure. The significance of the institutional structure is primarily related to the social democratic compromise and an extensive welfare state in a situation of low unemployment that has led to (a) the trade unions' acceptance of technological innovation in a situation of very low unemployment and the strength to insist on labor's share in profits, and (b) employers incentive to build high trust and competence building relations with workers. Within the economic structure, the dominance of a large firm, export oriented, high R&D engineering *development block* has strongly promoted the early adoption of advanced technology. The result has been that

> the whole institutional set up, the whole innovation system, seems to have become geared towards one specific model of development (towards the movement along one trajectory) and the ease with which advances have been made in this direction may have weakened the incentives to take on the more difficult task of developing the technological infra-structural basis necessary for radically new product areas (Edquist and Lundvall, 1989, p. 48).

The argument of Edquist and Glimmel (1989) concerning sociotechnical CIM trajectories is that in this context Swedish firms have tended to introduce enlarged jobs and achieve reductions in the vertical division of labor as a key factor associated with the introduction of technology—in a labor market context characterized by a severe labor shortage and strong trade union organization. These twin factors have led to management attempting to make jobs more interesting in order to reduce absenteeism and turnover, and attract workers. However, this has not been intimately related to the requirements of major product development and diffusion, nor to the strength of the country's vocational and educational system—factors that are more significant in understanding the decentralized CIM strategies characteristic of West German engineering industry. It is no surprise, therefore, that Sweden is one of the world leaders in both the development of user-oriented CIM design methodologies and the focus upon the retention and enhancement of craft skills in new software designs. As emphasized by Edquist and Glimmel, however, the weaknesses in product innovation, a degree of management conservatism, and the absence of a level of craft training equivalent in strength to the German system, may all pose serious problems for the continuation of the Swedish model. Despite the presence of exemplary projects, such as that of the Volvo plant at Uddevalla, they pointedly observe that "its a long way to Uddevalla, even for those working next door" (Edquist and Glimell, 1989, p. 39).

13.4 CONCLUSION

Any review of international perspectives on CIM is immediately faced by the variety of different disciplines that address themselves to the understanding and explanation of CIM. Among these are, of course, the different branches of engineering and computer science creating CIM components and designs, and human factors practitioners involved in the design and implementation of human–machine interfaces, usable systems, and so forth. In addition, however: management scientists and consultants offer a variety of different approaches to managing the selection and implementation of CIM; economists of innovation have provided elaborate models and detailed empirical evidence explaining the diffusion of CIM components; innovation policy analysts have been concerned with prescriptions over how diffusion processes may be speeded up by government intervention; education and training and personnel experts as well as human factors researchers have detailed the importance of new forms of skills and training and changing attitudes toward the importance of the human factor in production; and industrial sociologists or industrial relations researchers have investigated the forms of skills and work organization resulting from the introduction of CIM systems, and examined the influences upon and influence of different CIM strategies upon these forms.

As revealed by the approaches reviewed in this chapter, there are a number of different perspectives focusing particularly on variations in CIM systems, with specific concentration on the conditions that promote more decentralized, flexible, and skill-enhancing forms of CIM. An understanding of these influences is of crucial importance in any attempt to extend the impact of human factors insights and ideals in the design and implementation of CIM systems. At one level this can be understood as the application of sociotechnical systems (STS) principles to CIM. However, the contemporary approaches outlined here go far beyond the traditional limitations of STS to take into account systems of management–labor conflict, the institutional and economic factors shaping technological trajectories, and the limitations of traditional participatory approaches to system design.

As indicated by the societal effects and national systems of innovation approaches, a key issue in CIM design and implementation is the *nature and form of technical knowledge and its role within the production process*. The incremental, ad hoc, and tacit character of much effective production knowledge means that the final shape of effective CIM systems will not only be influenced by this knowledge but should act to support and enhance it. The form taken by such knowledge will, however, differ between firms, sectors, and nations. The consequent shape of CIM systems must also vary to take this into account. In order to understand and influence this process attention is required to both the *economic structure or set of interfirm linkages* within development blocs and the *broader institutional structure* constituted by national patterns of skill formation and conditions shaping enterprise calculation.

Yet, how should these conditions be influenced? What direction of CIM is most desirable? An attempt to answer these questions has been provided by the set of anti-Taylorist or antitechnocentric perspectives on CIM. They have been successful in elaborating some system design principles, techniques for system design processes, and establishing working exemplary models of more decentralized and skill-enhancing CIM structures. These have, however, emerged from a European context and have been strongly influenced by German industrial traditions, particularly craft traditions within the machine tool industry, and Scandinavian experiments in craft-oriented system design (Salzman, 1991). Less attention has been paid to desirable forms of system design appropriate to nations or firms with lower levels of skills, occupations traditionally defined as lower skilled or part-time and more traditionally occupied by women, industries more characterized by mass production or process production systems, and cultures less influenced by craft traditions or strong work orientations. Moreover, the success of the anti-Taylorist CIM approaches in grappling with some key principles in systems design has not been accompanied by much sophisticated examination of the broader institutional influences on CIM system design that should inform the efforts of any movement to effectively direct CIM development. This latter enterprise may be assisted in the near future by the societal effects and national systems of innovation approaches, perspectives that have traditionally reflected little on how desirable CIM trajectories should be defined and operationalized.

The conclusion that should emerge from this review of international perspectives on CIM is not, however, that there is a clear understanding of the range of variation in CIM systems, the factors influencing this variation, or the nature of more desirable forms. Once general statements about CIM have been made, there is a substantial degree of uncertainty in our understanding in all these areas. As recently illustrated in models developed of the Japanese lean production system, new versions are being offered of best practice production in manufacturing (Womack et al., 1990). Yet already the universal character of the lean production system, and its long-term sustainability in both Japan and foreign production cultures has been brought into question by evidence of cultural antagonism and labor protest both within and outside Japan (Berggren, 1992; Dankbaar, 1991). The purpose of this chapter will have been fulfilled, however, if these uncertainties are perceived as a major challenge for investigation rather than a problem to be avoided; more attention consequently paid in the development of CIM models to *systematic* variations in the context and purposes for which CIM is employed, and more research directed to integrating international perspectives on CIM to address this issue.

ACKNOWLEDGMENTS

The author wishes to thank the German von Humboldt Foundation for its generosity in funding an 18-month period of research in Europe that made it possible to gain access to the divergent international perspectives on CIM. Thanks are also due to the Australian Department of Industry, Technology and Commerce for funding a number of visits to Europe to participate in European Commission networks fostering new approaches to the diffusion of human factors in CIM.

REFERENCES

ACTU (1987). *Australia Reconstructed*, Australian Government Publishing Service, Canberra.

Aicholzer, G., and Schienstock, Eds. (in press). *Technology Policy: Towards an Integration of Social and Ecological Concerns*, De Gruyter, Berlin.

Ayres, R., and Ranta, J. (1989). Factors governing the evolution and diffusion of CIM, in *Trends and Impacts of Computer Integrated Manufacturing*, J. Ranta, Ed., International Institute for Applied Systems Analysis, Laxenburg, Austria.

Badham, R. (1986). Theories of Industrial Society, London.

Badham, R., Ed. (1991). Technology and culture, Special Issue of AI and Society, **5**(4), Oct.–Dec.

Badham, R., Ed. (1992). Systems, networks and configurations: Inside the implementation process, Special Issue of *International Journal of Human Factors in Manufacturing*, **3**, 1.

Badham, R. (Ed.) (1993). Systems, networks, and configurations inside the implementation process, Special Issue of *International Journal of Human Factors in Manufacturing*, **3**, 1.

Badham, R., and Mathews, J. (1989). The new production systems debate, *Labour and Industry*, **2**(2), 194–247.

Badham, R., and Schallock, B. (1991a). Human factors in CIM development: A human centred view from Europe, *International Journal on Human Factors in Manufacturing*, **1**, 2.

Badham, R., and Schallock, B. (1991b). The human shaping of CIM, in *Information Technology and People*, R. A. Peacock, et al., Eds., Open University/Peter Pereguinnes, Milton Keynes, UK.

Bandemer, S., Hennig, J., and Hilbert, J. (1991). *Prospects of Anthropocentric Production Systems in West Germany*, CEC FAST-MONITOR, APS Research paper Series, FOP 251, Brussels.

Berggren, C. (1992). Lean Production: The End of History, STA Working Paper Series, STS Department, Wollongong University, Wollongong 2500, Australia.

Bessant, J., and Grunt, M. (1985). *Management and Manufacturing Innovation in the UK and West Germany*, Gower, London.

BMFT, (1990). Research and Development Program "Work and Technology, Project Management Team "Work and Technology" on behalf of the Federal Minister for Research and Technology, Bonn, April.

Boguslaw, R. (1969). *The New Utopians*, Prentice Hall, New York.

Brödner, P., (1988). Options for CIM: "Unmanned factory" versus skill-based manufacturing, *Computer Integrated Manufacturing Systems*, **1**(2), 67–74.

Brödner, P. (1989). In search of the computer aided craftsman, *AI and Society*, **3**, 39–46.

Brödner, P. (1990). *The Shape of Future Technology*, Springer- Verlag, London.

Bruno, S., Cohendet, P., Desmartin, F., Llerena, D., Llerena, P., and Sorge, A. (1991). *Modes of Usage and Diffusion of New Technologies and New Knowledge: A Synthesis Report*, CEC Monitor-FAST Programme, Prospective Dossier No. 1, Vol. 2, European Commission, Brussels.

Buchanan, D., and Boddy, B. (1983). *Organisations in the Computer Age: Technological Imperatives and Strategic Choice*, Gower, Aldershot.

Child, J. (1987). Organizational design for advanced manufacturing technology, in *The Human Side of Advanced Manufacturing Technology*, T. Wall, et al., Eds., Wiley, Chichester, London.

Clegg, C., Ravden, S., Corbett, J. M., and Johnson, G. (1989). Allocating functions in computer integrated manufacturing: A review and a new method, *Behaviour and Information Technology*, **8**, 175–190.

Clegg, C., and Symon, G. 1989. A Review of Human-Centred Manufacturing Technology and a Framework for Its Design and Evaluation, MRC/ESRC Social and Applied Psychology Unit, University of Sheffield, Memo No. 1036.

Corbett, J. M. (1987). Human work design criteria and the design process: The devil in the detail, in *Skill-Based Automated Manufacturing*, P. Brödner, Ed., Pergamon Press, Oxford.

Corbett, J., Rauner, F., Rasmussen, L., and Corbett, J. M. (1990). *Crossing the Border: The Social and Engineering Design of Computer Integrated Manufacturing Systems*, Springer, London.

Dalum, B. (1990). National Systems of Innovation and Technology Policy: The Case of Denmark, Paper presented to the workshop on Socially Oriented Technology Policy, Institute for Advanced Studies, Vienna. April 3–4.

Daniel, W. W. (1987). *Workplace Industrial Relations and Technical Change*, Frances Pinter, London.

Dankbaar, B. (1991). Presentation and analysis of "Toyotism," in *Man, Technology: Developments and Perspectives of New Production Systems*, Atlantech, CEC FAST-Monitor, FOP225, Brussels.

Economist (1987). Factory of the future, *Economist*, May 30, pp. 35–59.

Edquist, C. and Glimell, H. (1989). Swedish Frontiers of Change—A Guide to the Impact of New Technologies, Work Designs and Management Practices, Report prepared for the Directorate for Social Affairs, Manpower and Education, OECD, Mimeo, ADD-TECH Program, Linkoeping University, Sweden.

Edquist, C., and Jacobsson, S. (1988). *Flexible Automation: The Global Diffusion of New Technology in the Engineering Industry*, Basil Blackwell, Oxford.

Edquist, C., and Lundvall, B.-A. (1989). Comparing Small Nordic Systems of Innovation, Paper presented to the Maastricht seminar on National Systems Supporting Technical Progress, ADD-TECH Program, Linkoeping University, November, Sweden.

Ehn, P. (1988). *Work Oriented Design of Computer Artifacts*, Arbetslivcentrum, Stockholm.

Fleck, J. (1988). The development of information integration: Beyond CIM?, in *Trends and Impacts of Computer Integrated Manufacturing*, J. Ranta, Ed., International Institute for Applied Systems Analysis, Laxenburg, Austria.

Fleck, J., Webster, J., and Williams, R. (1989). The dynamics of IT implementation: A reassessment of paradigms and trajectories of development, Edinburgh PICT Working Paper Series, University of Edinburgh, Edinburgh, Scotland.

Forrester, R. (1987). *High Tech Society*, Blackwell, Oxford.

Fortune, (1983). The race to the automatic factory, *Fortune*,

Hagedoorn, J. (1989). *The Dynamic Analysis of Innovation and Diffusion: A Study in Process Control*, Pinter, London and New York.

Hayes, R., and Jaikumar, R. (1988). Manufacturing's crisis: New technologies, obsolete organisations, *Harvard Business Review*, September–October, 77–85.

Haywood, B., and Bessant, J. (1990). in *New Technology and Manufacturing Management*, Warner et al., Eds., Wiley/FAST, London.

Hildebrandt, E. (1989). The Social Constitution of the Firm, Paper presented to CAPIRN Workshop, Santa Cruz, November.

Ingersoll Engineers (1984). Proceedings of FMS-2, Ingersoll, London.

Kelly, J. E., (1982). *Scientific Management, Job Redesign and Work Performance*, Academic Press, New York.

Koetter, W., Gohde, H., and Weber, W. (1989). Technical and organizational options for skill based task design in a group technology project, *IFAC/IFIP/IMACS Symposium on Skill Based Automated Production*, Austrian Centre for Productivity and Efficiency, Wien, Austria, November.

Martin, T., Ulich, E., and Warnecke, H. (1987). Appropriate automation for flexible manufacturing, in *Selected Papers from Automatic Control World Congress*, R. Isermann, Ed., July, IFAC, Munich, pp. 27–31.

Martin, T. (1990). The need for human skills in production—the case of CIM, *Computers in Industry*, **14**, 205–211.

Matzner, E., and Streeck, W., Eds. (1991). *Beyond Keynesianism*, Elgar, Alderslot, UK.

McLoughlin, I. and Clark, J. (1988). *Technological Change at Work*, Open University, Milton Keynes.

Molina, A. H. (1989). The Transputer Constituency: Building up UK/European Capabilities in Information Technology, PICT Research Report Series No. 1, University of Edinburgh, Edinburgh, Scotland.

OECD, (1988a). *New Technologies in the 1990s: A Socio-Economic Strategy*, OECD, Paris.

OECD, (1988b). *Science and Technology Policy Outlook*, OECD, Paris.

Pavitt, K. (1990). What we know about the strategic management of technology, *California Management Review*, Spring, pp. 94–106.

Perez, C., (1983). Structural change and the assimilation of new technologies in the social and economic eystems, *Futures*, **7**, 357–375.

Perrow, C. (1983). The organizational context of human factors engineering, *Administrative Science Quarterly*, **28**, 521–541.

Ranta, J., Ed. (1989). *Trends and Impacts of Computer Integrated Manufacturing*, International Institute for Applied Systems Analysis, Laxenburg, Austria.

Rosenbrock, H., Ed. (1990). *Designing Human-Centred Technology: A Cross Disciplinary Project in Computer Aided Manufacturing*, Springer-Verlag, London.

Salzman, H. (1991). Engineering and technology design in the United States, in R. Badham

Schallock, B. (1991). Decentralised shop floor control, paper presented to *Automatica 1991*, November, Melbourne, Australia.

Schultz-Wild, R. (1990). Application of IT and the restructuring of production work in West Germany, paper presented to the New Technologies, *Institutional Changes and Working Life Seminar*, Holmenkollen, Oslo, June.

Senker, P., and Beesley, M. (1986). The need for skills in the factory of the future, *New Technology, Work and Employment*, **2**, 9–18.

Sorge, A., and Streeck, W. (1988). Industrial relations and technical change: The case for an extended perspective, in *New Technology and Industrial Relations*, R. Hyman and W. Streeck, Eds., Blackwell, Oxford.

Sorge, A., Hartmann, G., Warner, M., and Nicholas, I. (1983). *Microelectronics and Manpower in Manufacturing—Applications of Computer Numerical Control in Great Britain and West Germany*, Gower, Aldershot.

Streeck, W. (1987). Industrial Relations: Agenda for Change: The Federal Republic of Germany, Draft Report to the OECD Working Party on Industrial Relations, 18 August.

Susman, G., and Chase, R. (1986). A socio-technical analysis of the integrated factory, *Journal of Applied Behavioral Science*, **22**, 257–270.

Teicholz, E., and J. Orr, Eds. (1987). *Computer-Integrated Manufacturing Handbook*, McGraw-Hill, New York.

Teresko, J. (1987). CIM: Much more than adding computers, *Industry Week*, February 9, pp. 10–12.

Walton, R., and Susman, G. (1987). People policy for the new machines, *Harvard Business Review*, March–April, 98–106.

Walton, R., and Susman G. (1987). People policies for the new machines, *Harvard Business Review*, March–April.

Warnecke, H.-J. (1989). Integration of information and material flow in a pilot plant—development and experiences, in *Trends and Impacts of Computer-Integrated Manufacturing*, J. Ranta, Ed., International Institute for Applied Systems Analysis, Laxenburg, Austria.

Warner, M., Wobbe, W., and Broedner, P., Eds. (1990). *New Technology and Manufacturing Management*, Wiley/FAST, London.

Weber, M. (1974). *Protestant Ethic and the Spirit of Capitalism*, Allen and Unwin, London.

Weiner, E., and Brown, A. (1989). Human factors: The gap between humans and machines, *The Futurist*, May–June, pp. 9–11.

Westong, F. C. (1988). Computer integrated manufacturing systems: Factor or fantasy? *Business Horizons*, **31**, 4.

Williams, K., Williams, J., and Thomas, J. (1984). *Why Are the British Bad at Manufacturing?* Routledge and Kegan Paul, London.

Wobbe, W., (1989). Computer and integrated manufacturing (CHIM), *IFIP—Conference Information Systems, Work and Organization Design*, Humboldt University, Berlin, July 10–13.

Womack, J., Jones, D., Roos, P. (1990). *The Machine that Changed the World*, Macmillan, New York.

Zuboff, S. (1988). *In the Age of the Smart Machine*, Heinemann, New York.

Zylstra (1989). (10-20 - possibly similar to Havn in Human Factors)

Zysman, J., et al. (1991). Globalisation and Production, Berkeley Roundtable in the International Economy (BRIE) Working Paper, Berkeley, California.

CHAPTER 14

ECONOMIC JUSTIFICATION

S. KOLLI and H. PARSAEI
Department of Industrial Engineering
University of Louisville
Louisville, Kentucky

D. H. LILES
Department of Industrial Engineering
University of Texas At Arlington
Arlington, Texas

14.1 INTRODUCTION

Advanced manufacturing systems refer to computer-oriented technologies used in product design, manufacturing, service operations, and decision support. Robots, numerically controlled (NC) machines, flexible manufacturing systems (FMS), computer-aided design/computer-aided manufacturing (CAD/CAM), and computer-integrated manufacturing (CIM) are components of advanced manufacturing systems. The productivity of a manufacturing organization is dependent on the abilities, skills, and physical and mental characteristics of the human element in the system.

The implementation of advanced manufacturing systems is capital intensive. Estimation of costs and the identification of benefits to be gained from their implementation are time consuming. Nevertheless, the evaluation and

Organization and Management of Advanced Manufacturing. Edited by Waldemar Karwowski and Gavriel Salvendy.
ISBN 0-471-55508-8

selection of an advanced manufacturing system from a set of alternatives requires the use of a economic justification methodology.

Advanced manufacturing systems are briefly discussed in Section 14.2. The relevance of human factors in advanced manufacturing systems is discussed in Sections 14.3 and 14.4. Section 14.5 discusses the justification for advanced manufacturing systems. Various justification methodologies are presented in section 14.6.

14.2 ADVANCED MANUFACTURING SYSTEMS

Advanced manufacturing systems refer to automated, normally computer-oriented technology used in design, manufacturing, service operation, and decision support (Canada and Sullivan, 1989).

A detailed discussion of the advanced manufacturing systems is presented in Groover (1987). Advanced manufacturing systems provide various benefits such as competitiveness in the market, flexibility, low costs, and efficient throughput time. Some of the important benefits due to advanced manufacturing systems over those of less automated systems are shown in Table 14.1. A survey has shown that advanced manufacturing systems have reduced overall manufacturing cost by 30 percent (Meredith, 1987).

The performance of the manufacturing system is dependent on human operators. Therefore, changes in the manufacturing environment due to the implementation of computer-supported technology also necessitate renewed study and analysis of human aspects. The next sections briefly discuss human factors for advanced manufacturing systems.

Evaluation of the multitude of costs and benefits, both qualitative and quantitative, and the analysis of trade-offs is very complex. Methodologies for justifying advanced manufacturing systems are presented in a later section.

14.3 HUMAN-RELATED PROBLEMS IN MANUFACTURING

Some of the common human-related problems in the traditional manufacturing environment are back pain, fatigue, occupational stress, boredom, and diseases due to exposure of workers to uncomfortable or dangerous environmental conditions. These human-related problems lead to a number of problems for an organization. For example, the presence of these problems in an organization results in reduced productivity, quality, competitiveness in the market, increased scrap rate, lead times, work-in-process inventory, absenteeism, loss of work hours, and so forth.

The performance of a manufacturing system is very much dependent on its design. The design of a manufacturing system should provide a comfortable environment for work. Any change in the equipment or machine tools in the system should have a corresponding ergonomic change, otherwise the envi-

TABLE 14.1 Benefits of Advanced Manufacturing Systems

Strategic
Better competitive advantage
Ability to adjust to varied product life cycles
Development of engineering and management expertise
Lower exposure to labor unrest
Ability to introduce products faster in the market
Being viewed as the leader in the use of new technologies
Improvement in return on equity
Improvement in overall productivity
Tactical
Lower work-in-process inventories
Reduced setup time
Reduced throughput time
Reduced variability in throughput time
Lower fixtures and jigs costs
Improved manufacturing control
Improved quality
Reduced scrap rate
Reduction of floor space used
Reduced labor costs
Reduced tooling costs
Reduced rework
Better status monitor of machines, tools, and material handling devices
Improved data management
Improved control of operations
Improved control of parts
Improved response time to demand variations
Improved ability to adjust to machine breakdowns
Improved working conditions
Improved ability to respond to design or process change

From Choobineh (1986).

ronment becomes uncomfortable for work. For example, back pain may result from work involving loading and unloading of machines, storing and retrieving tools and parts, or in work involving a prolonged sitting or standing posture.

Fatigue is another common problem in the manufacturing industry. It is caused by uncomfortable environmental conditions, prolonged physical and/or mental effort, physical problems, nutrition, illnesses, and so forth. This state causes a loss of efficiency and disinclination for any kind of effort. Occupational stress results from the inability of a worker to cope with the demands of a job.

Another human-related problem in manufacturing is boredom, which normally results from prolonged repetitive work or prolonged monotonous supervisory work.

These human-related problems due to improper or poor ergonomic design of systems lead to reduced productivity, quality, increased down time, scrap rate, medical costs, and compensation—all of which are undesirable to a manufacturing organization. There are several examples in the literature that deal with the losses due to improper ergonomic design of human–machine systems (Simpson, 1988).

14.4 HUMAN FACTORS FOR ADVANCED MANUFACTURING SYSTEMS

Changes in the manufacturing environment due to the implementation of advanced manufacturing systems necessitate study of the effect of these systems on the human element and the design of an appropriate human–machine system. Various studies indicate that advanced manufacturing systems eliminate heavy and dangerous work, but the work is dissatisfying and stressful to the worker. Since these manufacturing systems are supported by a computer, a low level of worker input is required, and the tasks include monitoring, programming, and performing some support activities. Boredom, monotony, and stress were reported to have resulted from these tasks (Edwards, 1989). Tasks such as programming and maintenance offer some challenge that may have a positive effect on the worker.

Effective use of human factors engineering in manufacturing leads to various benefits such as improved health and safety of the employees, increased productivity, improved quality, reduced lead time, and reduced scrap rate (Alexander, 1985). Changes in human–machine environment, work design, material handling, and other manufacturing activities that occur because of human factors engineering can improve the productivity of manufacturing operations (Priest, 1985).

The important tasks in a manufacturing environment with computer-supported technology are programming, monitoring, and support activities. Human factors engineering should be utilized for work design and the design of the human–machine systems, so that productivity could be increased and the employee's well-being is improved.

14.5 JUSTIFICATION OF ADVANCED MANUFACTURING SYSTEMS

Various costs should be estimated and the qualitative and quantitative benefits identified. These costs and benefits are used to evaluate, rank, and select a project from a set of alternatives. Implementation of advanced manufacturing systems results in both tactical and strategic benefits. The justification should be based on both the short-term and long-term benefits.

Various methodologies are available for the economic justification of advanced manufacturing systems. The greatest barrier to the implementation of advanced manufacturing systems is the inability of traditional financial methods and capital budgeting techniques to measure the effectiveness of new technology (Canada and Sullivan, 1989, 1990; Kolli and Parsaei, 1992; Meredith and Suresh, 1986; Parsaei and Wilhelm, 1989). Most of the benefits due to the implementation of advanced manufacturing systems are intangible or nonmonetary and are hard to quantify and translate into values that can be used for evaluation.

Since there are a number of criteria associated with the implementation of advanced manufacturing systems, the justification of these systems should be based on these criteria. Multiple-criteria decision-making methods justify projects based on multiple, conflicting criteria. The decision maker can also specify a weight of importance to a criterion. Due to the uncertainty associated with the problem and imprecise information available, it is beneficial to conduct sensitivity analysis (a "what-if" kind of analysis) with the justification methodology used (Sullivan, 1986).

14.6 JUSTIFICATION METHODOLOGIES

This section discusses the various justification methodologies, both traditional financial methods and nontraditional methods. Justification methodologies can be divided into two groups based on the number of objectives they optimize. They are single-criterion methods and multiple-criteria methods.

Single-criterion methods optimize a single criterion such as net present worth. Multiple-criteria methods optimize various, conflicting criteria associated with the justification of advanced manufacturing systems. Each one of the groups contain both deterministic and nondeterministic methods. Figure 14.1 illustrates the methodologies for the justification of advanced manufacturing systems. These methods are discussed later with examples provided for some methods.

Single-Criterion Methods

The justification methodologies falling in this category employ a single economic objective to justify investment in advanced manufacturing technology. Traditional financial methods including net present value (NPV), internal rate of return (IRR), benefit/cost ratio (B/C), payback period, mathematical programming, and minimal annual revenue requirement (MARR) belong to this category. This category also includes certain optimization methods and the productivity model as well as nondeterministic methods such as sensitivity analysis, decision trees, optimistic/pessimistic analysis, and Monte Carlo simulation.

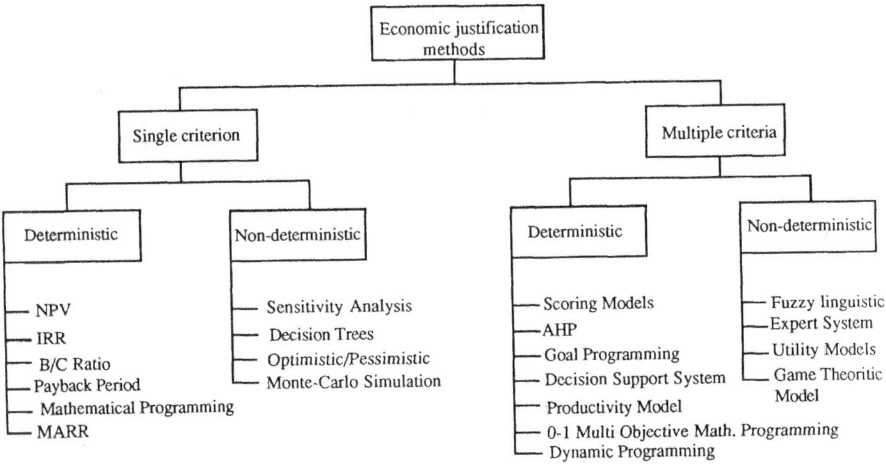

Fig. 14.1. A proposed grouping of economics justification methods.

Single-Criterion Deterministic Methods

Net Present Value The net present value is the sum of the net cash flows discounted at the minimum acceptable rate of return, to the present time or time zero (Stevens, 1989). Theoretically, an investment is economically justified if the NPV $\geq$ 0. The benefits of investment in advanced automated manufacturing projects are generally realized over a long period of time. This has the effect that when a high hurdle rate (discount rate) and short recovery period are used in the analysis to compute the NPV, as is often done in justifying investments in traditional economic analyses, high-technology projects are many times rejected. Therefore, the proper choice of a discount rate and recovery period is very important. The net present value can be calculated using the following formula:

$$\text{NPV} = \sum_{j=0}^{n} \frac{X_j}{(1 + k)^j}$$

where NPV is the net present value, X_j is the net cash flow in year j, n is the number of years of cash flow, and k is the minimum acceptable rate of return.

An example illustrating the effect of high hurdle rates on the selection of a project using NPV is presented in Table 14.2. Project 1 requires low initial cost and has a life of 6 years; whereas, project 2 is a capital-intensive automated manufacturing project with a project life of 10 years. For a discount rate of 15 percent, NPV analysis indicates that the automated manufacturing project is the more desirable alternative. But, for a higher hurdle rate of 30 percent, the automated manufacturing project is not desirable. The results

TABLE 14.2 Net Present Values for Two Proposed Projects at MARR = 15 and 30%

Project 1 Low Initial Cost		Project 2 Automation Project	
End of Year	Annual Cash Flows	End of Year	Annual Cash Flows
0	−600,000	0	−2,500,000
1	250,000	1	200,000
2	250,000	2	200,000
3	300,000	3	400,000
4	350,000	4	500,000
5	100,000	5	800,000
6	50,000	6	1,000,000
		7	1,000,000
		8	1,200,000
		9	1,200,000
		10	1,500,000
NPV (15%)	275,165	NPV (15%)	759,880
NPV (30%)	36,610	NPV (30%)	−860,450

of the example indicate that the choice of the discount rate should be made carefully.

Internal Rate of Return The interest rate, which makes the sum of the discounted cash flows equal to zero, is called the internal rate of return (Stevens, 1989). Investment in a project considered to be economically acceptable if the internal rate of return is greater than, or equal to, a selected hurdle rate (minimum attractive rate of return, MARR). This method of economic evaluation of projects is similar to the net present value method and has similar benefits and shortcomings in that, since the acceptability of a project depends on MARR, it is important to choose an appropriate value of MARR based on certain considerations such as risk of the project.

$$\sum_{j=0}^{n} \frac{X_j}{(1+i)^j} = 0$$

where X_j is the net cash flow in year j, n is the number of years of cash flow, j is the year in which cash flow X_j occurs and, i is the internal rate of return.

Benefit/Cost Ratio Investment in a project is considered economically justifiable if the benefit to cost ratio is greater than or equal to 1. The definition of benefits and costs is important in this method. This method is widely used in government organizations and public utilities to measure the economic effectiveness of an investment in a project that will benefit some portion of the public. Benefit/cost ratio is also called the "savings investment ratio" by some government organizations. This method is highly dependent on the proper definition of benefits and costs and provides no advantage over the net present value of the internal rate of return methods (Stevens, 1989).

Benefit/cost ratio can be calculated using one of the following methods (Canada and Sullivan, 1989):

$$\mathrm{B/C} = \frac{B}{\mathrm{CR} + (O + M)}$$

$$\mathrm{B/C} = \frac{B - (O + M)}{\mathrm{CR}}$$

where B/C is the benefit/cost ratio, B is the annual worth of benefits to the user, O is the equivalent uniform annual operating cost, M is the equivalent uniform annual maintenance cost and CR is the capital recovery cost (equivalent annual cost of the initial investment).

Example: Consider a project that requires an initial investment of \$40,000. The minimum acceptable rate of return specified is 12 percent and the project life is 6 years. The annual receipts as result of the project are \$25,000 and the annual operating and maintenance costs are \$2000. The benefit/cost ratio is calculated as follows:

$$\mathrm{CR} = \$40{,}000(A/P\ 12\%,\ 6) = \$40{,}000 \times 0.2432 = \$9728.00$$

$$B = \$12{,}000\ (O + M) = \$2000$$

where $(A/P\ 12\%, 6)$ is the capital recovery factor for 12 percent over 6 years.

$$\text{B/C ratio based on conventional method} = 12000/\ (9728 + 2000)$$
$$= 1.023\ (> 1.0)$$

$$\text{B/C ratio based on the modified method} = (12000 - 2000)/\ 9728$$
$$= 1.028\ (> 1.0)$$

Since the benefit/cost ratio is greater than 1.0, investment in the project is acceptable.

Payback Period The payback period is the time required to recover the initial investment, or, in other words, to balance the incoming cash flows and the cash outflows. A project is considered acceptable if the payback period is equal to, or less than, a predetermined payback period. The simplicity of this analysis tool has made it the most prevalent method used in industry to justify investments. The payback period can be determined from the following equation:

$$\sum_{j=0}^{p} X_j = 0$$

where p is the payback period and X_j is the net cash flow in year j.

There are certain disadvantages with this method. A project judged acceptable based on the payback period may not be acceptable by the NPV or IRR methods, and vice versa. This is because the payback period does not consider the time value of money (interest). This method also fails to incorporate the benefits resulting from strategic issues involved in the investment.

Modified Minimum Annual Revenue Requirement The minimum annual revenue requirement is simply the minimum income required to cover all incremental costs, including capital recovery on the incremental investment, return on the incremental investment, incremental costs of goods sold, and the incremental taxes inherent in an investment alternative. This approach was modified by Soni et al. (1988) to consider the possibility of a firm facing a declining market share, and, consequently, declining gross incomes, in the event that an investment in advanced manufacturing is *not* undertaken.

This methodology compares a series of minimum annual incremental revenue requirements to a corresponding series of annual incremental gross incomes. If the discounted value of the difference between the incremental gross income and the incremental revenue requirements is greater than, or equal to, zero, the project is judged acceptable (Soni et al., 1992).

Mathematical Programming This approach attempts to use a mathematical programming formulation to choose a subset of interrelated projects for investment from a given set of projects. The objective function is to maximize the net present value subject to various constraints of interdependence, for example, mutual exclusivity, multiproject budget, and labor and material restrictions. One of the advantages of this approach is that it can streamline computational complexity. However, there are certain drawbacks with this approach (Soni et al., 1992):

1. It ignores a potential cross correlation of cash flows between the projects.
2. The discount rate must be estimated by the analyst under an assumption of perfect and complete markets. The analyst must also provide all the information regarding future cash flows, interdependency, and so forth (Reeve and Sullivan, 1988).

Single-Criterion Nondeterministic Methods

Sensitivity Analysis Sensitivity analysis is one of the most important parts of any decision-making process. In the economic evaluation of investment in projects, a decision maker must consider uncertainty and risk. A sensitivity analysis can be performed to handle the issue of uncertainty and risk. This analysis is done when there may be inherent errors in the estimation of the values of parameters in the problem, or when preferences associated with the various projects change. Graphical and tabular methods are available to aid in the performance of sensitivity analyses.

Decision Trees The drafting of a decision tree is a diagrammatic procedure useful in modeling complicated problems involving risk, sequential decisions, and outcomes. The decision tree approach is appropriate in justifying automated manufacturing technology for situations wherein the analyst is making several similar decisions over a period of time. A number of commercial software packages are available to assist in decision tree analysis, such as SUPERTREE, Decision Programming Language, and so forth. All offer graphical capabilities to reduce the otherwise time-consuming calculations required.

A related approach to decision trees is the influence diagram approach (Call and Miller, 1990). A problem modeled as a decision tree can be converted to an influence diagram. A combination of these two approaches is computationally efficient.

Optimistic/Pessimistic Analysis If an economic evaluation problem involves uncertainty, the decision maker has no knowledge of the chances, or probabilities, of occurrence of an outcome. The optimistic/pessimistic analysis approach is used to find optimistic and pessimistic estimates of all decision variables and compare the available alternatives based on some measure of effectiveness (Canada and Sullivan, 1989).

Monte Carlo Simulation This technique, also called the "method of statistical trials" is used to analyze situations involving risk. This method is useful when other analytical approaches are difficult or not feasible to use. An outcome for each variable of interest (element) is randomly selected from a probability distribution assumed to represent each criterion of interest, and these outcomes are then combined. A certain number of trials are conducted and the combined outcomes are checked for some prescribed degree of accuracy. The important requirement of the technique is that the outcomes of all variables of interest be randomly selected (Canada and Sullivan, 1989; Sullivan and Orr, 1982). this method can also be used for problems with more than one criterion.

Multicriteria Methods

Like many real-world problems, the decision of investing in advanced manufacturing technology frequently involves multiple and conflicting objectives, for example, minimizing costs, maximizing flexibility, minimizing machine downtimes, maximizing efficiency, and so forth. The decision maker has to consider a number of criteria and may assign certain priorities for each criterion so that the most important criterion is satisfied first. Many multicriteria techniques have been developed and applied to the problem of justifying advanced manufacturing technology. Some of the deterministic methods for solving multicriteria problems are multiobjective mathematical programming, scoring models, the analytic hierarchy process (AHP), outranking methods and goal programming. Nondeterministic methods include fuzzy techniques, expert systems, game-theoretic methods, and multiattribute utility and value analysis.

Multicriteria Deterministic Methods

Multicriteria 0–1 Integer Programming This is a simple approach to modeling and solving the problem of selecting the best project from a given set of projects. Given a set of projects $X = \{x_1, x_2, \ldots, x_n\}$ where x_i, $i = 1, 2, \ldots, n$ represents the project and

$$x_i = \begin{cases} 1, & \text{if the } i\text{th project is selected} \\ 0, & \text{otherwise} \end{cases}$$

It is assumed that the minimum acceptable values of each attribute, such as NPV, payback period, work-in-process inventory, flexibility, competitive advantage, and so forth, is known and available. If qualitative factors are considered in the problem, a weight on a scale from 0 to 1 is used for each factor. The values of each of these attributes for each project form the coefficients of the decision variables in the constraints and the objective function of the model. The main advantage of this method is that it can solve a problem with a large number of alternatives in a short time.

A drawback of the problem is that the decision maker must know, or be able to determine, the b_j values in the constraints (the minimum acceptable values of each attribute). Then, a multicriteria 0–1 integer programming problem can be formulated as

$$\text{Minimize or maximize } \{f_1(x), \ldots f_k(x)\}$$

$$\text{Subject to}$$

$$\sum a_{ij}x_i \geq b_j$$

$$\sum x_i = 1$$

$$x_i \text{ is 0 or 1, } i = 1, 2, \ldots, n \text{ and } j = 1, 2, \ldots, m$$

Here, the $f(x)$ refer to the various objective functions such as minimizing payback period, work-in-process inventories, and so forth and maximizing net present value, competitiveness, and so forth. Various multicriteria decision-making techniques are available in the literature to solve these models, many of which have been incorporated into computer software to run interactively and/or noninteractively.

Linear Programming Chandra and Schall (1988) developed a linear programming and Leontief input–output model for the justification of flexible manufacturing systems. Several performance measures of the FMS are obtained through the Leontief input–output model and are then used as inputs in the analytic hierarchy process for the economic justification step.

Stam and Kuula (1991) formulated the problem of selecting a flexible manufacturing system as a multiobjective linear program. The problem of selecting appropriate technology is solved in a two-phase approach, the first phase using analytic hierarchy process and the second solving the linear program.

Dynamic Programming Dynamic programming is an appropriate tool for use in the justification problem since the decision on the level of investment for automated manufacturing depends not only on the current realizations of costs and revenues but also on previous decisions regarding investment in automated manufacturing systems. The justification of advanced manufacturing can be considered as a sequence of strategic investment decisions over time, where at each stage a decision concerning additional investment must be made. Therefore, in many instances, the investment process may be separated into various states at each stage, making dynamic programming suitable to handle this type of justification problem.

Goal Programming Goal programming has been used to model investment decision in a flexible manufacturing system from a multiobjective context. It has been shown capable of providing an acceptable solution in reasonable time.

This mathematical programming tool features the ability to analyze multiple conflicting goals. Long-term planning can be done in the same time as that required to solve a short-term problem. The length of the planning horizon does not effect the time for finding the solution.

A major drawback of a goal programming model is that it does not consider the dynamic nature of the decision environment. Certain assumptions must be made in modeling an FMS using goal programming, such as:

- Each part type must pass through one or more operations.
- There are one or more machines that can process each operation.
- Each machine can process one operation at a time.
- Only one unit of a special tool is available.

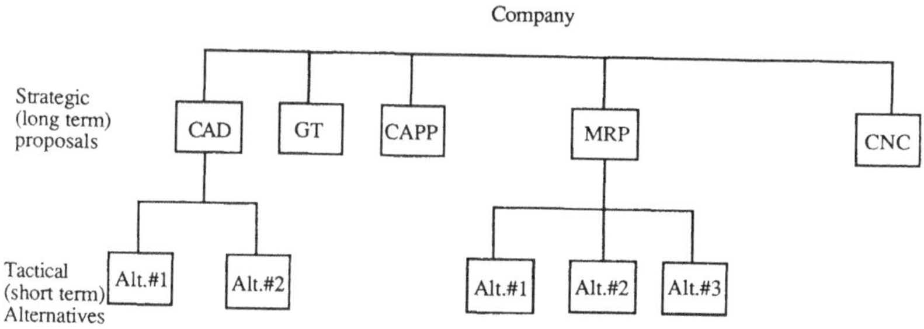

Fig. 14.2. The graphical representation of strategic proposals and tactical alternatives.

- The part-moving time is assumed not to affect the lead time.

Also, this modeling approach cannot take into consideration some important qualitative factors.

Scoring Models Scoring models feature the ability to accommodate the consideration of intangible, or economically nonquantifiable, elements involved in an investment decision in an analytical fashion. Scoring models can be classified into:

- Unweighted 0–1 factor models
- Unweighted factor scoring models
- Weighted scoring models
- Constrained weighted factor scoring models

Input data for scoring models are subjective "quesses" rather than sophisticated forms of the same quess required by other models.

Among common scoring models, the linear additive model is widely used by researchers for scoring competing alternatives. Parsaei and Wilhelm proposed a methodology based on linear additive model developed by Klee (1971). This methodology aids decision makers in evaluating the desirability of a firm's long-term and short-term advanced automated manufacturing technologies.

The methodology is implemented in two phases. Phase 1 examines the desirability of various strategic (long-term) automation proposals available, while phase 2 evaluates each tactical (short-term) alternative for implementing the strategic proposal selected as most desirable in phase 1. Figure 14.2 illustrates the use of the two phases of this decision methodology.

The phased approach was chosen to emphasize the bifurcated nature of the actual decision process. To implement the methodology, two sets of ordinal scale weights must be developed by decision makers (Tables 14.3 and

TABLE 14.3 Ordinal Scale Weights to Rank the Importance of Strategic and Tactical Decisions

Very important	1.00
Important	0.75
Necessary	0.50
Unimportant	0.25

14.4); they are then employed in the linear additive models in each of the two phases of the methodology (Parsaei and Wilhelm, 1989).

The linear additive model utilized in the two phases of this methodology can be presented by the following relationship:

$$\text{Max } D_j = \Sigma\, W_i X_j \qquad j = 1, 2, \ldots, m$$

where D_j is the score earned by the jth decision alternative, W_i is the weight assigned to the ith attribute, which reflects the relative importance of that attribute in the decision process, and X_{ij} is the expected performance of the jth decision alternative with respect to the ith attribute.

An illustrative example of this methodology is presented in Tables 14.5 and 14.6 (Parsaei and Wilhelm, 1989).

Total Productivity Model This method is used to assess the impact of the proposed equipment investment on profitability from the standpoint of productivity. If a project has a total productivity level greater than, or equal to, a predetermined productivity level, it is accepted. This approach is appealing to manufacturers because one of the prime reasons for considering investment in advanced automated manufacturing systems is the potential payoff in increased productivity.

The total productivity model suggested by Sumanth and Pino (1986) considers five productivity measures related to personnel, material, capital, energy, and other measures. Projects are evaluated based on their ability to score at acceptable levels on these productivity measures. This method is

TABLE 14.4 Ordinal Scale Weights Utilized to Measure the Performance of Each Strategic and Tactical Decision with Respect to Each Attribute

Superior	1.00
Good	0.80
Above average	0.60
Average	0.40
Below average	0.20
Poor	0.00

TABLE 14.5 Phase 1: Selection of the Best Strategic Automation Proposal

			Strategic Proposal 1		Strategic Proposal 2		Strategic Proposal 3	
i	A_i	Attribute	X_{i1}	A_iX_{i1}	X_{i2}	A_iX_{i2}	X_{i3}	A_iX_{i3}
1	0.125	Flexibility in production mix	0.60	0.075	0.80	0.100	0.80	0.100
2	0.187	Flexibility in scheduling	0.80	0.149	0.80	0.149	0.80	0.149
3	0.25	Cost reduction and savings	1.00	0.250	0.80	0.200	0.60	0.150
4	0.25	Product quality improvement	0.60	0.150	0.80	0.200	0.80	0.200
5	0.181	Reduction in WIP	0.40	0.075	0.40	0.075	0.60	0.112
		K_j		0.699		0.724		0.711
		Normalized K_j		0.965		1.00[a]		0.980

[a]The most desirable strategic proposal (Max $K_j = \Sigma A_iX_{ij}$).

K_j = the total score earned by the jth strategic proposal in satisfying the corporate goals for factory automation,

A_i = a ratio scale variable created by dividing the weight, A_i (ordinal weight from Table 14.3) assigned to the ith strategic decision attribute, which expresses the relative importance of the corporate automation goals achieved by implementing each proposal, by the sum of the weights assigned, assuming $1 \leq i \leq m$,

X_{ij} = the weight (ordinal value) assigned to the ith strategic decision attribute, which reflects the anticipated performance of the jth strategic proposal with respect to strategic decision attribute i.

computerized, and the software package offers features such as easy modification and the ability to conduct what-if analyses.

Analytic Hierarchy Process The analytic hierarchy process (AHP) was developed by Saaty (1980). It uses a hierarchical representation of factors (elements) influencing a system and makes pairwise comparisons among the factors in order to rank alternatives to solve a decision problem. It, in effect, attempts to analyze the impacts of alternatives, or elements, at the lowest level on the overall objective, or focus of a hierarchy.

In the last two decades, this method has been applied to numerous decision problems such as transportation problems, corporate planning, marketing strategy, energy policy, project selection, and budget allocation. However, many of the applications are related to socioeconomic decision making with only a few applications having been reported in the area of manufacturing/production systems.

The strength of the AHP method lies in its ability to structure a complex, multiattribute, and multiperiod problem hierarchically. It provides a comprehensive structure that represents a combination of one's rational, irrational, and intuitive judgments in the decision process. These judgments need not be consistent or transitive. The degree of consistency of the judgments is calculated as a by-product of the analytical process.

This analytical process basically consists of three stages of problem solving.

TABLE 14.6 Phase 2: Selection of the Best Tactical Alternatives

			Tactical Alternative 1		Tactical Alternative 2		Tactical Alternative 3	
i	B_i	Attribute	Y_{i1}	B_iY_{i1}	Y_{i2}	B_iY_{i2}	Y_{i3}	B_iY_{i3}
1	0.266	Capital recovery cost	1.00	0.266	0.80	0.212	1.00	0.266
2	0.133	Floor space required	0.40	0.053	0.60	0.079	0.80	0.106
3	0.133	Expandability of system	0.60	0.079	1.00	0.133	0.80	0.106
4	0.266	Compatibility	0.80	0.212	1.00	0.266	0.60	0.159
5	0.202	Ease of operation	0.40	0.080	0.80	0.161	0.60	0.121
		L_j		0.690		0.851		0.758
		Normalized L_j		0.810		1.00[a]		0.890

[a]The best available alternative (Max $L_j = \Sigma B_iY_{ij}$).

L_j = the total score representing the ability of the jth tactical alternative to achieve the goals of the strategic proposal,

B_i = a ratio scale weight computed by dividing the ordinal scale weight, B_i, assigned to the ith characteristic (or attribute) to reflect its importance in the strategic proposal by the total of the weights assigned to all attributes, assuming $1 \leq i \leq p$,

Y_{ij} = the ordinal scale weight assigned to the ith characteristic, which reflects the anticipated performance of tactical alternative j with respect to the ith characteristic (or attribute).

They are decomposition, comparative judgments, and synthesis of priorities. The process could be described as consisting of the following steps:

- Organizing the factors (elements) that affect the overall objective into a hierarchy. The elements of a hierarchy normally become more specific as one moves toward the bottom of the hierarchy. The lowest level of the hierarchy usually contains the alternatives under evaluation.
- Constructing a set of pairwise comparison matrices for each of the lower levels. The elements in the lower level are compared to each other based on their effect on each element in the higher level. These judgments are expressed in integers. This step results in a square matrix of judgments.
- Determine local priorities using the eigenvalue method. The higher the local priority for an element, the more important is the element with respect to the others.
- Once the pairwise comparisons are made, the consistency is computed using the eigenvalue, resulting in consistency index and consistency ratio.
- After the local priorities for all the factors in the hierarchy have been determined, a hierarchical composition is used on the eigenvectors, which results in the scoring of the alternatives (elements at the lowest level of the hierarchy).
- The consistency of the entire hierarchy is computed by multiplying each

consistency ratio by the priority of the factor and adding them together. The consistency ratio should be less than 10 percent to be acceptable.

The AHP is widely used due to the following advantages.

- The method is easy to apply.
- The method requires structuring a hierarchy of attributes and the objective, which facilitates understanding of the problem and determination of solution.
- The method provides functions to evaluate inconsistencies in decision maker's judgments.

Varney et al. (1985), Wabalickis (1988), and Walsh (1988) solve the justification problem using the analytic hierarchy process. Stam and Kuula (1991) use a two-phase approach to solve the problem of selecting an FMS. The first phase involves a screening of the alternatives using the analytic hierarchy process implemented in the EXPERT CHOICE software.

Deterministic Outranking Methods Outranking methods are used in multicriteria decision making to rank various alternatives based on the decision maker's preferences. These techniques for decision analysis were developed and are in wide use in Europe. Therefore this theory may be considered a "European school of thought."

The outranking methods in multicriteria decision making were first introduced in the mid-60's. Various methods such as ELECTRE, ORESTE, PROMETHEE, and MELCHIOR belong to this family.

The advantages of these methods are simplicity, clarity, and stability. They have been extended to include specialized features such as fuzzy set aspects.

These methods have been successfully applied in making public policy decisions and other decisions involving large capital investments. The applicability of outranking methods for the multiattribute evaluation of advanced manufacturing systems is suggested by Kolli and Parsaei (1992). Siskos et al. (1983) provide a bibliography on outranking methods.

Multicriteria Nondeterministic Methods

Game-Theoretic Models Game-theoretic models can also be used to analyze investments in advanced automated manufacturing systems. They feature the ability to handle the element of strategic interdependence between firms, which is an important factor because the industries that invest in advanced manufacturing technology exhibit significant strategic interdependence between competitors (Tombak, 1990).

Fuzzy Linguistic Methods The use of fuzzy linguistic variables to evaluate investment in advanced automation technology was proposed by Wilhelm and

Parsaei (1991). This method involves the use of linguistic variables from the theory of fuzzy sets to support the phased implementation of a computer-integrated manufacturing (CIM) strategy. Fuzzy linguistic models permit the translation of verbal expressions into numerical ones, thereby dealing quantitatively with imprecision in the expression of the importance of each strategic goal (e.g., enhanced competitiveness, high product quality, low product cost, etc.) and the enabling technology involved in implementing CIM systems (e.g., CAD, CAM, group technology, computer-aided process planning, etc.).

Wilhelm and Parsaei (1991), in their proposed methodology for the evaluation of CIM system investments, defined two fuzzy linguistic variables: importance and capability. The use of these linguistic variables allows the analyst to specify the importance associated with each of a set of strategic goals that impact the CIM strategy, and the tactical capability of each available CIM technology to meet the strategic CIM goals of the organization. Their study developed a heuristic algorithm based on the use of fuzzy linguistic variables to characterize the capability of available enabling technologies, and to select the best technology at each phase in the implementation of the CIM plan (Wilhelm and Parsaei, 1988, 1989).

Stochastic Programming Stochastic programming has been used to analyze cost–flexibility trade-offs involved in investment in product-flexible manufacturing capacity. Fine and Freund (1990) present a two-stage stochastic programming formulation of a product-flexible manufacturing capacity investment model. Advantages of this approach are that the decision maker understands the cost–flexibility trade-offs better and closely participates in the decision-making process. The value of flexibility can be obtained through this method. A stochastic dynamic program is presented by Kulatilaka (1988) for the analysis of flexible automation.

Expert Systems An expert system is a computer program that has the capability of solving complex problems by certain rules and logical reasoning mechanism. These rules represent the problem-solving approaches of experts. The main feature of expert systems is the ability to handle problems with inexact data (Canada and Sullivan, 1989).

An expert system consists of a knowledge base, inference engine, control mechanism, cache or working memory, and the user interface. Expert systems can be developed with various computer languages and expert system shells. They can also interact with other computer software.

Expert systems have been developed for handling the problem of evaluating investment in advanced automated manufacturing systems. Sullivan and LeClair (1985) developed an expert system for IBM PCs called XVENTURE. XVENTURE is used for solving investment justification problems quickly and consistently. XVENTURE sets company survival as its primary objective during the decision process. The decisions from the expert system could be:

GO—Venture is justifiable; DEFER—do not venture now; NO GO—reject/abandon the venture (Sullivan and Reeve, 1988).

Another expert system developed by Padmanabhan (1989) combines the knowledge-based system with analytical techniques. This expert system is developed using the PCPLUS expert system shell and can be used for the justification of flexible manufacturing systems. Zhang and Lu (1990) developed an expert system framework for economic evaluation of machining operation planning. The operation plan derived from the integrated manufacturing and management system ensures that the required specifications are met and economically utilizes labor, material, and working capacity. Boubekri et al. (1991) developed an expert system designed to assist management in the selection of industrial robots based on functional, organizational, and economic criteria. Fisher and Nof (1987) developed a knowledge-based expert system for the selection and replacement of manufacturing facilities. This expert system combines the heuristic decision-making techniques of human expert and economic analysts in quantitative models.

Utility Models Multiattribute utility models consider the decision maker's preferences expressed in the form of utility functions in order to select the most satisfying alternative (Keeney and Raiffa, 1976). These utility functions can be expressed as either an additive or a multiplicative function, provided that each pair of attributes is preferentially independent of its complement. The utility function varies from one decision maker to another. One of the difficulties of this method is concerned with the trade-off questions to be answered during the process of constructing utility models. The overall utility is determined by combining the utility obtained from each attribute and the alternative with the highest utility is selected.

ADDITIVE UTILITY MODEL The utility $U(x)$ for the attributes $X_1, X_2, \ldots, X_n$ can be expressed as

$$U(x) = \sum_{i=1}^{n} K_i * U_i\,(x_i)$$

where k_i is the weight for the ith attribute, and $U_i(k_i)$ is the utility of the outcome x_i for the ith attribute.

MULTIPLICATIVE UTILITY MODEL The utility $U(x)$ for the attributes $X_1, X_2, \ldots, X_n$ can be expressed as

$$U(x) = \frac{\prod_{i=1}^{n}[Kk_i * U_i(x_i) + 1]}{K}$$

where k_i is the weight for the ith attribute, $U_i(k_i)$ is the utility of the outcome x_i for the ith attribute, and K is a scaling constant that is a solution to

$$1 + K = \prod_{i=1}^{n}(1 + K \cdot k_i)$$

The shape of the utility function is purely preferential and it varies from one decision maker to the other. The main disadvantage of the MAUM is concerned with the trade-off questions to be answered during the process of constructing utility models. The MAUM is often perceived as highly subjective, counterintuitive, and frustrating.

In this method the overall utility of the advanced manufacturing system is determined by combining the utility obtained from each attribute. The alternative with the highest utility is selected (Bunn, 1984; Keeney and Raiffa, 1976). Sloggy (1984) explains the procedure to develop a multiattribute utility model for advanced manufacturing systems and how to evaluate various proposals.

Decision Support System A decision support system refers to a computerized approach to establishing an information system for managerial analysis. It typically includes a model that may be a set of financial, simulation, and optimization techniques, a database, and a terminal through which the user interacts with the model for solutions to a managerial problem. A decision support system (DSS) focuses on a decision process rather than data processing. Decision support systems are developed for middle and top management. A what-if kind of analysis is possible with the DSS, which provides the manager with necessary information for making better decisions.

Monahan and Smunt (1987) developed a multilevel DSS for evaluating the overall performance of automated flexible manufacturing systems. Kassicieh and Shultz (1987) also introduce a decision support framework for evaluating computer-integrated manufacturing systems.

The application of traditional economic evaluation methods, simulation, mathematical programming, or accounting techniques for the evaluation of advanced manufacturing systems suffer from limitations when used alone, but good analysis is possible when all the techniques are combined. With a combined decision support system several alternatives can be studied in less time. Also, modifications are possible during the decision process, and sensitivity analysis can be performed. Stam and Kuula (1991) use a visually interactive decision support framework involving multicriteria analysis designed to aid the decision maker, particularly top management in selecting the most appropriate technology. Badiru et al. (1991) present a spreadsheet model containing modules for AHP, utility model, and system value model developed with Lotus 1-2-3 macro language. This spreadsheet model is useful in handling multiple attributes encountered in the economic evaluation of advanced automated manufacturing technology.

Nondeterministic Outranking Methods The outranking methods mentioned before have been extended to incorporate special cases such as fuzziness and uncertainty. D'Avignon and Vincke, (1988) present an outranking method to incorporate uncertainty of consequences. Martel et al. (1986) present a fuzzy outranking relation for multicriteria decision making. This fuzzy outranking relation is characterized by a degree of credibility obtained from a confidence index and a doubt index.

14.7 CONCLUSIONS

Due to increased competitiveness in the market, companies are investing in advanced manufacturing systems. These advanced manufacturing systems provide significant benefits, such as increased productivity, improved quality, reduced lead time, and improved employee's well-being. Economic aspects are important in the implementation of advanced manufacturing systems. This chapter discussed techniques for justifying investment in advanced manufacturing technology.

REFERENCES

Alexander, David C. (1985). Making ergonomics pay: Adopting a business approach, *Industrial Engineering*, July, 32–39.

Badiru, Adedeji B., Foote, Bob C., and Chetupuzha, Joseph (1991). A multiattribute spreadsheet model for manufacturing technology justification, *Computers and Industrial Engineering*, **21**(1–4), 29–33.

Boubekri, Nourredine, Sahoui, Majid, and Lakrib, Chabane. (1991). Development of an expert system for industrial robot selection. *Computers and Industrial Engineering*, **20**(1), 119–127.

Bunn, Derek W. (1984). *Applied Decision Analysis*, McGraw-Hill, New York.

Call, Hollis J., and Miller, William A. (1990). A comparison of approaches and implementation for automating decision analysis, *Reliability Engineering and System Safety*, 115–162.

Canada, J. R., and Sullivan, W. G. (1989). *Economic and Multiattribute Evaluation of Advanced Manufacturing Systems*, Prentice-Hall, Englewood Cliffs, NJ.

Canada, John R., and Sullivan, W. G. (1990). Persistent pitfalls and applicable approaches for justification of advanced manufacturing systems, *Engineering Costs and Production Economics*, **18**, 247–253.

Chandra, Jeya, and Schall, Susan O. (1988). Economic justification of manufacturing systems using the Leontief input-output model, *Engineering Economist*, **34**(1), 27–45.

Choobineh, Fred (1986). Justification of flexible manufacturing systems, in *Flexible Manufacturing Systems—Current Issues and Models*, Fred Choobineh and Rajan Suri, Eds., Industrial Engineering Press, Norcross, GA., pp. 269–281.

D'Avignon, Gilles R., and Vincke, Ph. (1988). An outranking method under uncertainty, *European Journal of Operations Research*, **36**, 311–321.

Edwards, Jeffrey, R. (1989). Computer aided manufacturing and worker well being: A review of research, *Behavior and Information Technology*, **8**(3), 157–174.

Fisher, Edward L., and Nof, Shimon Y. (1987). Knowledge-based economic analysis of manufacturing systems, *Journal of Manufacturing Systems*, **6**(2), 137–150.

Groover, Mikell P. (1987). *Automation, Production Systems, and Computer Integrated Manufacturing*, Prentice-Hall, Englewood Cliffs, NJ.

Kassicieh, Suleiman, and Shultz, Carl R. (1987). Decision support flexible manufacturing systems, *OMEGA Int. J. of Mgmt. Sci*, **15**(6), 495–502.

Keeney, R. L., and Raiffa, H. (1976). *Decisions with Multiple Objectives: Preferences and Value Tradeoffs*, Wiley, New York.

Klee, A. J. (1971). The role of decision models in the evaluation of competing environmental health alternatives, *Management Science*, **18**(2), B52–B67.

Kolli, Sai, and Parsaei, Hamid R. (1992). Multi-Criteria Analysis in the Evaluation of Advanced Manufacturing Technology using PROMETHEE, 14th Conference of Computers and Industrial Engineering, Cocoa Beach, Florida.

Kulatilaka, N. (1988). Valuing the flexibility of flexible manufacturing systems, *IEEE Transactions on Engineering Management*, **35**(4), 250–257.

Martel, Jean-Marc, D'Avignon, Gilles R., and Couillard, Jean (1986). A fuzzy outranking relation in multi-criteria decision making, *European Journal of Operations Research*, **25**, 258–271.

Meredith, Jack R., and Suresh, Nallan C. (1986). Justification Techniques for Advanced Manufacturing Technologies, *International Journal of Production Research*, **24**(5), 1043–1058.

Meredith, Jack R. (1987). Implementing the automated factory, *Journal of Manufacturing Systems*, **6**(1), 1–13.

Monahan, George E., and Timothy L. Smunt (1987). A multilevel decision support system for the financial justification of automated flexible manufacturing systems, *Interfaces*, **17**(6), 29–40.

Padmanabhan, Srikanth (1989). A tandem expert support system as justification for a flexible manufacturing system, *Journal of Manufacturing Systems*, **8**(3), 27–39.

Parsaei, Hamid R., and Wilhelm, Mickey R. (1989). A justification methodology for automated manufacturing technologies, *Computers and Industrial Engineering*, **19**(3), 363–373.

Priest, John W. (1985). Ergonomic changes in work place can improve the productivity of production operations, *Industrial Engineering*, July, 40–43.

Reeve, J. M., and Sullivan, William G. (1988). Strategic Evaluation of Interrelated Investment Projects in Manufacturing Companies, Cost Management Systems Research Report Submitted to CAM-I, International.

Saaty, T. L. (1980). *The Analytic Hierarchy Process*, McGraw-Hill, New York.

Simpson, G. C. (1988). The economic justification for ergonomics, *International Journal for Industrial Ergonomics*, **2**, 157–163.

Siskos, J., Wascher, G., and Winkler, H. M. (1983). *A Bibliography on Outranking Approaches (1966–1982)*, Cahier du LAMSADE No. 45, Universite de Paris–Dauphine, Paris.

Sloggy, John E. (1984). How to justify the cost of an FMS, *Tooling and Production*, December.

Soni, R., Liles, D. H., and Stevens, Jr., G. T. (1988). The Economic Justification of Investment in Advanced Manufacturing, Proceedings of the International Industrial Engineering Conference, St. Louis, Missouri.

Soni, Ramesh G., Parsaei, Hamid, and Liles, Donald H. (1992). *Economic and Financial Justification Methods for Advanced Automated Manufacturing: An Overview, Economics of Advanced Manufacturing Systems*, Hamid R. Parsaei and Anil Mital Eds., Chapman and Hall, London, pp. 3–19.

Stam, Antonie, and Kuula, Markku (1991). Selecting a flexible manufacturing system using multi criteria analysis, *International Journal of Production Research*, **29**(4), 803–822.

Stevens, G. T. (1989). *The Analysis of Capital Expenditures for Managers and Engineers*, Ginn Press, Needham Heights, MA.

Sullivan, William G., and Reeve, James M. (1988). XVENTURE: Expert systems to the rescue—They can help justify investments in new technology, *Management Accounting*, 51–58.

Sullivan, William G. (March, 1986). Models IEs can use for strategic, non-monetary factors in automation decisions, *Industrial Engineering*, 42–50.

Sullivan, William G., and LeClair, S. R. (1985). Justification of flexible manufacturing systems using expert systems technology, Conference Proceedings—AUTOFACT 1985, Society of Manufacturing Engineers, November, pp. 7-1 to 7-13.

Sullivan, William G., and Orr, R. Gordon (1982). Monte-Carlo simulation analyzes alternatives in an uncertain economy, *Industrial Engineering*, 43–49.

Sumanth, David J., and Del Pino, Lourdes (1986). Equipment Selection and Justification through Total Productivity Model, Proceedings of the 8th Annual Conference on Computers and Industrial Engineering, pp. 91–94.

Tombak, Mihkel M. (1990). A strategic analysis of flexible manufacturing systems, *European Journal of Operations Research*, **47**, 225–238.

Varney, Mark S., Sullivan, William G., and Cochran, Jeffrey K. (1985). Justification of Flexible Manufacturing Systems with the Analytical Hierarchy Process, Proceedings of the 1985 Annual IIE Conference, pp. 181–190.

Wabalickis, Roger N. (1988). Justification of FMS with the analytical hierarchy process, *Journal of Manufacturing Systems*, **7**(3), 175–182.

Walsh, Angela Jean (1988). A Methodology for the Economic Justification of Automated Manufacturing Systems, M. Engg. Thesis, Department of Industrial Engineering, University of Louisville.

Wilhelm, Mickey R., and Parsaei, Hamid R. (1991). A fuzzy linguistic approach to implementing a strategy for computer integrated manufacturing, *Fuzzy Sets and Systems*, **42**, 191–204.

Zhang, Guangming, and Lu, Stephan C. (1990). An expert system framework for economic evaluation of machine operation planning, *Journal of Operational Research Society*, **41**(5), 391–404.

CHAPTER 15

COMPUTER-INTEGRATED BUSINESS SYSTEMS IN COMPANIES OF THE FUTURE

H.-J. BULLINGER, K.-P. FÄHNRICH, and J. NIEMEIER
Fraunhofer-Institut für Arbeitswirtschaft und Organisation
Stuttgart, Germany

15.1 NECESSITY OF COMPUTER-INTEGRATED BUSINESS SYSTEMS

Companies today are situated in a complex network of relationships between each other and the market. Decisive influence factors have certainly been, among others, the change from a buyers market to a sellers market as well as market internationalization and competition, which are, in consequence, rapidly increasing. Simple "if–then relationships" have long ceased to be able to represent market–customer relationships, as the interdependencies and connections are too complex.

Integration as Reaction to the Modified Demands on Companies

It is now a common view that computer-integrated information processing is of strategic importance within companies. The arguments are that computer-integrated information processing:

Organization and Management of Advanced Manufacturing. Edited by Waldemar Karwowski and Gavriel Salvendy.
ISBN 0-471-55508-8

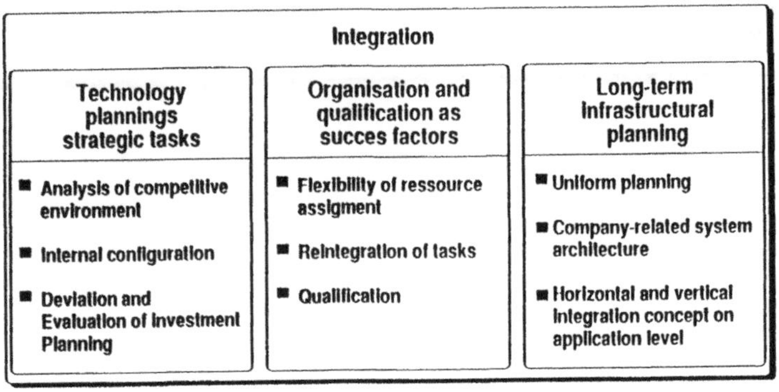

Fig. 15.1. Companies' reaction to changing competitive requirements.

- Creates new relationships to products and markets.
- Facilitates the implementation of new business strategies and taking new competition chances.
- Changes organization and management tasks.

It is considered vital to bring about a relevant management of computer-integrated information processing.

In the area of computer-integrated manufacturing (CIM) or office rationalization projects, approaches can be increasingly found that constitute a reaction to changed company targets, and that make an integration of information systems, existing or planned, appear advisable.

Technology Planning as a Strategic Task

On the CIM sector, as well as in the office automation area, it has been realized, first in theory and then in practice, that technology planning constitutes a long-term strategy approach for companies in changing competitive environments. There is a demand that all investments—be they in research and development, in distribution, or in organization—must be oriented toward the critical success factors in future competition. All investments in information and telecommunication technology must therefore be derived as well as assessed and judged from the point of view of the company's competitive environment (see Fig. 15.1).

Organization and Qualification

In the past it has frequently been tried to rationalize the technical and administrational office by means of a marked division of labor (Taylorism) in connection with a special (technology) support in order to reduce costs. There are a number of disadvantages as well as advantages resulting from high labor division and specialization, for example, time-intensive coordination and communication tasks, long information processing and information storing times. Much information has to be retrieved, stored, processed, and transformed several times. Computer performance does not have to be centralized any more. This brings about the opportunity of re-integrating tasks and of changing work contents by means of suspending exaggerated division of labor. As a consequence, staff members can be given back more responsibility and a greater share in the overall process.

15.2 NATURE OF INFORMATION IN A COMPANY

Information Technology: A Tool for Raising Productivity or a Strategic Weapon

Information as a resource, that is, potential for securing corporate success, or information as a production factor, give rise to questions concerning the conceptual and operative design of corporate divisions with respect to the factor "information."

On the one hand information is produced continually in the work process as a describer of the status of the work objects and respective work situation or of their change. Hence the nature of the information is descriptive, though its function for the work process is a controlling one. These aspects—the roots of the frequently cited diversity of information—must be carefully kept apart. Minimizing the descriptive component while ensuring the component necessary for controlling the work process is more than just a question of system efficiency. Questions of user acceptance and aspects of the adequacy of system solutions can give rise, however, to a higher proportion of descriptive information than would otherwise be necessary for control purposes.

Information as a New Production Factor

A new catch phrase in the discussion is "information as a new production factor." Although companies have learned to plan and organize with the traditional production factors and to control and supervise in quantifiable and nonquantifiable parameters, there is a distinct deficit when it comes to the production factor "information." Simply claiming that "information is the most valuable resource in our company" does not achieve anything in itself for the company. Appropriate management concepts must first be developed to manage the production factor information. The functional aspects must be clarified along with institutional questions such as the organization of information management.

Unfortunately the necessity of strategy-directed innovations is still a seldom met corporate principle because all too often productivity considerations conceal this deficit of action. In view of the economic environment, innovations at the right moment in the right direction are an entrepreneurial task.

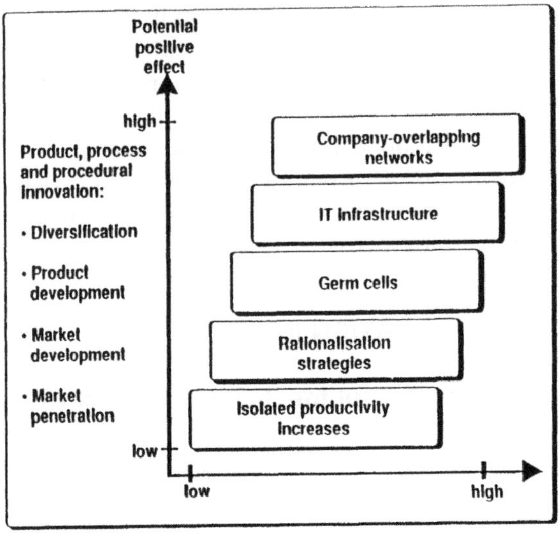

Fig. 15.2. Closing the strategic gap with information technologies.

In addition, information technology has recently been discussed in this connection as a "strategic weapon" in the hands of information management. Figure 15.2 demonstrates this concept.

An improvement of operative sequences, meaning isolated productivity increases and rationalization efforts, can close only a small part of the strategic gap, that is, the difference between "carrying on as before" and the desired development. They are reflected first and foremost in cost benefits when rendering corporate services. In addition, it is possible to discern four directions for expanding information and telecommunication systems. Information management is thus responsible for:

- Creating the focal points for information and telecommunication systems (forming germ cells) in the administrative, technical, and productive sectors
- Developing the general information technology infrastructure
- Installing company-overlapping information and telecommunication systems
- Exploiting information systems services to produce original product, process, and procedural innovations

Laying down the procedural sequence for shaping information management is not only a question of quick cost benefits and productivity increases. One possible approach, for example, is to create focal points parallel to taking infrastructural measures. Assuming that aspired growth is possible only through controlled growth, management must have commensurate experience in mastering the task of this parallel strategy successfully. By comparison, creating germ cells for information and telecommunication systems affords the advantage of being able to accumulate knowledge and experience in the company on the basis of these test type subsolutions, thus enabling companywide projects to be tackled with assurance.

The same applies, although with some qualification, to the installation of company-overlapping computer networks. A key aspect here is the type of link with other companies and its consequences for the development of internal and external standards that can lead to new dependencies, for example, on a component supplier. The logical coordination of the measures is one of the rules of all-round management, that is, information management can be understood only as a component of the corporate strategies such as market penetration, market and product development, and diversifications.

Discussion of Oversimplified Assumptions

The discussed requirements must be converted into effective concepts; thus, competitive advantages can be achieved and secured. Patent recipes are available in diverse forms for this purpose. Often, however, they entail oversimplified assumptions that tend to conceal a suitable approach rather than indicate new perspectives of action.

More Intensive Use of Information and Telecommunication Systems

Corporate success is not conditioned on the number of information technologies used, investment costs per workplace, use of the most advanced technology systems such as local area networks, or the use of a management information system (MIS). Doubtless they make a considerable contribution, but they are not in themselves indicators of success. Parameters such as the turnover and growth of earnings, increase of market shares, earning power of employed capital, and factors such as a company's standing are decisive in the long run for corporate success. These are the indicators that characterize the economic success of a company. (See Fig. 15.3.)

Patent recipes like "the use of information improves corporate success" hold too much disregard for the difficulties in developing applications and in the corporate implementation of information technology, not to mention the difficulty of proving the efficiency of these investments.

The challenges facing a company, be they the special conditions on the labor market or in its particular branch of industry, or be they substitute products and innovative competitors, can generally be tackled with structural personnel or technocratic adaptation strategies. All three adaptation strategies are supported by information strategies. In the past few years companies have relied increasingly on technocratic adaptation strategies, so that on this global level it is not possible to discern any difference as regards corporate success given different adaptation strategies and emphases. The intensity with which

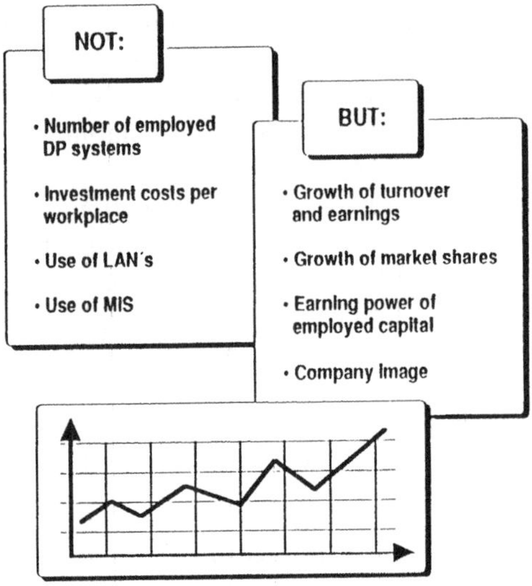

Fig. 15.3. Assessment criteria for corporate success.

these adaptation strategies are used is practically the same in successful and unsuccessful companies. This is clearly demonstrated in Fig. 15.4.

To conclude that each adaptation strategy is equally successful, that is, equivalent, would not be apt given the generalizing level of this cross-sectional view and the range of effect of information technologies. Closer differentiation of the findings reveals quite a variety of successful strategies with corresponding strategic policies. There are certainly some environmental situations in which a company would be advised to pursue a rather more conservative policy. In other markets, on the other hand, an active and innovative approach would be of great importance for the company's success.

Is Introduction of Information Technologies an Urgent Need? During the dominant general economic crisis of the past decade, the worst of which now seems to have been overcome through new growth impetus, information technologies (ITs) have undergone further maturing. More computer power is available at lower prices; more and better telecommunication systems are being installed. Well past the concept phase already are advances toward knowledge processing (decision support systems), and artificial intelligence (AI) systems or developments in the field of new system-based services in the area of distributed computer architecture, filing and retrieval of multimodal documents, and the ensuing considerations of organizational structures.

A critical area—now as ever—is the development of software, which lags far behind user needs and requirements. Important aspects here are such things as user convenience, modularity, and modifiability of software program packages. Judging by the mass of software in current use, it is apparent that

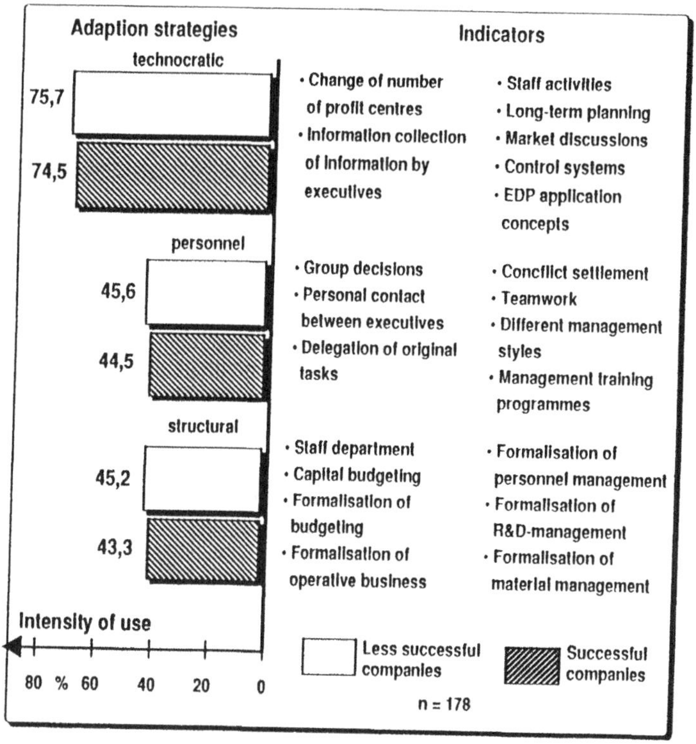

Fig. 15.4. Effects of different adaptation strategies on corporate success in large German companies.

it far from reflects the present state of software engineering, although the first movements in this direction are visible.

It is obvious that decisions are often far too closely tied to operative problems and first-stage solutions. Competitive advantages are in view, albeit still not clearly enough discernible for many. This route is risk, as is shown by the fact that it has not been possible to integrate any installed information system in existing companies as a pure service resource, nonselective of value and organization. Often alterations to the organizational structure and work sequences are a consequence and are not always welcome.

Economic crisis is overcome by applying innovations, as is proved by companies' demonstrated actions. Economic growth is increasingly becoming a function of companies' innovative activity. What is needed, therefore, is a change of companies' purely crisis-related innovative behavior. A study by Arthur D. Little (1985) has succeeded in giving a precise picture of the significance of innovations for the corporate scene in the near future. Of company directors surveyed in Europe, 87 percent rated the importance of innovations for the future of their companies as higher or much higher than today. The question remains, however, whether this attitude is reflected in corporate programs of activities and actual innovations or whether there are worlds separating the desired behavior and corporate enforceability that must first be bridged.

Role of the Information Manager Information management is the control of value creation processes and opens up opportunities for supporting a competitive strategy. This cannot only entail creating a new field of activity and its hierarchical arrangement. Claims like "the information manager will solve our problems" touch only the surface of the problem. Closer inspection of an information manager's necessary fields of competence raises the issue of the extent to which these requirements can be met by a single person. The picture of an information manager—the subject of years of discussion—has grown so much through the continual further development and continual expansion of these fields of activity that the solution can no longer be found in individual persons.

Understanding information management to be a new management strategy in the sense of a new management task, a communication factor, and a new controlling function not only calls for rethinking the structure but also gives rise to new economic dimensions in the strategic, structural, and managerial philosophy. The organizational incorporation of information management touches all spheres of activity of a reconstructional organization: functions and assignment of duties, management system, and hierarchy. As becomes increasingly clear, it is necessary to combine the basic functions of information management with functions that are performed in specific divisions by line superiors and with dedicated backup functions.

Information has moved increasingly into the focus of a conscious organization, and the task perspectives of information management have changed accordingly. Figure 15.5 illustrates this development for the public administration sector in the United States.

Let us return to the catch phrase "information as corporate resource." Properly used, it is reflected in a pertinent supply of information. The often-cited diversity of information is overwhelming.

Admid the diversity of information it is all too often forgotten that the information content of telecommunication is closely bound up with its form of presentation. Data on sales fluctuations in sterile columns of figures read differently than in neatly prepared diagrams. The visualization of information is one of the conditions of being quickly informed. These remarks touch only the surface of the problem, however. The task of information management embraces the planning, organization, regulation, and control of:

- The diversity of information
- The diversity of technology
- Organization and personnel components

Figure 15.6 illustrates this more closely. In future these tasks are to be

Period	Stage	Explanation
1900 → 1960	Paperwork management	■ Material control of the paperwork so as to improve efficiency
1960 → 1970	Management of automated technology (management of ADP)	■ Automation of manual techniques; automated and manual information management are considered to be fundamentally different
1970 → 1980	Information ressource management (IRM)	■ Integration of manual and automated methods of data processing and management ■ Technological and functional coalesence of technnologies of earlier decades ■ Coordinated and planned attitude to all technologies
1980 → 1990	Knowledge management	■ Complete penetration and dependence of information technologies in the organisation ■ Emphasis on the management of the information

Fig. 15.5. Development of the tasks of information management in the USA according to Mundhenke (1986).

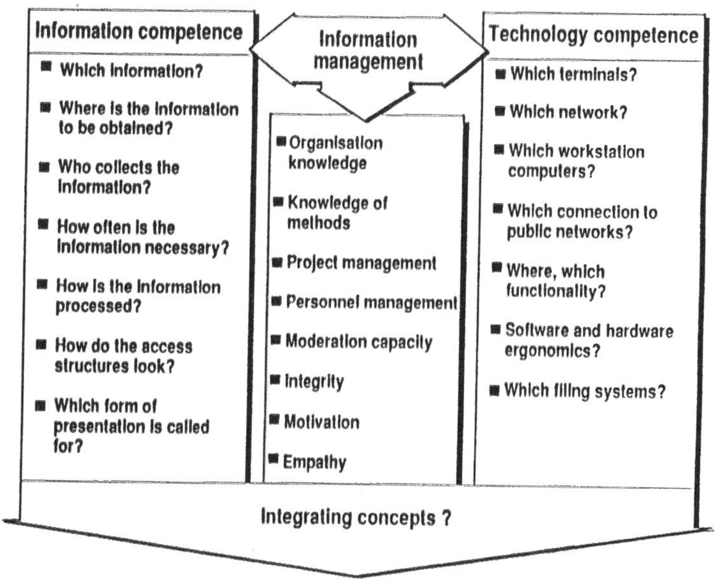

Fig. 15.6. Fields of competence of an information manager.

integrated in the plans of many companies by an information manager. Closer examination of the fields of competence of such an information manager raises the issue of to what extent these requirements can be met by real people.

A highly qualified team of subject specialists is required to be able to pursue this route. The primary objective is more a matter of preparing channels and "putting solutions on the tracks" and less one of initiating a "top-heavy head" in the company. The long-term objective is service carried by the specialist divisions' own dynamism. This objective can be achieved only through a successive transfer of responsibility to these specialist divisions. At all events care must be taken to avoid separating the problem-solving function completely from the operative handling factors or else, as experience shows, the problem solvers will quickly lose touch with the problem.

15.3 CIB: A COMPANY-SPECIFIC INFORMATION PROCESSING CONCEPT

Range of Computerized CIB Components

In general, the following types of technical computer and telecommunications systems are important for a computer-integrated business (CIB) concept.

Technical Computer and Telecommunications Systems These are mainly systems used at the manufacturing level such as computer-aided manufacturing (CAM) systems to control and monitor the production, processing, transport, and storing devices and computer aided quality (CAQ) systems with the tasks of planning tests, programming, and quality analysis. Computer aided planning (CAP) systems are used at the planning level to plan manufacture and installation for numerical control (NC) programming. Computer aided design (CAD) and computer aided engineering (CAE) systems are also used to support the design process and for prototypes of products.

Business Information and Telecommunications Systems Manufacturing requirements planning (MRP) is the central element in the business information and telecommunications system. At the planning level it performs the tasks of planning primary demand, material management, scheduling and capacity planning, and order monitoring. In the manufacturing sector MRP systems include production data acquisition and manufacturing control with the monitoring of volumes, time, and costs. Commercial information systems, such as cost accounting, financial accounting, and wages and salaries are situated at the administrative level in a company. Management information systems, decision support systems (DSS), and knowledge-based decision support systems are situated at the information systems level in the narrower sense.

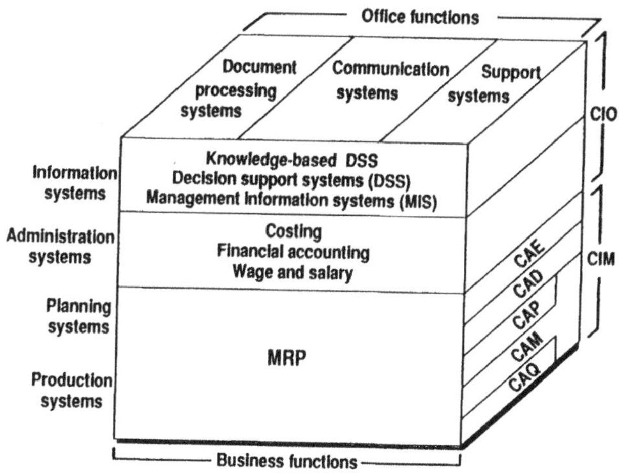

Fig. 15.7. The full range of technical support systems.

Office Automation (OA) Systems Information systems to support office procedures consist of document processing systems for preparing text and graphics, electronic mail systems, and various support systems such as appointments management, dictionaries, calculation programs, and project management. The main function of these systems is to support document processing, telecommunication, planning, and information functions in the office sector. The tool-like nature of these systems is in the foreground, that is, applications cannot be prestructured in detail. The full range of technical support systems is shown in Fig. 15.7.

The Concept of CIB

Against this background it is now possible to define what is meant by a CIB system, that is, an information and telecommunications systems that:

- Is derived from the company strategy.
- Is based on a unified infrastructure.
- Is not designed as a best technical solution but as the optimal exploitation of synergies.

Relation to Corporate Strategy Computer support in the production sector, both in the technical and the administrative office, has to be implemented at a situational level. This means that a large number of influencing factors, often of strategic importance, must be considered when organizing the information and telecommunications system. Basically, several directions of impact can be conceived with respect to the design of the information and telecommunications system. It can be configured for operative, strategic, or structural flexibility to meet the requirements of the situation. Operative flexibility means the ability to quantitatively adapt the basic business system to fluctuations in the demand volume. A company has strategic flexibility if it has the ability, over and above quantitative adaptation, to react to qualitative demands. Structural flexibility is needed particularly in technology-intensive businesses whose mix of product and services frequently changes and whose products and/or services are relatively short-lived. Organizational changes can therefore be carried out without disrupting the operating process.

Establishment of a Unified Infrastructure Here, both general trends in standardization in the OSI field (Open Systems Interconnection, e.g., MAP at Boeing and TOP by GM) and individual companies' efforts at standardization play an important role. Above all, the latter concern the basic technical systems sector (workstations, telecommunications systems, database, and archiving systems), the basic software tools, applications systems, and human–machine communication. Implementation tends toward distributed systems. The need to maintain technical standards in the fields of telecommunications, operating systems, and data management is underlined by the fact that the integration of the business administration sector with technical systems is ensured primarily by access to common databases and telecommunication between distributed systems. This therefore makes common and compatible planning of the technical infrastructure essential.

Exploitation of Synergies by Optimal Design of the Technical Solution An increase in efficiency and effectiveness can only be expected if the technical solution is in tune with long-term aims and organizational factors, the qualification structure of the employees is not overstretched, and an appropriate modus operandi is chosen. Often the best technical solution is not the most suitable one.

Single-function organization in terms of manufacturing technologies seen in many manufacturing companies will be expanded systematically in the context of a CIB concept and on the basis of the company strategy to encompass office automation technologies. Synergy effects are achieved by integrated technology planning of the information systems in the office and manufacturing derived from competitive strategies. This demands an overall consideration of the market, product, and manufacturing technology. The following section describes the way in which a computer-integrated business can be set up.

15.4 INTEGRATION STEPS IN SETTING UP A CIB

Information and telecommunications systems of a company will, in future, be a self-evident component of company strategy. The planning and design of a company-specific CIB system will increasingly be seen as a management task and requires in-depth thinking about long-term company strategy, the structure of the organization, management concepts, and personnel development (Bullinger, Niemer, and Huber, 1987; Bullinger, Niemer, and Schäfer, 1987). A company-specific CIB system is developed in several stages, all of which, for reasons of organization or qualification, have to be discussed to some extent. Generally, there are four stages, outlined as follows. Development goes from "stand-alone" information systems, through integrated sector and interdisciplinary solutions, to companywide information systems. The majority of practical examples are situated at stages 1 and 2.

Stage 1: Stand-Alone Information Systems

In this stage stand-alone data processing systems for dedicated tasks, such as financial accounting, order processing, word processing, financial planning, and so forth, are primarily used in systems to perform administrative and forward planning tasks. While small businesses usually feature single-workstation systems, larger companies generally have expandable multiworkstation systems. In the design and manufacturing sector these are task-specific systems such as CAD systems, robots, NC machines, and so forth, which are not integrated into a higher-level information system. The use of stand-alone, workstation-oriented technologies improves input–output relations in the local environment, shortens processing times, and upgrades the quality of work. The advantages of introducing stand-alone systems can be seen in that the existing organizational structure is only slightly affected and clear profitability statements can be made with respect to the investments involved. On the other hand, the synergy effects possible through higher-level information links cannot be achieved with existing media breaks. The characteristics of stand-alone information systems are summarized in Fig. 15.8.

The transition to the following stage of integration is prompted by a perception of the obstacles posed by the interruptions to data flow (media breaks). More and more integrated software packages based on the central computer concept are used to solve these problems. Data exchange between formerly discrete software functions is therefore made possible by integration at the software level.

Stage 2: Sector-integrated Information Systems (CAx Concepts)

The sector-integrated solution (cf. Fig. 15.9) is the second stage in integrating information and telecommunications systems. On the business administration side, MRP is included at this stage of integration. At best, only the rudiments of organizational and systems integration between the business administration/planning function and design and manufacturing processes, that is, between MRP and CAx systems, are available. As yet, integrated, interdisciplinary office automation systems based on distributed information systems are not

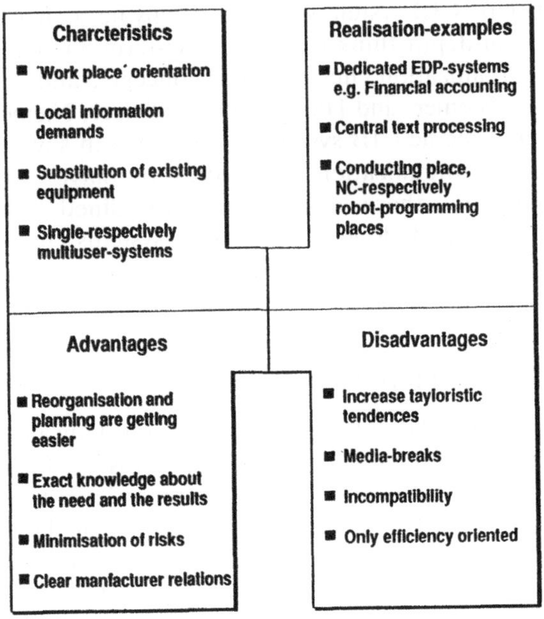

Fig. 15.8. The characteristics of stand-alone information systems.

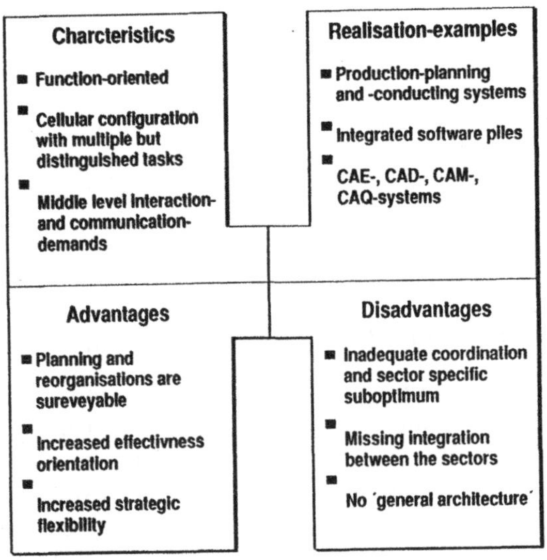

Fig. 15.9. Characteristics of sector-integrated information systems.

used. In the manufacturing sector flexible manufacturing systems, computer-aided manufacturing, and assembly islands and flexible transfer lines can, for example, be used.

In companies in this stage of integration there is a strong emphasis on sector-oriented division of labor, and the organizational processes are highly function-oriented. Stage 2 interdisciplinary integrated systems have the advantage of being more flexible than stage 1 systems, and company procedures are more transparent. The individual improvements achieved as the sectors expand do not, however, correspond to the optimum for the company as a whole. The reason for this is that, as in stage 1 of integration, the potentially available synergy effects cannot be achieved. The technical infrastructure of stage 2 integration is almost exclusively in the form of centralized computer concepts.

The disadvantages existing at this level of integration can be obviated by successively networking the computer resources between the individual sectors. For this purpose an interdisciplinary approach needs to be developed. At the organizational level a process-related attitude is important. At the technical support system level more comprehensive knowledge is required about all the components in a distributed system and their interrelations. The development process to the next stage of integration can be considerably simplified by using decision support systems, tools, and aids that permit the modeling of alternative architectures and the simulation of working processes based on these models. Computer-assisted modeling and design tools allow all the major components of the information system to be planned and designed graphically and interactively. They increase understanding of the information system and can be used as the operational basis for telecommunication on different subjects between management staff. Simulators that allow evaluations to be made on the basis of the models produced are used to optimize the architecture of the distributed information system. They enable the effects of alternative solutions to be displayed interactively. The use of such tools, which are currently being developed in the context of the ESPRIT project 2071 (COMANDOS—Construction and Management of Distributed Open Systems), promises increased planning flexibility and considerable support in system design. An adaptive approach is urgently needed for the design of information systems since their overall architecture has to be continuously adapted to changes in the company structure (Iivari, 1982).

Stage 3: Interdisciplinary Information Systems (CAx/CAx Concepts)

In the third stage of integration, the "automated islands" of the previous stage are linked through computer networks. Departments from various company sectors are now strongly affected by the required changes, with even the structures of the organization itself having to be modified in many instances. Stage 3 integration is the linking of different insular solutions, like the CAD/CAM pairing for example, the common management of basic data,

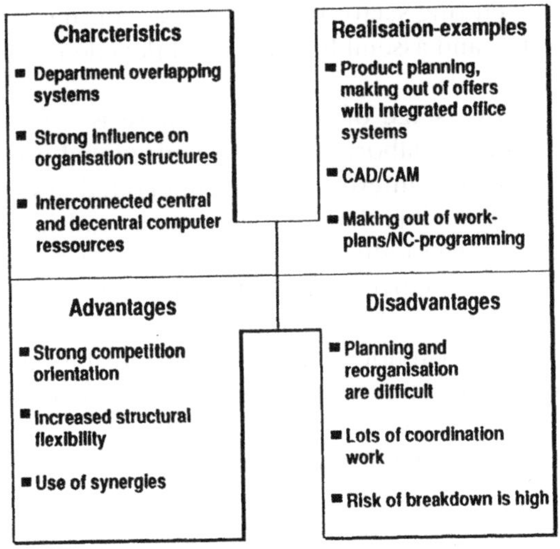

Fig. 15.10. Characteristics of interdisciplinary information systems.

achieved through automatic generation of parts lists, and transmission of data structures to an MRP system. The linking of working plan production and NC programming is also conceivable.

At this stage in integration the strong influence of technology on organizational structures is to be observed. The obvious aim of this technology is to permit stronger competitive orientation, to improve the structural flexibility of the company, and to exploit synergies between the sectors. This is linked to greater expenditure on reorganization and planning, since the distributed information systems that will come into use are not only bigger but also increasingly heterogeneous. Dependence on these systems also increases. There is a greater risk of failure in the technical parts of the systems. The characteristics of interdisciplinary information systems are shown in Fig. 15.10.

With the increasing complexity of the distributed systems used, it now becomes more and more expedient to use computer-assisted modeling and design tools and requisite simulation tools to assist in planning and reorganization. Security and risk management tools are offered to tackle the increased risk of systems failure. For example, the management of user authorizations or the estimation of the risks of systems failures can be assisted and simplified with interactive graphics tools. Furthermore, interactive graphics tools can be used for the management of user catalogues in order to reduce coordination work to a minimum.

Stage 4: Companywide Information Systems (CIB Concepts)

Computer-integrated business is based on the linking of rudiments from the CIM and office automation sectors and classical business data processing

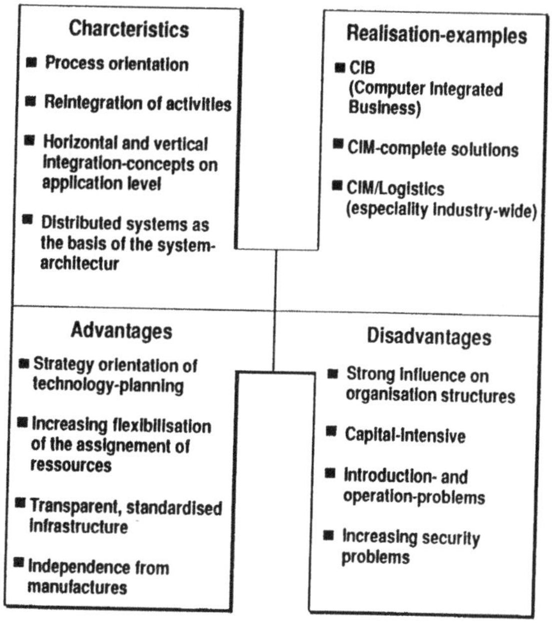

Fig. 15.11. Characteristics of companywide information systems.

systems. Interdisciplinary, transparent information networks are the basic technologies of stage 4 integration. The integrative nature of information and communication technologies makes it possible to construct universal, all-embracing information infrastructures. The processes to be supported can now be integrated at all levels of the hierarchy by information technology. Potential for integration at stage 4 results from the linking of technical information and telecommunication systems, business information and telecommunication systems, and office automation systems.

Linked processes such as order processing often have a central position in the company, since they affect almost all departments and extend from planning through production to the operation of machinery. Preparing quotes on the basis of a companywide information and telecommunication system that integrates both design and cost accounting on the organizational side and the CAD system and MRP on the technical side can, for example, considerably speed up and improve the quality of the quotation service. (See Fig. 15.11). This requires a profound change in the organization vis-à-vis the preceding stages of integration. The combination of company functions and the transition from working procedures organized on the basis of division of labor to process-oriented procedures require technology-oriented procedure sequences based on interdisciplinary databases. Strong process orientation including the reintegration of activities in some cases is needed in planning and design at this stage of integration.

The advantages of increased company flexibility, greater local presence, and a reduction in throughput times must be seen against the disadvantages of increasingly complex technology, high levels of tied capital and dynamically changing organization and coordination structures (cf. Fig. 15.10). The organizational and technical demands on the management of distributed information systems have therefore risen. Requisite organizational and technical measures must now be provided to ensure that the advantages gained by introducing computer-integrated information processing are not weakened by the dynamically changing organization and coordination structures. To do this, the information system has to be adapted to varying demands in close cooperation with technical and organizational departments. It must therefore be ensured that an adaptive approach is implemented so that the entire technical and organizational system can be continuously configured according to existing aims and tasks. This requires CIB-oriented integration of all the technical and organizational functions in a company. The tools introduced at stage 3 to manage distributed information systems are absolutely necessary at this stage of integration for the administration of the distributed information system.

While companies can become more independent of the suppliers of the technology, there is also the danger that, due to cross-business communications networks, they will become more deeply involved in their major clients' information processing. The exchange of CAD data, job planning directly in the supplier's computer, and electronic ordering systems are the first signs of trends developing in this direction. Management in the years to come is likely to be confronted by a major challenge in this respect: Will small businesses in particular be able to maintain the necessary independence and flexibility, and will it be possible to establish so-called virtual companies, that is, a flexible link between independent companies, to fulfil complex customer requirements instead of dependent firms in the information chain?

15.5 PRECONDITIONS FOR SUCCESSFUL MANAGEMENT OF DISTRIBUTED INFORMATION SYSTEMS

The success of CIB in the future depends decisively on the ability to confront the weaknesses mentioned earlier. The following four preconditions are particularly important for using distributed information systems.

Task-Specific, Technical Design Methods: Applications Management

The following knowledge is to be considered as the basis for the technical structure of distributed information system implementation (Horn, Ness, and Reim, 1988):

- Knowledge of the available information system components

- Knowledge of the relationship between the individual information system components
- Knowledge of the capabilities of individual information system components and of their interaction
- Knowledge of the system behavior during partial failure

The evolutionary development of information systems is suggested as a method of procedure by Floyd (1984), Gilb (1987), and Iivari (1982). This is based on the assumption that the system environment and demands change unpredictably. Up to 70 percent of the cost of information systems arises during a phase-oriented approach in the so-called maintenance phase. Therefore, a coercive demand is that the system be capable of modification. This leads to an iterative procedure in developing the system, with repeated runs through a design cycle and the appearance of consecutive versions of the system. It is no longer possible to separate this into a design, operating, and maintenance phase (Floyd, 1984; Iivari, 1982; Horn, 1988).

Evolutionary development leads to frequent changes in a software program. These are reflected either in modified software structures or in additional new or adapted existing software modules. Kramer, Magee, and Ng (1989) introduced the term "configuration programming" for this. Software evolution consists of configuring specific application classes based on prefabricated software modules. Support in software evolution consists of the following steps: task analysis, selection of the software modules needed from a library of reusable modules, configuration of the selected modules, and updating of the module library.

Task-Specific, Organizational Design Methods: Organizational Management

The following knowledge is to be considered as the basis for the organizational structure of distributed information system implementation (cf. also Horn, Ness, and Reim, 1988):

- Knowledge of the organizational units and their tasks
- Knowledge of the organizational units' need for support
- Knowledge of the relationship between organizational units
- An idea of the intensity of use and the location of support functions, resulting in demands on the type of information system components provided
- Knowledge of the required degree of local autonomy

This results in the need for new cooperative structures between the specialist world, the divisional organizer, and the technical world (applications developer, systems manager).

Tools to support this interdisciplinary process are needed. The ability to infer application development and systems management requirements is important, particularly in the interplay between technical design of the distributed information system. However, existing attempts at organizational management often fail to include the conceptual orientation in information systems.

Tools specially used to support the requirements specification are often restricted to graphic representations when investigating and documenting requirements. Since the information gained is not kept in a knowledge database, it can hardly be used for software development or system configuration. It is therefore difficult to link the partial functions of applications management and systems management.

The new cooperative structures between the specialist world, the divisional organizer, and the technical world lead to the organizer having new, additional computer-assisted aids at his or her disposal in order to achieve the integration of available information systems into organizational processes for restructuring the organization.

Possibility of Adapting and Developing the Whole System Further: Systems Management

Systems management includes a series of operative tasks. Bacon et al. (1988) distinguish five related tasks: configuration management, fault management, security management, accounting management, and performance management. Configuration management is very important in the context of adaptation and further development and has to be observed and modified so as to enable a running distributed information system to be adapted to new requirements and to be further developed. In this way optimal functioning is guaranteed in every situation.

The following knowledge and capabilities are considered as the basis for the adaptation and further development of distributed information system implementation (cf. also Horn, Ness, and Reim, 1988):

- Knowledge of existing functions and applications
- Knowledge of the possibilities of integrating existing or new applications
- Ability to quickly change the system for adaptation to new requirements
- Openness to new technologies and procedures for expansion of the information system

In this respect systems management is seen as an intrinsic part of a distributed information system, since it is only this which makes its use possible in business. Linking to the application and the organizational structure is necessary to achieve integration into business processes.

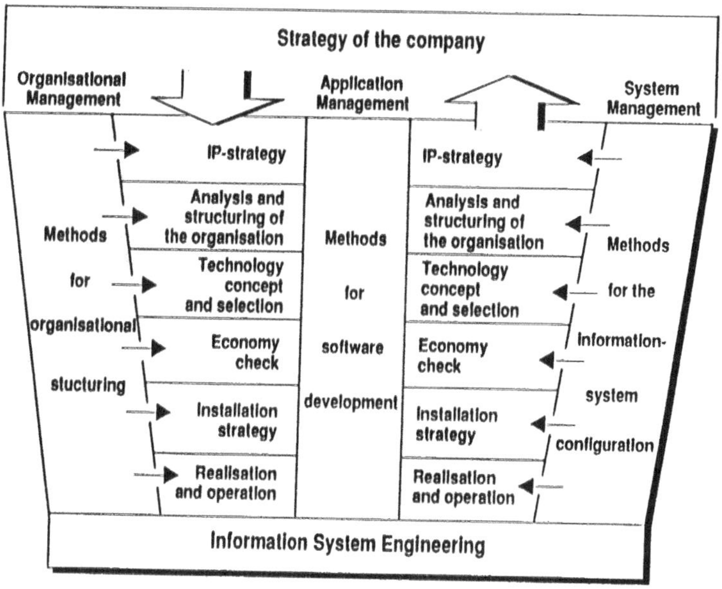

Fig. 15.12. Information system engineering as part of CIB.

Support in the Integrated View of Technical and Organizational Design in a Constantly Changing System: Information System Engineering

These three preconditions will merge. In future, there will no longer be independent design processes for subtasks. A new type of all-embracing design process will come into being: information system engineering. This is the integrated view of application management, organizational management, and system management (Fig. 15.12).

Information system management is contingent on the following:

- A clear understanding of the relationship between organizational units and information system components
- A clear understanding of the relationship between the individual information system components

At the organizational level CIB means the integration of the company organization, application, and system management.

REFERENCES

Bacon, J. M., Horn, C., Langsford, A., Mullender, S. J., and Zimmer, W. (1988). MANDIS: Architectural basis for management, in *Research into Networks and Distributed Applications*, R. Speth, Ed., North-Holland, Amsterdam, pp. 795–809.

Bullinger, H.-J., Niemeier, J., and Huber, H. (1987). Computer integrated business (CIB)-systeme, *CIM Management*, **3**, 12–19.

Bullinger, H.-J., Niemeier, J., and Schäfer, M. (1987). Aufbauorganisation des Informationsmanagements, *Nachr. Dok.*, **38**, 193–199.

Floyd, C. (1984). A systematic look at prototyping, in *Approaches to Prototyping*, R. Budde, K. Kuhlenkamp, L. Mathiassen, and H. Züllighoven., Eds., Springer, Berlin, pp. 1–18.

Gilb, T. (1987). Evolutionäres Entwickeln—Eine alternative Methode des Software-Engineering, *Computer Magazine*, **16**(1/2), 17–19.

Horn, C., Ness, A., and Reim, F. (1988). Construction and management of distributed office systems, in *EURINFO '88, Concepts for Increased Competitiveness*, H.-J. Bullinger, E. N. Protonotarios, D. Bouwhuis, and F. Reim, Eds., First European Conference on Information Technology for Organizational Systems, pp. 378–385.

Iivari, J. (1982). Taxonomy of the experimental and evolutionary approaches to systemeering, in *Evolutionary Information Systems*, J. Hawgood, Ed., North-Holland, Amsterdam, pp. 101–119.

Kramer, J., Magee, J., and Ng, K. (1989). Graphical configuration programming, *Computer*, **22**(10), 53–65.

Little, A. D. (1985). *Management der Geschäfte von morgen*, Gabler, Wiesbaden.

Lorin, H. (1988). *Aspects of Distributed Computer Systems*, 2nd ed., Wiley, New York.

Mundhenke, E. (1986). Informationsmanagement. Erfolgreiche Einführung in US–Verwaltungen, *Office Management*, **9**, 846–862.

CHAPTER 16

HUMAN FACTORS IN ADVANCED MANUFACTURING: A REVIEW AND REAPPRAISAL

J. R. WILSON
Department of Manufacturing Engineering and Operations Management
University of Nottingham
Nottingham, United Kingdom

R. J. KOUBEK and G. SALVENDY
School of Industrial Engineering
Purdue University
West Lafayette, Indiana

J. SHARIT
Department of Industrial Engineering
School of Engineering and Applied Sciences
State University of New York–Buffalo
Buffalo, New York

W. KARWOWSKI
Center for Industrial Ergonomics
Department of Industrial Engineering
University of Louisville
Louisville, Kentucky

Organization and Management of Advanced Manufacturing. Edited by Waldemar Karwowski and Gavriel Salvendy.
ISBN 0-471-55508-8

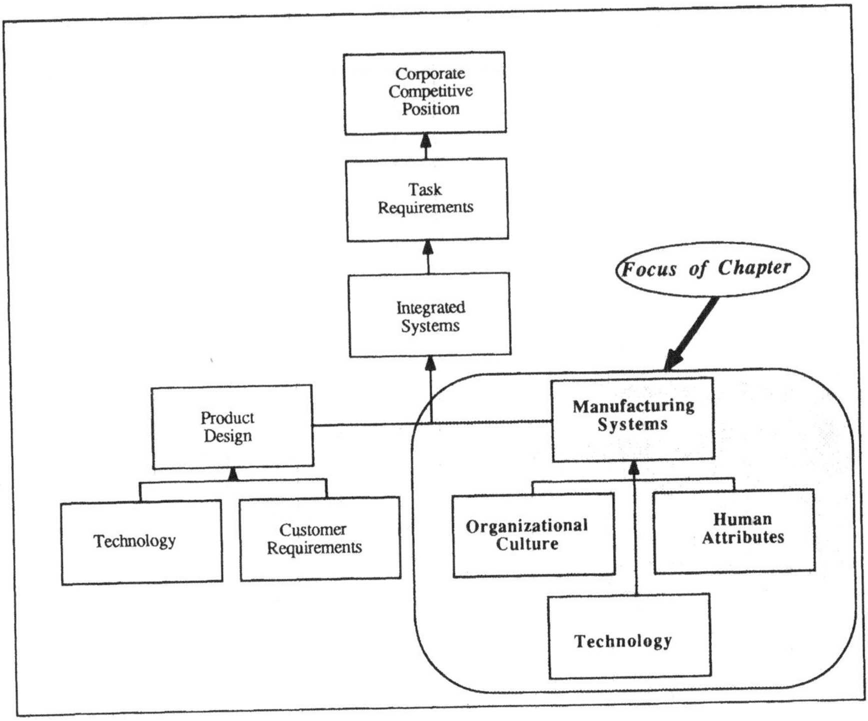

Fig. 16.1. A Conceptual framework for human aspects of AMT.

16.1 INTRODUCTION

The basis for competition in the world marketplace has changed significantly. The traditional cost-based approach, once the essence of competitive strategy, has been augmented by quality and product differentiation. Companies are now facing a fourth dimension, time (Itoh, 1991). Here, a company's ability to adapt quickly and with agility to market pressures becomes a measure of its survival.

In today's marketplace corporate survival depends on constant improvement of a company's competitive position. Advanced manufacturing technology (AMT) implementation must consider the competitive factors of cost, quality, product differentiation, and time to justify technology expenditures. This multifaceted competitive strategy justification must also be used when considering the human element in the production system.

Majchrzak (in press), in reviewing 154 publications based on empirical evidence, confirms that focusing on technology alone during modernization is insufficient for advancing significantly in all four competitive dimensions outlined by Itoh (1991). She concludes that the desired benefits of AMT are achieved as a result of "preplanning that includes an understanding of a) the likely organization and people infrastructure needed to support the technology change, and b) the inputs that will contain options for managing a positive infrastructure. . . . Hence, the desired benefits of AMT are achieved as a result of decisions made (by the planners of the technology), not as a result of the technology itself." Also, Brodner (1991) states, "Consequently, technology, work organization, and skill profiles must be well suited to each other and must be designed congruently."

Figure 16.1 presents a general model and conceptual framework for the interaction of technology, organization, and people (TOP) factors in the development of manufacturing systems. While discussion of all aspects in this framework is beyond the scope of this chapter, the material presented here focuses primarily on the bottom-right elements of the figure: technology, organization, and people. Since technology cannot be discussed in isolation of its respective impact on the organization and people, technology-related factors are presented in the context of the organization and people issues. To begin, however, a discussion of research methodology issues is provided in order to lay the foundation for understanding later sections. Following this, organization and people issues will be presented in turn.

16.2 RESEARCH METHODOLOGY ISSUES IN AMT

Research Paradigms

Aside from on-site real-time data collection, the human factors professional has two primary means for studying behavior in AMT settings: laboratory studies and the use of simulators. Each of these paradigms is discussed below.

Laboratory Research For tasks in the manufacturing systems domain containing significant rule- and knowledge-based components—such as supervisory control and scheduling tasks—human factors researchers have been adept at identifying human capabilities and limitations as well as at formulating frameworks for decision support that attempt to both compensate for these limitations and maximize human capabilities (e.g., Ammons et al., 1988; Dunkler et al., 1988; Nakamura and Salvendy, 1988; Sanderson, 1991; Sharit and Elhence, 1990; Sharit et al., 1988). These frameworks, however, depend on the functional and conceptual analyses of the human tasks under consideration, and despite all attempts to achieve a "workplace validity," it is questionable whether laboratory studies achieve anything more than a general understanding of human performance in this task domain.

The answer does not appear to lie in new experimental design paradigms as suggested by Simon (1987), although this approach to capturing the multifactorial complexity of the real world is somewhat consistent with the advice of Chapanis (1967) concerning the pitfalls of laboratory experimentation for solving real-world problems. More reasonable, although less conventional with respect to traditional human factors research methodology, are the approaches suggested by Rouse (1987) and Rasmussen (1988). Rouse espouses a system design perspective that is more concerned with the "why, how, and what if, rather than how much," while Rasmussen emphasizes a more elaborate analysis of the work domain for the analysis of cognitive work for system design:

> For this kind of analysis, observation and recording of overt activities is insufficient. This method must be supplemented by an analysis of the semantic content of the work based on a technical analysis by domain experts together with interviews and discussions with people actually performing the work. This kind of knowledge is no simple matter because it requires trained interviewers with an intimate knowledge of the actual work domain. (p. 2)

In general, industries employing advanced manufacturing systems will almost always appear reluctant to accept or integrate experimental research findings into their operations, probably because of a general distrust for data derived from scenarios other than those found in the workplace. While some of the general lessons learned from such research may prove valuable, the face validity and overall utility is likely to be perceived as low.

Simulators Training simulators have been used mostly when to operate under real-system conditions is inefficient, expensive, hazardous, inconvenient, or impossible (Flexman and Stark, 1987). Although the use of simulators in the training of operators of industrial systems has become more common in recent years, their application has been primarily in the power industry (e.g., electrical power generating systems and nuclear power plants) and for chemical processing and oil refinery systems. One factor that has likely impeded the development of training simulators for advanced manufacturing systems is the discrete nature of these systems relative to process control systems. Manufacturing systems are often undergoing changes that their discrete makeup allows them to accommodate. Given the expense involved in developing simulators, manufacturers may not consider them worthwhile investments.

There are a number of reasons why human factors researchers should pursue the prospect of employing training simulators in manufacturing systems. First, computer-integrated manufacturing systems can represent a confusing environment for learning complex skills. Moreover, it is not convenient to have the operator experiment by manipulating parameters under real operating conditions. Second, the simulator can expose the operator to many more situations than would normally be encountered, thereby potentially increasing the efficiency of operators during both training and real operating conditions. The interplay of factors such as (1) the hierarchical nature of computer-integrated manufacturing systems; (2) the large number and variety of discrete operations with variable processing times; (3) complex spatial configurations that can encompass machine centers, buffer areas, material handling systems, and storage facilities; and (4) system performance objectives that can be applied to individual parts, part families, and batches argues in favor of a training program that can allow the operator to systematically exercise some degree of control of the variables associated with manufacturing operations. Training simulators can also support the identification of mental models (Wilson and Rutherford, 1989) or other relevant notions of knowledge representation for situations involving contingencies and interruptions, which frequently occur in manufacturing operations. Related to this are studies of operator trust and performance strategies (e.g., Moray and Lee, 1990). Finally, by simulating the complex network of information flows that exists in computer-integrated manufacturing systems, training simulators can help operators and supervisors learn to integrate both vertical and horizontal information (Sharit et al., 1988), that is, hierarchical information roughly representing levels of progressively shorter planning horizons and information at a similar operating level but spanning across loosely interrelated subsystems (such as automated guidance vehicles traveling between flexible manufacturing systems to an automated storage facility).

The challenge for human factors researchers is to build simulators that can accommodate these characteristics of advanced manufacturing systems. By utilizing the simulator's inherent capacity for allowing variables to be systematically manipulated, the error taxonomies and other approaches to human behavior analysis, together with fundamental knowledge concerning human abilities and limitations, can then provide the basis for decision support requirements and therefore the basis for decision support system design.

Human factors researchers also have the opportunity to apply some relatively new methodologies that can embellish manufacturing systems. Knowledge-based simulation methods (Barr et al., 1989) serve to strengthen the system model by allowing system states to be inferred from rules rather than requiring the user to describe the details of each system state. Because many manufacturing problems, especially in scheduling and control (Sanderson, 1989), are handled more effectively through heuristics, more powerful system models can be built by integrating the numerical features of simulation with knowledge-based components. In addition, advantage can be taken of the ability for knowledge-based programs to explain the reasoning process that led to a particular result, providing further support for operator training.

An AI-based simulation tool that deserves consideration is the cognitive environment simulation (Woods et al., 1988). This tool was developed to explore how operators in nuclear power plants decide on what actions are appropriate in emergency situations. A variety of different assumptions concerning both the knowledge the human has available and also the types of cognitive processing that can characterize the human's response to the demands, reflecting the ongoing events in the system, can be systematically manipulated. In the advanced manufacturing environment this type of tool can assist in design problems such as establishing worker capacity requirements by evaluating the cognitive workload that can potentially arise across a number of demanding production control scenarios.

Field Studies An alternative to laboratory-based experiments and the use of simulators is to perform research in the real-world environment. For studying human factors in manufacturing, field research is presently the most frequent paradigm used. In this approach, data are collected from personnel while performing their jobs in the workplace. Sackett and Larson (1990) provide an excellent review of the strengths and weaknesses associated with this approach, which are summarized below.

The first strength of field studies identified by Sackett and Larson (1990) is the length of time between manipulation of the independent variable and measurement of the dependent variable. For example, in a study to test the impact of reward systems on worker productivity, field studies provide the opportunity to implement the systems and measure their impact on performance several months, or even years, after personnel are performing under the particular system. This provides for long-term stability and reliability of the research findings. Such a time lapse is not feasible in laboratory or simulation settings.

Since in field studies personnel are performing their normal work procedures, the activity has meaning to the participants. Compared to traditional laboratory studies, where subjects are typically paid a stipend to perform a task that is independent and unrelated to their normal daily life, participants in field studies typically are more involved in the task. Such involvement can produce more reliable and realistic data. In addition, Sackett and Larson (1990) point out that potential demand effects from the experimental setting are reduced by task performance in the normal work environment. Subject responses are less likely to be influenced by the experimental setting. Finally, field studies allow manipulation of the independent variable in a manner similar to that normally found in the workplace. In laboratory settings, often the levels at which the independent variables are manipulated must be extreme to produce an effect in a limited time period. This can result in a difficulty transferring results to real-world settings. Field studies avoid the artificiality by imposing realistic variable manipulations.

When deciding to use a field study, the above benefits must be weighed against its primary weakness: lack of control necessary to determine the exact effect of experimental manipulations on the dependent variable measures (Sackett and Larson, 1990). Difficulty can arise in controlling the independent, dependent, and confounding variables. In the example cited above where reward systems are manipulated to assess their impact on productivity, selection of the reward systems to be tested is not exclusively a decision of the experimenter. Company officials will most likely be involved to assure human resource policies and union agreements are followed. In addition, dependent variable measures often must be approved. Due to equipment recording capabilities, company procedures, and needs for confidentiality, the measures of productivity may be less precise and reliable than desired. Finally, given the longer time frame usually associated with field studies, a host of intervening, confounding variables may arise. The very environmental complexity that provides realism for the field study measures can interfere with reliability and validity of these same measures. Personnel changes, equipment failures, and changes in work flow are just a few of the innumerable potentially confounding variables that may arise and impact the dependent variable measures. With such a dynamic environment, it is quite difficult to control for all possible disruptions.

Given these strengths and limitations, those studying human issues in manufacturing must decide whether their question is best served by the laboratory, simulation, or a field study paradigm. At present, given the complexity of the manufacturing environment and the lack of clarity regarding the important human issues, it is a difficult task to provide realistic laboratory-based research that yields findings valid for the real-world environment. As such, field studies are seen most frequently in the literature. However, from these studies, principles are beginning to emerge that are suitable for testing in the more tightly controlled laboratory environment. As such, while field studies will most likely continue to be a primary means for exploration of these issues in the near future, the role of laboratory-based research should continue to grow as the issues become more well defined.

Concepts for Understanding Operator Behavior

While laboratories and simulators provide an environment in which to examine human aspects of AMT systems, they do not dictate what aspects are to be studied. Two features of human behavior often explored in complex task performance are errors and knowledge representation—particularly mental models. These concepts and their relation to AMT are discussed below.

Error Analysis Traditionally, the potential for humans to commit errors (Reason, 1990, Chapter 28) has been considered an important, if not critical, aspect of those industries in which human-initiated actions could lead to hazardous consequences. With respect to human decision and control in advanced manufacturing systems, human error has not been subjected to nearly as formal an analysis. Nevertheless, human error has important implications for this domain as well and can be loosely defined as human behavior that leads to degraded system performance.

Prabhu et al. (1992) proposed a framework for deriving human-centered information requirements based on categories of temporal errors, which are essentially errors arising predominantly as a result of temporal factors. These categories of errors related to time stress, time estimation, human memory processes, the reasoning process, the coordination of cognitive processes, and the error recovery systems.

Such errors can lead to actions performed at the wrong times and the omission or incorrect sequence of actions. Errors related to temporal delay could potentially arise when there are time delays between actions and between intention and action [what Reason (1990) refers to as reduced intentionality]. In advanced manufacturing systems, these delays could be caused by factors such as workstation configuration or material handling systems design.

Some categories of error pertain primarily to unidimensional processes related to human abilities and limitations; other categories, however, are more complex, and are consequently further broken down. For example, the coordination and integration of cognitive processes reflects the interactions between human cognitive processes and system events and is broken down into the subcategories of errors related to temporal scheduling and temporal execution control. The latter, for instance, relates to control of the execution of decisions over time and addresses those errors pertaining to people changing strategies over time and revising assumptions and knowledge of system dynamics.

By analyzing these temporal error categories within the frameworks for error analysis proposed by Rasmussen (1986) and Reason (1990), a taxonomy of temporal errors in computer-integrated manufacturing systems was formulated. This taxonomy was designed to help the analyst identify the most likely "error shaping factors" (i.e., factors that promote errors), which ultimately can be used to identify decision support system requirements. In addition to Prabhu et al., other types of error taxonomy are likely to appear (e.g., Sanders and Moray, 1990) and are encouraged in order to steer our thinking about human decision and control performance in this task environment in these terms.

Mental Models For nearly 50 years the notion that people form mental models of their environment has been a topic of psychological inquiry (e.g., Craik, 1943). For nearly 30 years researchers in human–machine systems have utilized the concept as being almost self-evident (e.g., papers in the collection of Edwards and Lees, 1974). Within a variety of litertaure we also have seen work in explaining parallel or identical notions such as naive theories, device models, mental maps, domain models, etc. Now, in the last few years, workers in human–computer interaction have picked up the notion and are attempting to define many different kinds of user models as well as models of the user. All this time the development of the mental model notion within cognitive psychology and human factors seem to have continued almost without any point of mutual contact.

Researchers in human factors continually seek theories and concepts from cognitive psychology in order to adapt and apply them to the specification and implementation of systems in offices, services, factories, leisure, and so on. Theories of knowledge representation and development of user models receive special attention, and the notion of mental models seems to hold particular interest, attractiveness, and potential utility for human factors workers.

It is easy to see why this is so. If we can predict or understand what internal model or conceptual model a new system user might hold about that system and its relevant domain or what model might be built by interaction with the system, then we can improve interface design, training, operating procedures, and so on. In Norman's (1983) view, by understanding the potential users' mental models and by adapting their own conceptual model accordingly, designers might develop a system "image" that better matches, sustains, and helps develop an appropriate user mental model.

For operators of AMT we might agree that mental models are internal representations of relevant technology, built on experience and observation of the world and of the particular system. Such models might constitute system structure, function, or state, and certain spatial, causal, or contingency relations. If we regarded them as qualitative simulations that are run by the operator to make inferences, then we can see how they might act to underpin understanding and behavior.

Methods have been suggested for identification and representation of mental models held by people using technical systems, which are appropriate to ergonomics and AMT rather than to laboratory-based experimental psychology (e.g., Payne, 1991; Rutherford and Wilson, 1991). However, if the notion is to be useful and usable within human factors, we need to be far clearer about relevant methods of data collection and analysis, procedures for conversion of data into a mental model, specification, test, and refinement of that model.

Within the world of human factors—whether applied to manufacturing environments or to the larger domain of human–machine systems—some fundamental questions must be answered about mental models. It is necessary to propose what a mental model might be, how it might be identified and represented, and how it might subsequently be applied in systems analysis and design. Perhaps most fundamentally we must address the question of whether we are talking about anything other than "operator/user knowledge" when we apply the term within the fields of human–computer interaction and human–machine systems. [See Rogers et al., (1991) for further discussion.]

Facilitating Application of Research Findings to AMT Environments

To benefit those industries employing advanced manufacturing systems, human factors researchers may need to rethink their strategies. Current pressures in this industry tend to make it unlikely that project directors, designers, and the like would take the time to develop, along with human factors professionals, research programs that could positively impact their operations; thus channels need to be found that make such links to human factors researchers attractive. This can be done through government support that establishes liaisons whose purpose it is to work on human factors problems considered worthwhile by both human factors researchers and manufacturing representatives.

Essentially, human factors researchers need to market the potential contributions of their expertise to the design and operation of computer-integrated manufacturing systems. The use of successful case studies that illustrate the benefits of such intervention, as well as other cogent arguments based on insights derived from a systems analysis of manufacturing operations, are examples of strategies that can be used. Human factors scientists will need to identify those interventions that would likely result in significant gains. The process must begin with a systems analysis of the target facility that could include both traditional task (Drury et al., 1987) and cognitive task analyses. Only after such an analysis could one ascertain whether the existing base of human factors insights and methods could be applied in a cost-effective manner. Human factors scientists usually do not have to confront such stringent criteria in highly regulated industries, such as the nuclear sector; in such industries, the most minimal insights into human error and system safety are welcomed.

Traditional links, typically through project directors at government research agencies that urge researchers not to work in a vacuum (i.e., to establish links with an appropriate industry for "validating" their research), do not ultimately translate into the acceptability criteria noted earlier. The key is to help establish a formal mechanism to ensure that these criteria are met. If this is achieved, manufacturers are less likely to be disillusioned and cynical with respect to the insights and arguments offered by human factors researchers. The contexts in which laboratory research would now appear most useful include developing tools that support manufacturing operations and testing designed to reveal the effectiveness with which humans can perform certain types of tasks.

16.3 ORGANIZATION AND MANAGEMENT ISSUES IN AMT

Using appropriate research methods as described above, analysis and design of human aspects in AMT need to consider the organization, people, and technology simultaneously to achieve a congruent design [Majchrzak and Gasser, in press (a)]. Congruency shifts the focus from optimizing each of these components to designing an appropriate "fit" between the components. In this section, several organizational issues necessary for congruent design will be discussed with reference to AMT. Discussion of the "people" aspects will be presented in Section 16.4.

Human Resources for AMT Operation

When changing the technology or organization of their manufacturing systems, companies are looking to improve the effectiveness of their operations and to enhance their competitive edge in the marketplace. How successfully they do this will be assessed, not just by increases in productivity (traditionally seen in terms of output quantity per unit of time, cost, or other resource), but also by improved production quality and reduced reworking, better flexibility and innovation in operations, and reduced indirect as well as direct costs. Ergonomists would argue that central to achieving all of these outcomes are human factors issues. The ways in which new technical or organizational systems are introduced will determine the success of their implementation, and the jobs and roles that are developed for staff at all levels will be critical to successful operation of the systems.

With any system involving people, individual differences mean that general solutions are hard to define, but recognition of such differences is a step in the right direction. Growth and social needs, highlighted by Cummings and Blumberg (1987) as contingency factors, relate in fact to two principal areas of human factors interest in manufacturing systems: group work and operator control. Control, at the heart of people's psychological needs, is a key issue in human-centered systems of the future. What is in no doubt is that modern systems of manufacturing, aiming at cooperation, flexibility, and total quality, will produce a premium on workers' social and cognitive skills and will entail work force cooperation and communication beyond what has gone before.

Total Quality Management and Total Involvement

Changed production philosophies and controlling technologies, specifically those in the "family" of just-in-time (JIT) approaches and derivatives, make a number of demands on the way manufacturing is conceived and organized. *Eliminating waste*—removing or eliminating buffer stocks and work in progress, *pull not push* of production by demand rather than availability, a *flexible work force*, and the *exposure of production problems* (rather than working around them or covering them up) are all ingredients of a JIT approach and will place new and very different demands upon industrial work forces and hence upon human factors researchers (Wilson, 1992). Many authors link a JIT philosophy to a policy of total quality control or management (TQC or TQM). This is built on quality of materials, parts, and products and also of the product design, the process of manufacture, and the people, operators, and management. Operator involvement in this, e.g., use of statistical process control (SPC) as explained by Klein (1989), may restrict operator freedom, expectations, and control.

That any radical production management change will require the total involvement of *all* employees is self-evident. All successful change needs input, support, communication, and involvement at all levels. Words and ideas are, though, easier than practice here. There is as yet no widely available comprehensive body of human factors knowledge regarding new manufacturing management approaches of total quality, flexible firms, and JIT philosophies. One of two views can be taken: that JIT and related policies and practices will be beneficial for shop-floor worker control and for good job redesign or that the changes in manufacturing policy will be harmful in terms of human factors in its broadest sense. Whichever of these views will prevail, it is undoubtedly true that changes in the future organization and control of manufacturing will have a tremendous impact on the jobs and roles of all those involved. In the most optimistic analysis, good job redesign and a human-centered manufacturing system design philosophy will be seen as necessary preconditions for successful change to modern manufacturing practices. With such a perspective, worker participation, facilitative coordination rather than supervision, and a flexible approach to production will all be necessary to prevent gains in production effectiveness being offset by impossible work roles and reluctant cooperation among the operators who are expected to make the system work. The job of the human factors community now is to provide evidence and tools for this.

Process of Change

Many different approaches might be taken in organizational redesign initiatives (see Eason, 1988, p. 159), but as with all change, implementation is the key to success. The overriding task in work redesign is to ensure that the process and content of change are both geared to sustain the process itself. Mechanisms to do this should be built from the outset, involving new company communication procedures that are group or team based. Considerable political and structural flexibility is needed in the process in order to be able to "change the change" if necessary. Only by setting up a robust, iterative, and adaptive process can enthusiasm for work redesign in the abstract be translated into practical developments contributed to and promoted by all parties involved (Wilson and Grey, 1990).

The challenges of managing an organizational redesign program are different from those experienced during the exploration and initial implementation stages. In particular, it is important to be aware how decisions made in the earlier stages of the process might ultimately affect the survival of the change. Trade-offs will need to be made between short-term benefits and longer term requirements. For example, trade-offs must usually be made between diffusing the process and spreading resources too thinly (Mansell and Rankin, 1983). In addition, a distinction is desirable between short-term direct work design requirements and longer term organizational support mechanisms. Important, too, is that the elements of redesign should be measurable wherever feasible as feedback is an essential element in maintaining long-term commitment.

There must be a clear perception of the role of the change agent—and of what will happen if and when that agent leaves. Change agents must have skills in project management. Shared goals must be established and the iterative nature of the process recognized, and there must be the freedom, power, and courage to give the process time. Underlying much of the above is that the change, its mechanisms, and development of its content should be participative. Thus, it must be stressed, change is not as simple as slogans would have us believe. Participation can be at many depths and be of many types, and it is a potential minefield of obvious and less obvious traps for the unwary (see Wilson, 1991a, for a review). Nonetheless, commitment to participation and to the practical mechanisms to underpin it are necessary to make changes in work design and in manufacturing management happen successfully.

16.4 PEOPLE ISSUES IN AMT: JOB DESIGN

Issues in AMT related particularly to the individual tend to focus on the type of work and the environment in which this work is performed. The type of work is primarily a function of the job design, while work environment issues are traditionally addressed through ergonomics analyses. The discussion below first presents general issues in job design for AMT environments and then briefly describes a few current topics, such as locus of control, knowledge requirements, and work groups. Then, Section 16.5 will present ergonomics issues in AMT.

So much has been written about job design or job redesign that it is sometimes difficult to know what is myth, what is wishful thinking, and what is the truth. The fact is, of course, that there is no definite "true path" of job redesign applicable in all cases. Each company, each situation, is different, a one-off. All that can be accepted is that there are certain approaches to investigating, redesigning, and implementing jobs that have better chances of success than others.

In broad terms, the approaches in the literature are work simplification, job rotation, job enlargement, job enrichment, and group work design. The dominant approach for the past 10 or 15 years, whether for individuals or groups, has emanated from job enrichment and expectancy theory of motivation. It has been to identify those characteristics of jobs that lead to critical (positive) psychological states, which in turn lead to outcomes of greater internal or external motivation, improved performance, and greater job satisfaction. The widely known job characteristics model posits autonomy, feedback, skill variety, task identity, and task significance as the key job characteristics (Hackman and Oldham, 1980). Other authors have added to these such characteristics as clarity (of role or goals), skill utilization, and others (Wilson, 1987). Many cases (although not always well documented) can be found in the literature showing, by and large, the benefits accruing to individuals and companies from job redesign. Often such change accompanies change in technology also (see Buchanan and Boddy, 1983; Butera and Thurman, 1984; Davis and Taylor, 1972, 1979; Work Research Unit, 1982, for early collections of cases).

In situations where technical systems or methods of controlling manufacturing are to change, the need to consider the design of people's jobs is even greater. To fail to do so can result in poor utilization of the new systems and perhaps even worse performance than was obtained before. This may be due to inappropriate consideration of the role of people, not making the most of existing personnel skills and knowledge, creating interfunction communication and coordination problems, or promoting fear of or resistance to change generally. Perhaps even more important are the benefits to be gained, in terms of greater work force commitment, flexibility, responsiveness, and better organization of operations through appropriate work redesign with changing technology.

As far as contemporary job design is concerned, the preferred route is to use as appropriate the established approaches in a form of work restructuring. We must be aware that jobs are more than just tasks; they comprise work content and distribution of work, work roles, and tasks (Clegg, 1984). Jobs may be seen as a prescribed work role, comprising tasks, methods, and organizational setting; for actual role behavior to match that prescribed, people have to be capable, willing, and have the opportunity to perform.

Majchrzak (1988) sees four dimensions of operators' jobs that will be affected by flexible factory automation: coordination and interdependence needs, information needs, human–machine redundancy, and discretion and responsibility (pp. 46–55). On the other hand, Wall et al. (1990) identify the critical dimensions of operator jobs with AMT as being control of timing, methods, and boundary; monitoring and problem-solving demands; production responsibility; and social interaction (social contact and social support). At a slightly wider level, the important issues implicit in human-centered work redesign for AMT can be seen as development of new information and control structures, including information interfaces on the shop floor; flexible allocation and separation of function; greater control and skill utilization at operator level; better coordination among the work force, probably through self-organizing groups or teams, support at local level of reliable operator performance and decision making, including diagnosis and intervention tasks and troubleshooting; and appropriate means of implementation of change and of sustaining and building on that change (Wilson, 1991b).

The case for good job design, moving away from work simplification, can be made in terms of better performance and better employee well-being and in terms of better new technology utilization and a stronger, more responsive, and flexible organization. The approach to be taken in job redesign must reflect the actual case conditions, but will draw upon both established theories of job and role characteristics along with the more operational and AMT-oriented concerns. From a research perspective we still have large gaps in our knowledge of how different factors of job design affect which outcomes, how to measure and assess these outcomes, how to predict effects elsewhere in the organization, and how to ensure any change is dynamic.

Local Control

Local or operator control has been proposed as being desirable from both performance and job quality viewpoints and may even be required with some forms of new manufacturing technology such as cell manufacture within JIT systems. The concept derives in part from the sociotechnical systems notion of controlling variances at or close to source. With one of the models for job redesign proposed above (Wall et al., 1990), gains measured in terms of performance and worker attitudes are reported as resulting from increasing operator control over maintenance, setup, and intervention as well as changing performance measurement criteria (Jackson and Wall, 1991; Wall et al., 1991).

One of the main arguments for increasing operator control has always been that to do otherwise is to underutilize a major company resource—the innate skills and knowledge (formal and informal) of its shop-floor staff (Leplat, 1990; Wood, 1989, 1990). Acknowledgment of such tacit skills and knowledge and the organization of work to allow use of them could have benefits for both work effectiveness and quality of work life, particularly given contemporary organizational aims of flexibility, adaptability, and innovation.

In a case study reported recently by Jackson and Wall (1991) and Wall et al. (1991), operators at computer numerical control (CNC) assembly machines were given a greater range of tasks, including: preventive maintenance; dealing with one-off jobs; resetting, calibration, and adjustment; and coping with inadequate parts kits. This was seen as a philosophy of controlling key variances closer to their source. The results indicate greater intrinsic job satisfaction, less feelings of job pressure, a reduction in demand on support staff, and a reduction in downtime for high (but not low) variance systems. The investigators feel they have shown that the last finding is due not to less waiting time for maintenance leading to less time per incident, but to better intervention by operators using their knowledge and skills to reduce the number of incidents through true preventive strategies.

Tacit Knowledge

Skills and knowledge in an organization are formally planned for and introduced through policies on training, job evaluation, and specification and by design of work systems generally. It is accepted that workers may also have "informal" skill and knowledge—attributes gained directly through experience and generally not understood or utilized by organizations. This may be termed tacit knowledge and craft skill. Many definitions of these concepts may be found, some of which at least reflect difficulties in distinguishing clearly between skills and knowledge. A working definition is that tacit skill and knowledge are skill and knowledge that are generally inarticulable and unacknowledged in the context in which they are present and are manifest in everyday application.

Recent work has highlighted the importance of tacit knowledge in supporting human skilled performance, reflecting a general current interest in worker abilities, and has identified the gulf between actual skill used by production workers on the shop floor and that officially decreed by planners and managers (Davids and Myers, 1991). Skills and abilities of workers are not well understood by management and are often taken for granted, and tacit skills do indeed exist in people performing even routine "unskilled" work (Jones, 1989; Wood, 1990). The facilitation of this tacit knowledge can reduce the disruptive impact of uncertainty in manufacturing systems through local control, and control of variances at the source through operator intervention can enhance AMT system utilization (Martin, 1990). Hirschorn (1984, p. 100) gives an example from the United States that "demonstrates the value of shop-floor knowledge and expertise in improving production methods."

The comparison is between the conjectures and beliefs of the planners and the reality of the shop floor. As Clegg and Corbett (1987) put it:

> "Many technologies and production systems were, and continue to be, designed on the premise that the human component . . . is inflexible and stupid, and that, where possible, the potential for human contribution should be minimized and designed out of the system . . . whereas, in practice, it is often people's flexibility and intelligence that allows reasonable levels of performance out of poorly designed systems." (p. 181)

During a shop-floor participative observation study in two small manufacturing enterprises, data were collected by observation, taking concurrent verbalizations of behavior and conducting interviews. The principal investigator also worked for a period on the shop floor carrying out many of the operators' tasks. Data were reduced and represented by a planned, rigorous procedure. "Scenarios" were constructed for cases where it appeared workers used skills or knowledge outside what might be expected of them in their formal job specification or in the role expected of them by management. These scenarios were then tested with a "planner" in the company—the production manager for instance—by asking them questions in the form of "what happens when?" or "who does this?" Comparisons have proved very instructive in terms of the knowledge that production workers have, where they get their information for the tasks they carry out, and the decisions they make (Darracott and Wilson, 1992).

At this time it would appear that the notion of tacit knowledge is a very real one and that techniques do exist to identify examples. Difficulties to be overcome include developing and formalizing techniques that are usable generally by company personnel in-house and finding ways to assist companies to utilize the evidence of tacit knowledge that is uncovered. The interest in the problems of formal and tacit skills, knowledge, and attitudes is aroused both by new technologies (Leplat, 1990) and cultural changes demanded by new ideas of managerial, organizational, and operational philosphies; Holberton (1991) quotes a senior U.S. production executive, in an article on the significance of TQM: "this sort of change is about attitudes . . . this is a knowledge-based culture—not a rules-based one."

Group Work

Improved job design, local control, and utilization of tacit knowledge can all be achieved when shop-floor staff operate as a team, as a group of workers with some degree of autonomy. Within a group structure it is possible to design in many of the desirable attributes of jobs, for instance, control of timing, methods, and boundary; optimum demands, responsibility, flexibility, and variety; social contact; and social support. Certainly a key element in successful autonomous work groups is that they provide for considerable self-direction and responsibility; the group has collective control not only over its work but also over distribution of roles, training, and recruiting new members. Perhaps most important, groups created early enough can be the medium and mechanism for development and implementation of the change itself, whether of technology, work organization, or both.

Early on, a decision must be made as to the (approximate) degree of autonomy to be vested in the group. This decision itself is not independent of decisions about, for example, tasks, skills, and relationships with support functions. Over two decades ago Gulowsen (1972) defined the following levels of group autonomy:

1. Group members determine individual production methods.
2. The group decides on questions of internal leadership.
3. The group decides on questions of recruitment.
4. The group determines the internal distribution of tasks.
5. The group decides on questions of production method.
6. The group decides when it will work.
7. The group decides what additional tasks to take on.
8. The group decides on questions of external leadership.
9. The group has influence over its quantitative goals.
10. The group has influence over its qualitative goals.

However, our research base is still not strong enough for us to recommend about appropriate levels with confidence. Essentially, both individual and group autonomy need to be considered in terms of work method autonomy, work schedule autonomy, and the degree of the group's autonomy in choosing work criteria.

In moving away from the traditional work design of low autonomy in shop-floor jobs there may be some iteration in the transition process (Grey, 1986). For example, in some instances the implementation of semiautonomous groups has been accompanied by formal guidelines specifying the boundaries of autonomy between groups and their associated support functions (O'Brien et al., 1987). Such a process has been found to have additional benefits not only in terms of group training but also due to the fact that it can highlight many broader systemic effects of group work redesign.

The broader context within which work redesign takes place can have a profound impact on the extent of significant change (e.g., Tack et al., 1986). Particular practical difficulties that have emerged in setting up a group working in general in the United Kingdom include redesigning the job of the first line manager, the incorporation of technical jobs within semiautonomous groups, and communication within and between groups (IDS, 1984). In addition, increasing moves toward groups working at the shop-floor level have been accompanied, particularly in the United States, by the development of pay-for-knowledge systems of compensation (e.g., Lawler and Ledford, 1985). A key job characteristic for consideration in the implementation of group performance is that of autonomy. It would seem that the transition toward the achievement of a desired level of autonomy needs to be accompanied by early specification of group boundaries both on a quantitative and qualitative basis.

While the present emphasis on movement toward work groups would suggest gains in all situations, recent research (Adler, 1991) has brought into question this assumption. Work groups do not necessarily bring about improvements in satisfaction or productivity. According to Majchrzak and Gasser [in press (b)], of importance is the *match* between the work group and work environment. Here the emphasis is shifted from designing good work groups to designing groups with features and objectives congruent with the environmental demands. From this, they have developed a framework for determining when work groups are needed, what type of work group is needed, and what systems are needed to support the work group. This framework has resulted in 14 propositions, as presented in Table 16.1. As can be seen, these propositions account for the level of autonomy provided for a work group, as discussed above, when determining what type of work group is needed.

16.5 PEOPLE ISSUES IN AMT: ERGONOMICS OF THE WORKPLACE

The final general area of human issues in AMT is the work environment. Recent attention to topics such as work groups and worker empowerment presented above should not be at the expense of fundamental ergonomic workplace design issues. The discussion below provides an overview of these considerations.

Worker Safety and Injury Control

As discussed by Karwowski (1990), safety and health hazards associated with complex manufacturing systems become one of the critical aspects of system design and operation. Although in the past manufacturing automation efforts have been studied from the standpoints of computer control requirements, communication protocol needs, electronics, mechanics, and kinematics, today they must also be considered from the human factors, reliability, and safety

TABLE 16.1 Propositions for Work Group Design

When is a Workgroup Needed?
Prop. 1: Work should be designed to be performed by a group of individuals when the work necessitates diverse perspectives.
Prop. 2: Work should be designed to be performed by a group of individuals when the work tasks have the following attributes: are reciprocally interdependent, create a workload greater than one person can handle, and are either so complex that specialized skills are needed or are so routine that boredom must be alleviated.
Prop. 3: Work should be designed to be performed by a group of individuals when flexibility is needed in the production environment, such that the workgroup is expected to handle multiple sources of variances on the job.
Prop. 4: Work should be designed to be performed by a group of individuals when the conditions specified in Propositions 1 through 3 have been met AND top management is not, in practice, opposed to team-based work design.
What Type of Workgroup is Needed?
Prop. 5: Individuals in workgroups will function more effectively performing non-overlapping tasks when work objectives are focused primarily on quantity of output (i.e., efficiency is most important), tasks are no more than moderately reciprocally interdependent, and the desire flexibility is low.
Prop. 6: Individuals in workgroups will function more effectively performing partially overlapping tasks when work objectives demand some need for shared problem definitions, task attributes are moderately complex with moderate reciprocal interdependencies, and desired flexibility is moderate.
Prop. 7: Individuals in workgroups will function more effectively performing completely overlapping tasks under two different situations: 1) Task complexity is very low and motivation is important (and therefore boredom is likely to result); or 2) Reciprocal interdependence is high, desired flexibility is high and task complexity is not more than moderate such that completely overlapped jobs are not only necessary but possible.
Prop. 8: If the values in practice of top managers oppose workgroup autonomy, the amount of autonomy experienced by the workgroup will be nonexistent.
Prop. 9: Workgroup members will perform their tasks more effectively with a high degree of autonomy over work methods when either creativity is needed, the tasks are complex, or flexibility to respond to uncertainties is desired.
Prop. 10: Workgroup members will perform their tasks more effectively with a moderate amount of autonomy over work methods when task complexity is low and work objectives focus exclusively on quantity and cost.
Prop. 11: Workgroup members will perform their tasks more effectively with a high degree of authority over both the purpose and work methods of the workgroup ONLY when a high degree of environmental flexibility is desired and work objectives of innovativeness are sought.
What Supportive Systems are Needed for Workgroup Effectiveness?
Prop. 12: Workgroup members will perform their tasks more effectively when rewards are matched to degree of task overlap and autonomy.
Prop. 13: Workgroup members will perform their tasks more effectively when training and selection practices are matched to degree of task overlap and autonomy.
Prop. 14: Workgroup members will perform their tasks more effectively when norms are matched to degree of autonomy.

After Majchrzak and Gasser (1992b).

perspectives (Yokomizo, 1987; Karwowski et al., 1988; Rahimi and Karwowski, 1990; Gray et al., 1992).

At present, very little is known about the extent of safety hazards due to AMT environments (Ayres and Miller, 1983; Mastersson, 1987). As automated production systems become more versatile and complex, the system disturbances, including start-up and shut-down problems and deviations in system control, pose serious safety hazards to the human operators. Ironically, in many abnormal situations operators usually rely on safe performance of the automated safety subsystems, which may increase the risk of severe injury (Kuivanen, 1991). A study of accidents related to malfunctions of technical control systems showed that about one-third of the accident sequences included a human intervention in the control loop of the disturbed system (Backström and Harms-Ringdahl, 1984).

According to the International Standards Organization (ISO/TC 184/WG4), the risks associated with hazards due to industrial automation vary with the types of industrial machines incorporated into the specific manufacturing systems and the application of such a system as to how it is installed, programmed, operated, maintained, and repaired. The ISO Committee on Industrial Automation Systems (Bloodgood, 1991) has recently proposed a new safety standard entitled "Safety of Integrated Manufacturing Systems." This new international standard, which was developed "in recognition of the particular hazards which exist in integrated manufacturing systems incorporating industrial machines and associated equipment," aims to "minimize the possibilities of injuries to personnel while working on or adjacent to an integrated manufacturing system."

As discussed by Karwowski (1990), there are strong indications from Europe, Japan, and the United States that the current needs for injury control and worker safety in integrated manufacturing systems are grossly underestimated (see research results reported by Aalto, 1990; Backström and Harms-Ringdahl, 1984; Etherton, 1988; Helander, 1990; Karwowski and Rahimi, 1990; Kuivanen, 1991; Nagamachi, 1988; Nicolaisen, 1987; Parsons, 1987; Sugimoto, 1985; Zimolong and Duda, 1992). Sugimoto (1987) has dispelled some of the common misunderstandings of the nature of automation, including the beliefs illustrated by the following false statements related to automation safety: (1) automation leads directly to increased safety, (2) microelectronic controllers are difficult to understand and are a kind of black box, (3) microelectronic controllers are highly reliable and thus assure safety, (4) machinery is made fail-safe by the safety functions of computer software, (5) safety devices sense a hazardous condition and function accordingly, and (6) as artificial intelligence advances further, robots will become safer.

Hazards Due to AMTs To date, only very limited statistics are available on the safety hazards and related accidents in emerging manufacturing technologies, especially in computerized automation and robotics systems (Karwowski, 1990). Therefore, it is difficult to fully assess the potential impact of these new manufacturing technologies on worker safety. Such an impact should be measured in terms of frequency and severity of related injuries. Injury control and accident prevention strategies for AMT need to be considered in view of the following characteristics of these systems (Kjellén, 1990):

1. Computer-integrated technology is associated with an increased complexity; that is, systems are made up of networks of closely related subsystems and their components.
2. Several accident problems emerge at interfaces between subsystems and when disturbances progress from one subsystem to the other (Leplat, 1984).
3. Traditional safety measures such as guards are in many cases not feasible.

4. It is difficult for the worker who moves inside the physical system to anticipate the possible energy flows, and this difficulty in anticipating the performance and effects of the system is enhanced in systems that utilize exotic physical processes.
5. Many advanced technological systems are characterized by the existence of substantial amounts of energy and, consequently, of a high potential severity of accidents.

The Occupational Health and Safety Rules of Japan (Sugimoto, 1987) assert that "industrial robots do not at present have reliable means of safety and workers cannot be protected from them unless their use is regulated." To identify the hazards caused by industrial robots and other components of AMT, past accidents need to be carefully analyzed. Nicolaisen (1985) states that there are not much data available on robot-related accidents, and there may be a high percentage of unrecorded accidents. Accidents involving robot use are difficult to identify in the classification of accidents caused by other machines (Karwowski et al., 1988; Sugimoto and Kawaguchi, 1985). Sugimoto (1987) reported results of the survey by the Japan Ministry of Labor (1985) of the accidents related to industrial robots. Across the 190 factories surveyed (with 4341 working robots), there were 300 robot-related disturbances, 37 cases of unsafe acts resulting in some near accidents (near hits), 9 injury-producing accidents, and 2 fatal accidents.

Nature of Human Operator Activities Worker activities in an AMT environment are basically those of supervisory control, monitoring, and maintenance. These activities may also be classified into four basic groups as follows: (1) programming tasks, that is, encoding information that guides and directs machinery operation; (2) monitoring of AMT production and control components; (3) maintenance of AMT components to prevent or alleviate machinery malfunctions; and (4) performing a variety of support tasks. Edwards (1989), based on extensive review of the impact of AMT on the worker well-being, concluded that although the utilization of AMT machinery may eliminate heavy and dangerous tasks, working with an AMT is often dissatisfying and stressful for the worker. Sources of stress included constant monitoring required in many AMT applications, limited scope of the activities typically allocated to workers, low level of worker interaction permitted by the system, and safety hazards associated with the unpredictable and uncontrollable nature of the equipment. Edwards (1989) also pointed out that workers who are involved in programming and maintenance activities feel the elements of challenge, which may have positive effects on the workers' well-being. However, these positive effects were often offset by the complex and demanding nature of these activities and by pressure exerted by management to complete these activities quickly.

Needs Assessment for Worker and Safety Health (Karwowski, 1990)

Work-related injuries in the United States have been described and characterized for a number of years by several different organizations, including the U.S. Department of Labor's Bureau of Labor Statistics (BLS), the National Center for Health Statistics (NCHS), the National Institute for Occupational Safety and Health (NIOSH), and the National Safety Council (NSC), a nonprofit, nongovernmental organization. Such information provides a basis for much of the epidemiologic, ergonomic, and technological research in NIOSH's Division of Safety Research.

Hazards Due to AMT Components and Their Interactions An accident is only one of the several outcomes of man–machine interactions under hazardous conditions; near accidents and damage incidents are much more common (Zimolong and Hale, 1989). Generally speaking, the occurrence of an error can lead to one of these consequences: (1) the error remains unnoticed, (2) the error can be compensated for by the system, (3) the error leads to a machine breakdown and/or system stoppage, or (4) the error leads to an accident. Since not every human error that results in a critical incident will cause an actual accident, the further distinction of the following outcome categories is appropriate as follows (Swain, 1985).

1. *Unsafe incidents*—Any unintentional occurrence that may or may not result in injury, damage, or loss.
2. *Accident*—Unsafe event resulting in injury, damage, or loss.
3. *Damage incident*—Unsafe event, which only resulted in some kind of material damage.
4. *Near accident*—Unsafe event in which injury, damage, or loss was fortuitously avoided despite a "close call."
5. *Accident potential*—Unsafe events, which could have resulted in injury, damage, or loss, but owing to circumstances, not even a close call was experienced.

Zimolong and Duda (1992) concluded that available data of reported incidents on AMT components in general and robot installations in particular are very poor and that only a few original reports on accidents, critical incidents, and abnormal stoppage cases have been reported up to now. There is also a lack of information on supplementary hazard data such as number of employees exposed and work hours of personnel. As a consequence, it is unclear how dangerous work with AMT really is with respect to such basic concepts as number of accidents and/or injuries per 1000 employees, per one million work hours, or per number of robots.

Risk Assessment and Safety Analysis In general, hazards can arise from the system itself, from its association with other equipment or the physical environment of the workplace, or from interactions of personnel with the system. Examples of main sources of hazards in AMT are given in Table 16.2. The NIOSH has indicated that the severe occupational traumatic injuries it is intent upon controlling include amputations, fractures, severe lacerations, and worker deaths that are the result of severe trauma incurred on the job. The potential for such trauma in integrated manufacturing systems is also very high (Etherton, 1988; Sugimoto, 1987; Zimolong and Duda, 1992). Etherton and Millard (1990) have utilized four different injury epidemiology measures of machine-related injury that are appropriate to this analysis: (1) number of fatalities, (2) number of amputations, (3) number of injuries, and (4) combined frequency and severity.

According to the ISO (1991), the risk assessment in AMT should be performed to minimize all risks and to serve as a basis for determining safety objectives and measures in the development of programs or plans to (1) create a safe working environment and (2) ensure safety and health of personnel. Each identified hazard should be assessed for its risk, and appropriate safety measures should be determined and implemented to minimize that risk. Hazards should also be ascertained for the single units, and interaction between single units, the operating sections of the system, and operation of the complete system for all intended operating modes and/or conditions including conditions where normal safeguarding means are suspended for such operations as programming, verification, troubleshooting, maintenance, or repair.

System Disturbance Methodology for Injury Control

In AMT, the human operators are needed for the purpose of controlling, programming, maintaining, presetting, servicing, or troubleshooting tasks. Disturbances in the system lead to situations that make it necessary for workers to enter the hazardous areas. In this respect, it can be assumed that disturbances remain the most important reason for human interference in AMT, because the systems will more often than not be programmed from outside the restricted areas (Anon, 1984). One of the most important issues for AMT safety is to prevent disturbances, since most risks occur in the troubleshooting phase of the system. The avoidance of disturbances is the common aim for both safety and cost-effectiveness (Kuivanen, 1990).

A classical approach in safety research has been to study the effects of the characteristics of the production system on human information processing and the probability of human errors. A current approach is to study human responses to deviations or disturbances in the system. Human actions in disturbed systems both prevent and contribute to the occurrence of accidents in the AMT environment. For example, a study of accidents related to malfunctions of technical control systems showed that about one-third of the accident sequences included a human intervention in the control loop of the disturbed system (Backström and Harms-Ringdahl, 1984).

In addition to productivity, the disturbances during the operation of a

TABLE 16.2 Main Sources of Hazards in AMT

a) Moving mechanical components in:
   1) normal operation either individually or in conjunction with other elements of the system or related equipment in the hazard zone,
   2) abnormal operation (e.g. falling, tipping).

b) Power sources

c) Stored energy

d) Interferences
   1) electrical (e.g. electromagnetic interference, electrostatic discharge (ESD), radio frequency interference (RFI),
   2) mechanical (e.g. vibration, shock).

e) Hazardous atmospheres or materials
   1) explosive or combustible,
   2) corrosive,
   3) radiation (e.g. ionization, thermal).

f) Failure or fault of:
   1) protective means including removal, disassembly, or defeating,
   2) components, devices, or circuits,
   3) power sources or means of power distribution including fluctuations or disturbances,
   4) information transmission.

g) Human error
   1) design, construction, or modification,
   2) operating systems, application software, and programming,
   3) application and implementation,
   4) set-up including work handling/holding and tooling,
   5) operation or use,
   6) maintenance and repair,
   7) documentation and training/instruction.

h) Ergonomic considerations
   1) lighting,
   2) vibrations,
   3) noise,
   4) climatic conditions,
   5) operator control stations design/layout

After ISO (1991).

flexible manufacturing system (FMS) have direct effect on the safety of the people involved in operating the system. The study reported by Kuivanen (1990) showed that about one-half of the disturbances in automated manufacturing decrease the safety of the workers. The main causes for disturbances were errors in system design (34 percent), system component failures (31 percent), human error (20 percent), and external factors (15 percent). Most machine failures were caused by the control system, and in the control system most failures occurred in sensors. An effective way to increase the safety of a FMS is to reduce the number of disturbances. This requires that the critical system components and functions are identified (Kuivanen, 1990) by using analysis methods, such as failure mode and effect analysis (FMEA), at the design stage. Risks and hazards can be identified by using, for example, an energy analysis method. Even better results can be obtained if these two methods are combined to identify critical system failures and disturbances.

The main research issues in AMT disturbance prevention are (Kuivanen, 1991) (1) the major causes of the disturbances, (2) unreliable components and functions, (3) the impact of the disturbances on safety, (4) the impact of the disturbances on the function of the system, (5) material damage, and (6) repairs. The following information is typically recorded in the log book for each system disturbance: (1) date, time, and name of the logger; (2) manufacturing data; (3) problems and disturbances (causes, effects, duration, measures); (4) notice for the next shift; and (5) testing and development measures. The intensive method of disturbance analysis consists of the minute-by-minute observation of one morning, evening, and night shift during two days. An example of the disturbance form and the classification of potential disturbances are shown in Tables 16.3 and 16.4, respectively.

Toikka et al. (1991) described the orientation model for evaluation of an operator's role in AMT disturbance control. The model includes five stages: (1) withdrawal (somebody else solves the problem), (2) routine disturbance handling (habitual, not concerned with the causes of disturbances), (3) unofficial developing activities (private log books for ideas and inventions), (4) official optimizing of the system (organizational forms to develop disturbance control), and (5) systematic way of working (developing the functional principles and organization of a FMS).

Safety Planning Guide and Checklist (Karwowski, 1990)

According to the ISO (1991), an integrated manufacturing system (IMS) is a group of two or more industrial machines working together in a coordinated manner normally interconnected with and operated by a supervisory controller or controllers capable of being reprogrammed for the manufacturing of discrete parts or assemblies. The design phase of the proposed ISO (1991) safety strategy includes (1) specification of the limits of parameters of the system, (2) application of a safety strategy, (3) identification of the hazards,

TABLE 16.3 AMT System Disturbance Data Collection

1-	identification information
2-	state of the system
3-	effect on production
4-	disturbance elimination measures
5-	development measures
6-	duration of the disturbance
7-	material damages
8-	production stoppage costs
9-	cause of disturbance
10-	effect on safety, and
11-	evaluation of disturbance orientation

After Kuivanen (1991).

(4) assessment of the associated risks, and (5) removal of the hazards or limitation of the risks as much as practicable.

The AMT safety specification (ISO, 1991) should include (1) description of functions, (2) layout and/or model, (3) survey about the interaction of different working processes and manual activities, (4) analysis of process sequences including manual interaction, (5) description of the interfaces with conveyor or transport lines, (6) process flow charts, (7) foundation plans, (8) plans for supply and disposal devices, (9) determination of the space required for supply and disposal of material, and (10) available accident records.

Integrated manufacturing system safety should be planned early at the design stage and be an integral part of the IMS technical planning. Safety planning can be achieved through suitable aids in the form of symbolic signs describing the IMS-related safety hazards. Recently, Nicolaisen (1990) presented a method that integrates safety and health aspects into layout planning (see Table 16.5). He also illustrated the use of the method on the planning of a FMS. The developed method is based on symbols representing "hazards" and "safety precautions" and on a "tasks/functions and persons/manufacturing devices" matrix.

The layout for the schematic (symbolic) representation of AMT depicts the most important elements (machines, manual working places, material flow, and control devices) distinguished by their function, type, and size and of the arrangement of these elements in the planned environment and of their future linkage (system structure) as far as it can be shown graphically. Nicolaisen (1990) suggested the use of symbolic signs (like road signs) for the planning of the protection of health and safety of workers in AMT (see Fig.

TABLE 16.4 Preliminary Disturbance Classification

1. Major causes of disturbances
   - design based errors
   - failures in components
   - user errors
   - external factors
   - unclear
2. Unreliable components and functions
3. The impact of disturbance on safety
   - risk exists
   - hazardous situation
   - minor accident
   - serious accident
   - catastrophe
   - effect exists/no effect
4. The impact of disturbances on production
   - difficulties to work
   - production slow down
   - one machine stops
   - cell stops
   - FMS stops
   - effect exists/no effect
5. Material damage and repairs made
6. Evaluation of disturbance control orientation
7. Relative time used for disturbance elimination

After Kuivanen (1990).

TABLE 16.5 Major Steps in FMS Safety Analysis

1	Hazard analysis,
2	Appropriate hazard symbols are transferred to graphic representation of layout ,
3	Sequences for individual dangerous tasks are determined and discussed, and put into the layout draft using circled numbers connected with lines, and
4	Safety are designed and marked in the layout using safety precautions symbols

After Nicolaisen (1987).

Tasks/ functions	Persons/manufacturing devices: Working team/ device operator	Maintenance staff	Transportation staff	Mfg. control/ job planning	Automatic system
Feeding and supply of material					
- raw material	○		○		(b)
- inserted parts	○		○		(b)
Part orientation, bending					
preassembly, heating inserting of parts					(pp) (ir)
Start/stop of process	○				(c)
Injection moulding					(im)
Handling of parts					(ir)
Quality control	○				(im)
Subsequent task					
- deburring					(ir)
- assembling, welding printing, processing					(p)
- palletizing, packing					(b) (ir)
Transport	○		○		(b)
Tool setting	○				
Programming	○				
Reprogramming	●				
optimization	●				
correction	●				
changing of program	●				

Fig. 16.2. Matrix of tasks/functions and groups or people/means of manufacturing in AMT (after Nicolaisen, 1990).

16.2). The following three questions were the starting points for the design of a layout according to safety aspects:

1. Which are the hazards in the system and where are they?
2. Which persons work at which places in the system (or in its environment) and is there any hazard for them?
3. Which safety precautions are to be taken at which place?

Tasks/ functions	Persons/manufacturing devices: Working team/ device operator	Maintenance staff	Transportation staff	Mfg. control/ job planning	Automatic system
System/process control	○				(c)
Maintenance/adjustment	○				
Troubleshooting	○				(c)
Repair	○	○			
Test	○	○			
manual operation in the case of severe breakdown	○				(ss)
Determination of the sequence of orders	○			○	
Determination of the working steps	○			○	
Control of material supply	○		○	○	
Monitoring of dates	○			○	
Forwarding of the orders	○		○	○	

○ = main task
○ = subtask
p = processing units
ir = industrial robot
b = buffer
im = injection moulding machine
c = control system
ss = subsystem
pp = pre-processing station

Fig. 16.2. (*Continued*)

This safety planning method was tested by several planning teams in practice, and the reactions were extremely positive according to Nicolaisen (1987). The advantage of the symbols library is that the designer does not have to be an expert in safety and health issues. The symbols library developed in this task will be used with the computer-aided layout planning method.

In accordance with the ISO (1991), all necessary requirements granting a safe operation need to be considered in the design of systematic safety planning procedure and the checklist. This includes all protective measures to effectively reduce the hazards and requires (1) integration of the man–machine interface, (2) early definition of the position of those working on the system (in time and space), (3) early consideration of ways of cutting down on isolated work, and (4) consideration of environmental aspects.

In accordance with the ISO (1991) standard, the safety planning procedure should address, among other aspects, the following safety issues of AMT:

1. *Selection of Operating Modes of System*. The control equipment should have provisions for at least the following operating modes:
   (a) Normal or production mode—all normal safeguards connected and operating
   (b) Operation with some of the normal safeguards suspended
   (c) Operation in which system or remote manual initiation of hazardous situations is prevented (e.g., local operation, isolation of power to, or mechanical blockage of hazardous conditions)
2. *Training, Installation, Commission, and Functional Testing*. When personnel are required to be in the hazard zone, the following safety measures should be provided in the control system:
   (a) Hold to run
   (b) Enabling device
   (c) Reduced speed
   (d) Reduced power
   (e) Movable emergency stop
3. *Safety in AMT Programming, Maintenance, and Repair*. During programming, only the programmer should be allowed in the safeguarded space. The system should have inspection and maintenance procedures to ensure continued intended operation of the system. The inspection and maintenance program should take into account the recommendations of the system supplier and those of suppliers of various elements of the systems. Personnel who perform maintenance or repairs on the system should be trained in the procedures necessary to perform the required tasks.
4. *Fault Elimination*. Where fault elimination is necessary from inside the safeguarded space, it should be performed after safe disconnection (if possible, lockout). Additional measures against erroneous initiation of hazardous situations should be taken. By means of instruction and warning signs, attention should be drawn to fault elimination at a system that cannot be observed completely.

16.6 CONCLUSIONS

Although it may appear that continued development and implementation of AMT should diminish human factors concerns, in reality, as evidenced by much of this book, the success of this technology requires that a larger number of human factors considerations be well understood. An understanding of the issues, however, is far from sufficient, and the true challenge will be to discover unique solutions for most effectively integrating humans into advanced manufacturing systems.

This chapter has not attempted to provide exhaustive coverage of all topics related to human aspects of AMT; this book as a whole forwards that goal. Rather, what is presented here is a eclectic sampling of significant issues related to the technology, organization, and people issues in the design and implementation of AMT. While the chapters in this book discuss each of these areas in greater detail, the rationale for presenting all three components in one chapter is to highlight that optimizing each component independent of the others will produce a suboptimal system.

Research methodologies will need to consider not only the variety of new computer-based technologies emerging in areas such as artificial intelligence and interface design but also how to construct tools that can be readily assimilated into the manufacturing environment. A host of organization, management, and job design issues will also need to be resolved. Also, as complexity of the work environment increases, skill requirements are changing to require tacit knowledge and decision-making capabilities. New organizational designs are also needed to address the group work environment and other job design features now made possible by AMT hardware and software. Special emphasis must be placed on workers' comfort and safety, and effective injury control in automated manufacturing environments must be developed. As pointed out by Sugimoto (1987), "as automation progresses and (manufacturing) systems grow in scale, their failure will have a far larger impact . . . and the accidents arising from automation reveal the difficult situation of blue-collar workers who are no longer able to keep up with the sophistication and complexity."

Choosing between a human-centered or technology-centered design is not viewed as the primary issue. Rather, goals for the "technology–organization–people" manufacturing system must be determined according to corporate values and strategy. Successful design of AMT systems will focus not only on the critical features of each component in the system but also on the relationships between these components for accomplishing the higher level corporate goals. Consideration of the relationships between components, while adding a higher level of complexity, increases the capability of obtaining a system design that achieves these goals. In conclusion, the definition of an AMT system must be comprehensive to include not only technology but also the organization and people. As in any true system design, attention should be focused not only on the individual components but also on the interaction of these components to assure congruency.

REFERENCES

Aalto, A. (1990). Accident and sickness absence rates of CNC-machinists and machine center operators in a Finnish machine stop. Unpublished Technical Report, Tampere, Finland.

Adler, P. S. (1991). Workers' assessments of three flexible manufacturing systems. *International Journal of Human Factors in Manufacturing*, **1**(1), 33–54.

Ammons, J. C., Govindaraj, T., and Mitchell, C. M. (1988). *IEEE Trans. on Systems, Man, and Cybernetics*, **18**, 744–756.

Anon (1984). Mechantronics: The policy ramifications, in *Mechantronics and Safety*, Asian Productivity Organization, Tokyo, pp. 93–183.

Ayres, R. U., and Miller, S. M. (1983). *Robotics: Applications & Social Implications*, Ballinger Publishing Company, Cambridge, MA.

Backström, T., and Harms-Ringdahl, L. (1984). A statistical study of control systems and accidents at work. *Journal of Occupational Accidents*, **6**, 201–210.

Barr, A., Cohen, P. R., and Geigenbaum, E. A. (1989). *The Handbook of Artificial Intelligence*, Vol IV, Addison-Wesley, New York.

Bloodgood, J. F. (1991). Comparison of Draft Standards ISO/TC 184/WG 4-WD and B11.20, NMTBA—The Association for Manufacturing Technology.

Brodner, P. (1991). Design of work and technology in manufacturing. *International Journal of Human Factors in Manufacturing*, **1**(1), 1–16.

Buchanan, D. A., and Boddy, D. (1983). *Organisations in the Computer Age*, Gower, Aldershot.

Butera, F., and Thurman, J. E. (1984). *Automation and Work Design*. North Holland, Amsterdam.

Carlsson, J. (1984). Robot Accidents in Sweden, National Board of Occupational Safety and Health, 5-171-84, Solna, Sweden.

Chapanis, A. (1967). The relationship of laboratory strokes to practical situations. *Ergonomics*, **10**, 557–577.

Clegg, C. W. (1984). The derivation of job designs. *Journal of Occupational Behavior*, **5**, 131–146.

Clegg, C. W., and Corbett, J. M. (1987). Research and development into "humanizing" advanced manufacturing technology, in *The Human Side of Advanced Manufacturing Technology*, T. D. Wall, C. W. Clegg, and N. J. Kemp, Eds., Wiley, Chichester.

Coleman, P. J. (1984). Descriptive epidemiology in job injury surveillance. *Journal of Occupational Accidents*, **6**, 135–146.

Craik, K. J. W. (1943). *The Nature of Explanation*, Cambridge University Press, Cambridge, England.

Cummings, T., and Blumberg, M. (1987). Advanced manufacturing technology and work design, in *The Human Side of Advanced Manufacturing Technology*, T. D. Wall, C. W. Clegg, and N. J. Kemp, Eds., Wiley, Chichester.

Darracott, N., and Wilson, J. R. (1992). Tacit Skills and Knowledge of Production Workers: An Overview and a Methodology. Report to the UK Department of Trade and Industry, Report of the IOE, Department of Manufacturing Engineering and Operations, University of Nottingham, UK.

Davids, K., and Myers, C. (1991). Role of tacit knowledge in human skill performance. *Journal of Human Movement Studies*, **19**, 273–288.

Davis, L. E., and Taylor, J. C. (1972). *Design of Jobs* (1st ed.), Penguin, Harmondsworth.

Davis, L. E., and Taylor, J. C. (1979). Design of Jobs (2nd ed.), Goodyear, Santa Monica, CA.

Dixon, J. R., Nanni, A. J., and Vollmann, T. E. (1990). The New Performance Challenge: *Measuring Operations for World Class Competition*, Dow Jones-Irwin, Homewood, IL.

Dunkler, O., Mitchell, C. M., Govindaraj, T., and Ammons, J. C. (1988). *IEEE Trans. Systems, Man, Cybernetics*, **18**, 223–237.

Drury, C. G., Barbara, P., Van Cott, H. P., Grey, S. M., and Corlett, E. N. (1987). Task analysis, in *Handbook of Human Factors*, G. Salvendy, Ed., Wiley, New York, pp. 370–401.

Eason, K. D. (1988). *Information Technology and Organisational Change*, Taylor & Francis, London.

Edwards, E., and Lees, F. P., Eds. (1974). *The Human Operator in Process Control*, Taylor & Francis, London.

Edwards, J. R. (1989). Computer aided manufacturing and worker well-being: A review of research. *Behaviour and Information Technology*, **8**(3), 157–174.

Etherton, J. (1988). Unexpected motion hazard exposures on a large robotic assembly system, in *Ergonomics of Hybrid Automated Systems*, Vol. I, W. Karwowski, H. R. Parsaei, and M. R. Wilhelm, Eds., Elsevier Science Publishing, Amsterdam, pp. 411–419.

Flexman, R. E., and Stark, E. A. (1987). Training simulators, in *Handbook of Human Factors*, G. Salvendy, Ed., Wiley, New York, pp. 1012–1038.

Gaudier, M. (1988). Worker's participation within the new industrial order: A review of literature. *Labour and Society*, **13**, 313–332.

Gray, S. V., Syan, C. S., and Wilson, J. R. (1992). Human control of robot motion: orientation, perception and compatibility, in *Human–Robot Interaction*, M. Rahim; and W. Karwowski, Eds., Taylor & Francis, London, pp. 48–64.

Gulowsen, J. (1972). A measure of workgroup autonomy, in *Design of Jobs*, L. E. Davis and J. C. Taylor, Eds., Penguin, Harmondsworth.

Hackman, J. R., and Oldham, G. (1980). *Work Redesign*, Addison-Wesley, Reading, MA.

Harms-Ringdahl, L. (1986). Experiences from safety analysis of automatic equipment. *Journal of Occupational Accidents*, **8**, 139–148.

Hayes, R. H., and Broadbent, D. (1988). Two modes of learning for interactive tasks. *Cognition*, **28**, 249–276.

Hayes, R. H., and Wheelwright, S. C. (1984). *Restoring Our Competitive Edge: Competing Through Manufacturing*, Wiley, New York.

Helander, M. G. (1990). Ergonomics and safety considerations in the design of robotics workplaces: A review and some priorities for research. *International Journal of Industrial Ergonomics*, **6**, 127–149.

Hirschorn, L. (1984). *Beyond Mechanisation*, MIT Press, Cambridge, MA.

Holberton, S. (1991). An idea whose time has not only come but will prevail. Quality is the crucial issue for companies in the 1990's. The significance of Total Quality Measurement is examined in the first of a series. *Financial Times*, Wednesday, March 20, p. 10.

IDS (1984). Group Working & Greenfield Sites. Study 314, Income Data Services, Ltd.

ISO/TC 184/WG 4 (1991). Industrial Automation Systems—Safety of Integrated Manufacturing Systems—Basic Requirements (CD 11161), Geneva, Switzerland.

Itoh, S. (1991). Customer-oriented manufacturing. *International Journal of Human Factors in Manufacturing*, **1**(4), 365–370.

Jackson, P. R., and Wall, T. D. (1991). How does operator control enhance performance of advanced manufacturing technology? *Ergonomics*, **34**, 1301–1311.

Japan Ministry of Labour (1985). Study on accidents involving industrial robots. Occupational Safety and Health Department, Labor Standards Bureau, Report No. 5, Japan.

Jones, B. (1989). When certainty fails: Inside the factory of the future, in *The Transformation of Human Work*, S. Wood, Ed., Hyman, London.

Karwowski, W., Parsaei, H. R., Nash, D. L., and Rahimi, M. (1988). Human perception of the work envelope of an industrial robot, in *Ergonomics of Hybrid Automated Systems*, W. Karwowski, H. R. Parsaei, and M. R. Wilhelm, Eds., Elsevier, Amsterdam, pp. 421–428.

Karwowski, W., and Rahimi, M., Eds. (1990). *Ergonomics of Advanced Manufacturing and Hybrid Automated Systems-II*. Elsevier, New York.

Karwowski, W. (1990). Injury control and worker safety in integrated manufacturing systems, Unpublished Technical Report, Center for Industrial Ergonomics, University of Louisville, Louisville, KY.

Klein, J. A. (1989). The human costs of manufacturing reform. *Harvard Business Review*, **67**, 60–61 and 64–66.

Kuivanen, R. (1991). The impact on safety of disturbances in flexible manufacturing systems, in *Ergonomics of Hybrid Automated Systems*, Vol. II, W. Karwowski and M. Rahimi, Eds., Elsevier, Amsterdam, pp. 951–956.

Lawler, E. E., III, and Ledford, G. E., Jr. (1985). Skill based pay: A concept that's catching on. *Personnel*, **62**(9), 30–37.

Leplat, J. (1984). Occupational accident research and systems approach. *Journal of Occupational Accidents*, **6**, 77–89.

Leplat, J. (1990). Skills and tacit skills: A psychological perspective. *Applied Psychology: An International Review*, **39**, 143–154.

Majchrzak, A. (1988). *The Human Side of Factory Automation: Managerial and Human Resource Strategies for Making Automation Succeed*, Jossey-Bass, San Francisco.

Majchrzak, A., and Gasser, L. (1992a). HITOP-A: A tool to facilitate interdisciplinary manufacturing systems design. *International Journal of Human Factors in Manufacturing*, **2**(3), 255–276.

Majchrzak, A., and Gasser, L. (1992b). Towards a conceptual framework for specifying manufacturing workgroups congruent with technological change. *International Journal of Computer-Integrated Manufacturing*, **5**(2), 118–131.

Mansell, J., and Rankin, T. (1983). Changing organizations: The quality of working life process. Occasional paper No. 4. Ontario Ministry of Labour/Ontario Quality of Working Life Centre.

Martin, T. (1990). A framework for studying the psychological effects of advanced manufacturing technology, in *Work Design in Practice*, C. Haslegrave, et al., Eds., Taylor & Francis, London.

Mastersson, L. (1987). Interaction between man and robots with some emphasis on "intelligent" robots, in *Occupational Health and Safety in Automation and Robotics*, K. Noro, Ed., Taylor & Francis, London, pp. 143–150.

Moray, N., and Lee, J. (1990). Trust, automation and performance in human-machine system, in *Ergonomics of Hybrid Automated Systems*, Vol. II, W. Karwowski and M. Rahimi, Eds., Elsevier, Amsterdam, pp. 669–673.

Myers, C., and Davids, K. (1991a). Tacit skill and performance at work: Integrating laboratory and field research. SAPU Memo No. 1242. MRC/ESRC Social and Applied Psychology Unit, University of Sheffield.

Myers, C., and Davids, K. (1991b). Knowing and doing: Different dimensions of skill at work. MRC/ESRC Social and Applied Psychology Unit, University of Sheffield.

Nagamachi, M. (1988). Ten fatal accidents due to robots in Japan, in *Ergonomics of Hybrid Automated Systems*, Vol. I, W. Karwowski, et al., Eds., Elsevier Science Publishers, Amsterdam, pp. 391–396.

Nakamura, N., and Salvendy, G. (1988). *International Journal of Production Research*, **26**, 576–583.

Nicolaisen, P. (1985). Occupational safety and industrial robots—present stage of discussions within the Tripartite Group on robotic safety, in *Robot Technology and Applications—Proceedings of the 1st Robotics Europe Conference, Brussels, June 27–28, 1984*, K. Rathmill, P. MacConaill, S. O'Leary, and J. Brown, Eds., Springer-Verlag, Berlin, pp. 74–89.

Nicolaisen, P. (1987). Safety problems related to robots. Seminar on Industrial Robotics, 1986, Brno, Czechoslovakia, United National Economic Commission for Europe.

Nicolaisen, P. (1990). Arbeitsschutz, in *Industrieroboter-Handbuch fur Industrie und Wissenschaft*, H. J. Warnecke and R. D. Schraft, Eds., Springer-Verlag, Berlin.

Norman, D. A. (1983). Some observations on mental models, in *Mental Models*, D. Gentner and A. L. Stevens, Eds., Erlbaum, Hillsdale, NJ, pp. 7–14.

O'Brien, C., Chalk, S., Grey, S. M., White, A., and Wormell, N. (1987). An application on the automative component industry, in *Just-in-Time Manufacturing*, C. Voss, Ed., IFS Publications, Springer-Verlag, Berlin, pp. 303–317.

Payne, S. (1991). A descriptive study of mental models. *Behaviour and Information Technology*, **10**, 3–21.

Parsons, H. M. (1987). Robotics and the health of workers, in *Occupational Safety and Health in Automation and Robotics*, K. Noro, Ed., Taylor & Francis, London.

Prabhu, P., Sharit, J., and Drury, C. (1992). Classification of temporal errors in CIM systems, development of a framework for declaring human-centred information requirements. *International Journal of Computer-Integrated Manufacturing*, **5**(2), 58–67.

Rahimi, M., and Karwowski, W. (1990). A research paradigm in human-robot interaction. *International Journal of Industrial Ergonomics*, **5**(1), 59–71.

Rasmussen, J. (1986). *Information-Processing and Human-Machine Interaction*, North–Holland, New York.

Rasmussen, J. (1988). Information technology: a challenge to the human factors society? *Human Factors Society Bulletin*, **31**(7), 1–3.

Reason, J. (1990). *Human Error*, Cambridge University Press, New York.

Rogers, Y., Rutherford, A., and Bibby, P. A. (1991). *Models in the Mind: Theory, Perspective and Application*, Academic Press, London.

Rutherford, A., and Wislon, J. R. (1991). Searching for the mental model in human-machine systems, in *Models in the Mind: Theory, Perspective and Application*, Y. Rogers, A. Rutherford, and P. A. Bibby, Eds., Academic Press, London, pp. 195–223.

Rouse, W. B. (1987). much ado about data. *Human Factors Society Bulletin*, **30**(9), 1–3.

Sackett, P. R., and Larson, J. R. (1990). Research strategies and tactics in industrial and organizational psychology, in *Handbook of Industrial and Organizational Psychology*, 2nd ed., Consulting Psychologists Press, Palo Alto, CA, pp. 381–418.

Sanders, J., and Moray, N. (1990). *Human Error*, Lawrence Erlbaum, Hillsdale, NJ.

Sanderson, P. M. (1989). The human planning and scheduling role in advanced manufacturing systems: An emerging human factors domain. *Human Factors*, **31**, 635–667.

Sanderson, P. M. (1991). Towards the model human scheduler. *International Journal of Human Factors in Manufacturing*, **1**, 195–8219.

Sharit, J., and Elhence, S. (1990). Allocation of tool-replacement precision-making responsibility in flexible manufacturing systems. *International Journal of Industrial Ergonomics*, **5**, 29–46.

Sharit, J., Eberts, R., and Salvendy, G. (1988). A proposed theoretical framework for the design of decision support systems: a cognitive approach. *International Journal of Production Research*, **26**, 1037–1063.

Simon, C. W. (1987). Will egg-sucking ever become a science? *Human Factors Society Bulletin*, **30**(6), 1–4.

Slocum, J. W., and Sims, H. P. (1980). A typology for integrating technology, organisation and job design. *Human Relations*, **33**, 193–212.

Sugimoto, N. (1985a). Safety measures for automated machines. *Safety Staff*, **12**, 4.

Sugimoto, N. (1985b). Systematic robot-related accidents and standardization of safety measures, in *Robot Safety*, M. C. Bonney and Y. Yong, Eds., Springer-Verlag, Berlin, pp. 23–29.

Sugimoto, N. (1987). Subjects and problems of robot safety technology, in *Occupational Safety and Health in Automation and Robotics*, K. Noro, Ed., Taylor & Francis, London, p. 175.

Sugimoto, N., and Kawaguchi, K. (1985). Fault-tree analysis of hazards created by robots, in *Robot Safety*, M. C. Bonney and Y. F. Yong, Eds., Springer-Verlag, KFS, Berlin, pp. 83–98.

Tack, P. J., Bolk, H., and Slagmolen, G. (1986). Decision making around flexible manufacturing systems (FMS): The problem of context and rationality, 3rd International Conference of Human Factors in Manufacturing, pp. 199–205.

Toikka, K., Norros, L., Hyötyläinen, R., and Kuivanen, R. (1991). Disturbance Control in Flexible Manufacturing, The Finnish Work Environment Fund, Publications A14, Tampere, Finland (in Finnish).

Wall, T. D., Corbett, J. M., Clegg, C. W., Jackson, P. R., and Martin, R. (1990). Advanced manufacturing technology: Towards a theoretical framework. *Journal of Organizational Behavior*, **11**, 201–219.

Wall, T. D., Corbett, J. M., Martin, R., Clegg, C. W., and Jackson, P. (1991). Advanced manufacturing technology, work design and performance: A change study. *Journal of Applied Psychology*, **75**, 691–697.

Wilson, J. R. (1987). Whose attitudes to which aspects of work?: Cool media for high participation. *Work and Stress*, **1**(4), 385–395.

Wilson, J. R. (1991a). Participation—a framework and a foundation for ergonomics? *Journal of Occupational Psychology*, **64**, 67–80.

Wilson, J. R. (1991b). Personal perspective: Critical human factors contributions in modern manufacturing. *International Journal of Human Factors in Manufacturing*, **1**(3), July, 281–297.

Wilson, J. R. (1992). Human resource issues in manufacturing systems: The effects of JIT/TQM initiatives, in *Human Factors in Design for Manufacturability and Process Planning*, M. Helander, Ed., Taylor & Francis, London.

Wilson, J. R., and Grey, S. M. (1990). But what are the issues in work redesign? in *Word Design in Practice*, C. Haslegrave, J. R. Wilson, and E. N. Corlett, Eds., Taylor & Francis, London, pp. 7–15.

Wilson, J. R., and Rutherford, A. (1989). Mental models: Theory and application in human factors. *Human Factors*, **31**(6), 617–634.

Wood, S., Ed. (1989). *The Transformation of Work?* Unwin Hyman, London.

Wood, S. (1990). Tacit skills: The Japanese model and new technology. *Applied Psychology: An International Review*, **39**, 169–190.

Woods, D. D., Roth, E. M., and Pople, H. (1988). *Reliability Engineering and System Safety*, **22**, 169–200.

Work Research Unit (1982). *Meeting the Challenge of Change: Case Studies*. Department of Employment, London.

Zimolong, B., and Duda, L. (1992). Human error reduction strategies in advanced manufacturing systems, in *Human-Robot Interaction*, M. Rahimi and W. Karwowski, Eds., Taylor & Francis, London, pp. 242–285.

Zimolong, B., and Hale, A. R. (1989). Arbeitssicherheit, in *Europaisches Handbuch der Arbeits und Organizationpsychologie*, S. Grief, W. Holling and N. Nicholson, Hrsg., Psychologie Verlagsunion, Munchen, pp. 126–131.

INDEX

Printed and bound by CPI Group (UK) Ltd, Croydon, CR0 4YY

16/09/2026

14963457-0006